# Improving Vegetatively Propagated Crops

# APPLIED BOTANY AND CROP SCIENCE

Series Editors: R. W. Snaydon, University of Reading, England;
J. M. Barnes, Co-operative State Research Service, United States Department of Agriculture, Washington, USA and F. L. Milthorpe[†], Macquarie University, New South Wales, Australia


Physiological Ecology of Forest Production   J. J. Landsberg
Weed Control Economics   B. A. Auld, K. M. Menz and C. A. Tisdell
Improving Vegetatively Propagated Crops   A. J. Abbott and R. K. Atkin (Eds)

[†]Deceased

# Improving Vegetatively Propagated Crops

Edited by
A. J. ABBOTT AND R. K. ATKIN

*University of Bristol, Department of Agricultural Sciences,
Long Ashton Research Station,
Long Ashton, Bristol BS18 9AF*

1987

ACADEMIC PRESS
Harcourt Brace Jovanovich, Publishers
London   San Diego   New York   Berkeley
Boston   Sydney   Tokyo   Toronto

ACADEMIC PRESS LIMITED
24–28 Oval Road,
London NW1 7DX


United States Edition published by
ACADEMIC PRESS INC.
San Diego, CA 92101


Copyright © 1987 by
ACADEMIC PRESS LIMITED


All rights reserved.
No part of this publication may be reproduced or transmitted in any
form or by any means, electronic or mechanical, including
photocopy, recording, or any information storage and retrieval
system, without permission in writing from the publisher.


**British Library Cataloguing in Publication Data**

Improving vegetatively propagated crops.—
(Applied botany and crop science).
1. Plant propagation
I. Abbott, A. J.  II. Atkin, R. K.  III. Series
631.5′3      SB119

ISBN 0-12-041410-4


Typeset by Photo·graphics, Honiton, Devon
and printed in Great Britain by
St. Edmundsbury Press, Bury St Edmunds, Suffolk

# Preface

Crop improvement has progressed from the collection, selection and cultivation of wild plants to scientifically-based plant breeding techniques which accelerate the production and release of new and better plant varieties. Crop improvement programmes, especially those for crops which are propagated vegetatively, have also benefited from the use of mutation breeding and tissue culture methods — advances that have further extended our ability to exploit genetic variation. These methods are now augmented by new techniques in genetic engineering which provide opportunities for manipulating the plant genotype in beneficial ways and may eventually provide greater genetic precision and diversity. However, we do not yet know how to apply them to the majority of plant breeding problems.

Although there has been commendable progress in improving many of the world's major crops, constraints remain which cause crops to fall short of the requirements of the farmer, processor or consumer. As the objectives for crop improvement widen, it is timely to consider how both orthodox and novel breeding methods might be used to hasten improvements in vegetatively-propagated crop plants — which, as a group, have longer generation times than most sexually propagated plants.

The book is divided into six parts. The first three review the selection strategies used to improve a range of (tropical) plantation and temperate crops, and forest trees. Traditional approaches and breeding objectives are being re-examined as new varieties are sought with specific improvements in yield, crop regularity or quality. These considerations lead to an appraisal of the options that the newer procedures may offer plant breeders.

Parts IV–VI survey the origins of genetic variation and the factors that affect the frequency of gene expression, methods for enhancing mutation rates and the opportunities this gives for creating new and better genotypes, and the origins and classification of chimeras. Chapters 22 and 23 consider the potential of tissue culture in improving forest trees and in the clonal propagation of plantation crops.

A recurring theme throughout this book is the paradox facing all plant breeders. We need to devise strategies that will allow us to release inherent genetic variability, yet, once improvement has been achieved, it must be captured, stabilised and multiplied in useful uniform variants. Traditional breeding methods must therefore continue to progress alongside modern molecular manipulation. The world still needs plant breeders, and breeders need a broad genetic base from which to select utilisable characters. The necessary genetic resources must be available, and well documented. This book therefore includes an account of various activities pertinent to the conservation of genetic resources, the value of which is widely recognised.

We have tried to bring together in one volume, specialist contributors, each distinguished in their particular field of study, to review the progress made in improving vegetatively-propagated crop plants. Such a volume cannot be fully comprehensive, but we hope that the topics covered may help in the breeding problems encountered in other crops. We intend this book to stimulate discussion and encourage co-operation among plant breeders and their collaborators in associated disciplines.

The passage of any book from manuscript through to the printed volume invariably involves encouragement and support from many persons, who together ease the task of the editors. This book is no exception, we are especially grateful to all the contributors for their patience and forbearance; their ready co-operation was greatly appreciated. We also thank Gina Fullerlove of Academic Press for sustaining our confidence in the scientific value of the book's subject, and our many colleagues who have given us practical support. In particular, Iain Campbell and Colin Lacey (now at the Plant Breeding Institute, Cambridge) for their advice, Ann Belcher for manuscript and proof checking and Dick Chenoweth for redrawing some of the diagrams.

*A J Abbott and R K Atkin*

# List of Contributors

| | |
|---|---|
| A. J. Abbott | University of Bristol, Department of Agricultural Sciences, Long Ashton Research Station, Long Ashton, Bristol BS18 9AF, UK |
| F. H. Alston | AFRC Institute of Horticultural Research (East Malling), Maidstone, Kent ME19 6BJ, UK |
| M. M. Anderson | Scottish Crop Research Institute, Invergowrie, Dundee DD2 5DA, UK |
| R. K. Atkin | University of Bristol, Department of Agricultural Sciences, Long Ashton Research Station, Long Ashton, Bristol BS18 9AF, UK |
| R. D. Barnes | Oxford Forestry Institute, Department of Plant Sciences, University of Oxford, South Parks Road, Oxford OX1 3RB, UK |
| M. G. Beech | AFRC Institute of Horticultural Research (East Malling), Maidstone, Kent ME19 6BJ, UK |
| R. M. Brennan | Scottish Crop Research Institute, Invergowrie, Dundee DD2 5DA, UK |
| C. Broertjes | Research Institute ITAL, P.O. Box 48, 6700 AA Wageningen, The Netherlands (Present address: Eykmanstraat 11, 6706 JT Wageningen, The Netherlands) |
| J. Burley | Oxford Forestry Institute, Department of Plant Sciences, University of Oxford, South Parks Road, Oxford OX1 3RB, UK |
| A. I. Campbell | University of Bristol, Department of Agricultural Sciences, Long Ashton Research Station, Long Ashton, Bristol BS18 9AF, UK |
| R. H. V. Corley | Unilever Plantations Group, P.O. Box 68, Unilever House, Blackfriars, London EC4P 4BQ, UK |
| W. Gottschalk | Institüt für Genetik, Universität der Bonn, Kirschallee 1, D-5300 Bonn, FRG |
| J. J. Hardon | Centre for Genetic Resources, Wageningen, The Netherlands |
| A. M. Van Harten | Department of Plant Breeding (IvP), Agricultural |

| | |
|---|---|
| | University, P.O. Box 386, 6700 AJ Wageningen, The Netherlands |
| J. G. Hawkes | School of Plant Biology, The University of Birmingham, P.O. Box 363, Birmingham B15 2TT, UK (Present address: Department of Geological Sciences, The University of Birmingham, P.O. Box 363, Birmingham B15 2TT, UK) |
| G. G. Henshaw | School of Biological Sciences, The University of Bath, Claverton Down, Bath BA2 7AY, UK |
| D. L. Jennings | Scottish Crop Research Institute, Invergowrie, Dundee DD2 5DA, UK |
| L. H. Jones | Unilever Research, Colworth Laboratory, Colworth House, Sharnbrook, Bedford MK44 1LQ, UK |
| R. P. Jones | AFRC Institute of Horticultural Research (East Malling), Maidstone, Kent ME19 6BJ, UK |
| J. Kleinschmit | Niedersächsische Forstliche Versuchsanstalt, Abteilung Forstpflanzenzüchtung, Staufenberg 6, OT 3513 Escherode, FRG |
| C. N. D. Lacey | AFRC Institute of Plant Science Research, Plant Breeding Institute, Maris Lane, Trumpington, Cambridge CB2 2LQ, UK |
| D. O. Ladipo | West African Hardwood Improvement Project, Forestry Research Institute of Nigeria, P.M.B. 5054, Ibadan, Nigeria |
| F. A. Langton | AFRC Institute of Horticultural Research (Littlehampton), Worthing Road, Littlehampton, West Sussex BN17 6LP, UK |
| R. R. B. Leakey | Institute of Terrestrial Ecology, Bush Estate, Penicuik, Midlothian EH26 OQB, UK |
| C. H. Lee | Harrisons and Crossfield, Oil Palm Research Station, Banting, Selangor, Malaysia |
| W. J. Libby | Department of Forestry and Resource Management, University of California, 145 Mulford Hall, Berkeley, CA 94720, USA |
| G. R. Mackay | Scottish Crop Research Institute (Pentlandfield), Pentlandfield, Roslin, Midlothian EH25 9RF, UK |
| S. A. Merkle | School of Forest Resource, The University of Georgia, Athens, GA 30602, USA |
| D. W. Simpson | AFRC Institute of Horticultural Research (East Malling), Maidstone, Kent ME19 6BJ, UK |

| | |
|---|---|
| H. E. Sommer | School of Forest Resources, The University of Georgia, Athens, GA 30602, USA |
| R. K. Soost | Department of Botany and Plant Sciences, University of California, Riverside, CA 92521, USA (Present address: P.O. Box 589, Inverness, CA 94937) |
| H. Tan | Plant Science Division, The Rubber Research Institute of Malaysia, P.O. Box 10150, Kuala Lumpur 01–02, Malaysia |
| R. A. E. Tilney-Bassett | Department of Genetics, School of Biological Sciences, University College of Swansea, Singleton Park, Swansea SA2 8PP, UK |
| D. I. T. Walker | West Indies Central Sugar Cane Breeding Station, Groves, St. George, Barbados, West Indies |
| P. F. Wareing | Department of Botany and Microbiology, School of Biological Sciences, The University College of Wales, Aberystwyth SY23 3DA, UK |
| H. Y. Wetzstein | Department of Horticulture and Adjunct School of Forest Resources, The University of Georgia, Athens, GA 30602, USA |
| D. Wilson | University of Bristol, Department of Agricultural Sciences, Long Ashton Research Station, Long Ashton, Bristol BS18 9AF, UK |

The need to review the specific problems encountered in vegetatively-propagated plants was conceived by our much respected colleague,

# Don Wilson

to whom we dedicate this book.

# Contents

**Preface** v
**List of contributors** vii

**PART I  IMPROVING PLANTATION CROPS**
**1  Trends in sugarcane breeding**
   D. I. T. Walker
   1.1  Genetic resources and genetic base broadening      3
       1.1.1  Base collections                              3
       1.1.2  Utilisation                                   5
       1.1.3  Commercial breeding programmes                9
       1.1.4  Working collections                           9
       1.1.5  Size of programmes                           11
       1.1.6  Selection of parents                         12
       1.1.7  Crossing systems                             15
       1.1.8  Selection systems                            21
   1.2  Other breeding methods                             24
   1.3  The future                                         24
   Further reading                                         24
   References                                              25

**2  Strategies in rubber tree breeding**
   H. Tan
   2.1  Problems associated with *Hevea* breeding          28
       2.1.1  Narrow genetic base                          29
       2.1.2. Long breeding and selection cycle            33
       2.1.3  Choice of parents                            33
       2.1.4  Selection of multiple characters             33
       2.1.5  Disease resistance                           34
       2.1.6  Genotype–environment interactions            34
       2.1.7  Seasonality, non-synchronisation of flowering   35
              and low fruit-set

|  |  |  |  |
|---|---|---|---|
| | 2.2 | Recent developments | 35 |
| | | 2.2.1 Broadening the genetic base | 36 |
| | | 2.2.2 Shortening the breeding and selection cycle | 39 |
| | | 2.2.3 Combining ability studies as aid in choice of parents | 45 |
| | | 2.2.4 Selection of advanced generation polycross progenies | 46 |
| | | 2.2.5 Breeding for disease resistance | 47 |
| | | 2.2.6 Use of crown budding for correction of secondary defects | 48 |
| | | 2.2.7 Improving planting recommendations | 48 |
| | | 2.2.8 Flower induction and improvement of fruit-set | 49 |
| | 2.3 | Breeding objectives and strategies | 50 |
| | 2.4 | International co-operation | 52 |
| | 2.5 | Conclusions | 53 |
| | References | | 54 |

## 3 Breeding and selecting the oil palm
*J. J. Hardon, R. H. V. Corley and C. H. Lee*

|  |  |  |
|---|---|---|
| 3.1 | Breeding for seed production | 64 |
| | 3.1.1 Selection progress | 64 |
| 3.2 | Selection for vegetative propagation | 68 |
| | 3.2.1 Expected yield improvement | 69 |
| | 3.2.2 Selection objectives | 71 |
| | 3.2.3 Secondary selection objectives | 73 |
| | 3.2.4 Selection populations | 75 |
| 3.3 | Clone trials | 78 |
| References | | 79 |

## 4 Breeding citrus—genetics and nucellar embryony
*R.K. Soost*

|  |  |  |
|---|---|---|
| 4.1 | Polyploidy | 83 |
| 4.2 | Incompatibility | 87 |
| 4.3 | Mutation breeding | 88 |
| 4.4 | Chimeras | 89 |
| 4.5 | Juvenility | 90 |
| 4.6 | Nucellar embryony | 92 |
| 4.7 | Specific characters | 96 |
| | 4.7.1 Cold-hardiness | 96 |
| | 4.7.2 Disease and pest resistance | 97 |
| | 4.7.3 Other characteristics | 98 |

Contents                                                                 xiii

    4.8  Rootstock breeding  98
           4.8.1  *Cold tolerance*  98
           4.8.2  *Salt tolerance*  99
           4.8.3  *Pests and diseases*  99
    4.9  *In vitro* culture  100
    4.10  Other techniques  102
    4.11  Crossability  102
    4.12  New cultivars  103
    4.13  Germplasm  104
    References  104

## PART II  IMPROVING TEMPERATE CROPS

### 5  Strategy for apple and pear breeding
*F. H. Alston*

    5.1  Methods used at East Malling  114
    5.2  Disease and pest resistance  114
           5.2.1  *Apple scab (Venturia inaequalis)*  114
           5.2.2  *Mildew (Podosphaera leucotricha)*  115
           5.2.3  *Rosy apple aphid (Dysaphis plantaginea)*  117
           5.2.4  *Fireblight (Erwinia amylovora)*  117
           5.2.5  *Resistance breeding policy*  117
    5.3  Yield  117
    5.4  Quality  118
    5.5  Breeding priorities  119
    5.6  Establishment of new cultivars  120
    5.7  Conclusions  122
    References  122

### 6  Plum breeding
*R. P. Jones and D. Wilson*

    6.1  Important plum species  125
    6.2  Breeding programmes  126
    6.3  Objectives of the breeding programme  126
           6.3.1  *Regular cropping*  127
           6.3.2  *Dessert-quality fruit*  129
           6.3.3  *Fruit size*  129
    6.4  Limitations on plum improvement  130
    6.5  Prospects  132
    References  133

## 7 Raspberry and blackcurrant breeding
D. L. Jennings, M. M. Anderson and R. M. Brennan

| | | |
|---|---|---|
| 7.1 | Raspberries | 135 |
| | 7.1.1 Yield | 135 |
| | 7.1.2 Season of ripening and autumn fruiting | 137 |
| | 7.1.3 Fruit quality | 137 |
| | 7.1.4 Resistance to diseases and pests | 138 |
| 7.2 | Blackcurrants | 139 |
| | 7.2.1 Yield | 141 |
| | 7.2.2 Juice quality | 142 |
| | 7.2.3 Resistance to diseases and pests | 142 |
| 7.3 | Conclusions | 144 |
| | References | 145 |

## 8 Strawberry breeding in the United Kingdom 149
D. W. Simpson and M. G. Beech

References 158

## 9 Breeding for improved ornamental plants
F. A. Langton

| | | |
|---|---|---|
| 9.1 | The generation of genetic variants | 161 |
| | 9.1.1 Interspecific hybridisation | 161 |
| | 9.1.2 Mutation breeding | 162 |
| | 9.1.3 Induced autopolyploidy | 166 |
| 9.2 | Breeding objectives | 168 |
| | 9.2.1 Seed propagation | 168 |
| | 9.2.2 Year-round production | 169 |
| | 9.2.3 Low-temperature tolerance | 171 |
| | 9.2.4 Pest and disease resistance | 172 |
| | 9.2.5 Longevity | 174 |
| | References | 175 |

## 10 Selecting and breeding for better potato cultivars
G. R. Mackay

| | |
|---|---|
| 10.1 Historical perspective and cytogenetic background | 183 |
| 10.2 General considerations | 184 |
| 10.3 A conventional breeding programme | 186 |
| 10.4 Non-conventional approaches | 190 |
| 10.5 Future strategies | 191 |
| 10.6 True potato seed | 193 |
| References | 193 |

*Contents*

## PART III  IMPROVING FOREST TREES

**11  Genetic resources and variation in forest trees**
W. J. Libby

| | |
|---|---|
| 11.1 Alternative strategies for forest trees | 200 |
|     11.1.1 *The status quo* | 200 |
|     11.1.2 *Seed production areas* | 201 |
|     11.1.3 *Hybrids* | 201 |
|     11.1.4 *Classical tree improvement: seed orchards* | 201 |
|     11.1.5 *The clonal option* | 203 |
| 11.2 Some particular characteristics of forest trees and tree breeding | 203 |
|     11.2.1 *Test extent and duration* | 203 |
|     11.2.2 *Levels of genetic variability* | 204 |
|     11.2.3 *Clonal-option versus seed-orchard selection strategies* | 205 |
|     11.2.4 *Developmental genetic variation* | 206 |
|     11.2.5 *Two theoretical clonal strategies* | 207 |
|     11.2.6 *The unit of management* | 207 |
|     11.2.7 *Genetic conservation* | 208 |
| References | 208 |

**12  Vegetative propagation for improved tropical forest trees**
R. D. Barnes and J. Burley

| | |
|---|---|
| 12.1 Vegetative propagation techniques | 212 |
| 12.2 Cyclophysis, topophysis and periphysis | 214 |
| 12.3 Genetic information | 215 |
| 12.4 Seed production | 216 |
| 12.5 Commercial vegetative propagation | 217 |
| 12.6 Conservation and the genetic base | 222 |
| 12.7 Experiments on environmental effects | 223 |
| 12.8 The future | 223 |
| References | 224 |

**13  Selection for improved tropical hardwoods**
R. R. B. Leakey and D. O. Ladipo

| | |
|---|---|
| 13.1 Clonal field trials of *Triplochiton scleroxylon* | 231 |
| 13.2 Prediction of branching habit | 236 |
|     13.2.1 *Application of the "predictive test"* | 238 |
|     13.2.2 *Possible limitations of the "predictive test"* | 239 |
| 13.3 Selection for wood quality | 240 |
| 13.4 Improvement of other tropical hardwoods | 241 |
| References | 241 |

## PART IV  SOURCES AND EXPLOITATION OF GENETIC VARIATION

### 14  Genetic variation in temperate forest trees
*J. Kleinschmit*

| | |
|---|---|
| 14.1 Levels of variation | 246 |
|     14.1.1 *Variation between populations (provenances)* | 248 |
|     14.1.2 *Variation between individuals within populations* | 249 |
|     14.1.3 *Degree of heterozygosity* | 250 |
| 14.2 Studies in Norway spruce (*Picea abies* Karst.) | 251 |
|     14.2.1 *Genetic variation in Norway spruce* | 252 |
|     14.2.2 *Correlations and grouping* | 252 |
|     14.2.3 *Selection and testing programme* | 257 |
| References | 257 |

### 15  Phase change and vegetative propagation
*P. F. Wareing*

| | |
|---|---|
| 15.1 The nature of phase change | 264 |
| 15.2 Phase change and vegetative propagation | 267 |
| References | 269 |

### 16  The chimeral problem
*R. A. E. Tilney-Bassett*

| | |
|---|---|
| 16.1 Chimera induction | 271 |
|     16.1.1 *Advantageous chimeras* | 273 |
|     16.1.2 *Disadvantageous chimeras* | 276 |
|     16.1.3 *Solid mutants* | 278 |
| References | 282 |

### 17  World strategies for collecting, preserving and using genetic resources
*J. G. Hawkes*

| | |
|---|---|
| 17.1 Genetic diversity | 285 |
|     17.1.1 *Genetic erosion* | 286 |
|     17.1.2 *Genetic resources development* | 286 |
|     17.1.3 *Genetic resources activities* | 287 |
| 17.2 Exploration | 287 |
|     17.2.1 *Seed sampling* | 287 |
|     17.2.2 *Sampling vegetatively propagated plants* | 288 |
| 17.3 Conservation | 290 |
|     17.3.1 *Orthodox seeds* | 290 |
|     17.3.2 *Recalcitrant seeds* | 291 |
|     17.3.3 *Vegetatively propagated crops* | 294 |

|  |  |  |
|---|---|---|
|  | 17.3.4 *In situ* conservation | 294 |
|  | 17.4 Evaluation | 295 |
|  |     17.4.1 *Categories of screening* | 295 |
|  |     17.4.2 *Elimination of duplicates* | 296 |
|  | 17.5 Data management | 297 |
|  | 17.6 Utilisation | 298 |
|  |     17.6.1 *Pre-breeding* | 298 |
|  | 17.7 Training | 298 |
|  | 17.8 International collaboration and funding | 299 |
|  | 17.9 Conclusions | 300 |
|  | References | 300 |
| **18** | **New techniques for germplasm storage** |  |
|  | G. G. Henshaw |  |
|  | 18.1 Germplasm storage *in vitro* | 304 |
|  | 18.2 Restricted-growth storage procedures | 307 |
|  | 18.3 Cryopreservation | 308 |
|  | 18.4 Conclusions | 311 |
|  | References | 312 |

## PART V  MUTATION BREEDING

|  |  |  |
|---|---|---|
| **19** | **The genetic basis of variation** |  |
|  | W. Gottschalk |  |
|  | 19.1 The selection value of mutant genes and its improvement through recombination | 317 |
|  | 19.2 Genotypes homozygous for many mutant genes | 319 |
|  | 19.3 The problem of desired mutations | 325 |
|  | 19.4 Pleiotropic gene action | 326 |
|  | 19.5 Problems of "genecology" | 329 |
|  | References | 332 |
| **20** | **Application of mutation breeding methods** |  |
|  | C. Broertjes and A. M. Van Harten |  |
|  | 20.1 When should mutation breeding be used? | 337 |
|  | 20.2 The genetic constitution of the parent variety | 340 |
|  | 20.3 Chimera formation | 341 |
|  | 20.4 *In vivo* adventitious bud techniques | 343 |
|  | 20.5 *In vitro* adventitious bud techniques | 344 |
|  | 20.6 Other propagation methods useful in mutation breeding | 344 |
|  | 20.7 The application of new techniques | 345 |
|  | References | 347 |

## 21 Selection, stability and propagation of mutant apples
C. N. D. Lacey and A. I. Campbell
21.1 Production and selection of mutant clones 351
21.2 Preliminary tests for periclinal chimerism 353
    21.2.1 *Conventional propagation* 353
    21.2.2 *Propagation after re-irradiation* 355
21.3 Tests for homohistant mutants through root cuttings 357
21.4 Conclusions 360
References 361

## PART VI  IMPROVEMENT THROUGH TISSUE CULTURE

## 22 Application of tissue culture techniques to forest trees
H. E. Sommer, H. Y. Wetzstein and S. A. Merkle
22.1 Criteria 365
22.2 Historical development 366
22.3 Tissue culture of hardwoods 366
22.4 Tissue culture of conifers 368
22.5 Acclimatisation and planting out 371
22.6 Conclusions 379
References 379

## 23 Clonal propagation of plantation crops
L. H. Jones
23.1 Uses of tissue culture 385
23.2 Some commercial considerations 386
23.3 Advantages of clones 387
23.4 Scaling-up the laboratory process 387
23.5 Application of propagation *in vitro* to plantation crops 388
    23.5.1 *Oil palm* (Elaeis guineensis *Jacq.*) 388
    23.5.2 *Coconut* (Cocos nucifera *L.*) 397
    23.5.3 *Date palm* (Phoenix dactylifera *L.*) 398
    23.5.4 *Rubber* (Hevea brasiliensis *Muell. Arg.*) 398
    23.5.5 *Cocoa* (Theobroma cacao *L.*) 399
    23.5.6 *Cassava* (Manihot esculenta *Crantz*) 401
    23.5.7 *Sugarcane* (Saccharum officinarum) 401
23.6 Conclusions 402
References 403

**Index** 407

# Part I Improving Plantation Crops

# 1 Trends in Sugarcane Breeding

D. I. T. WALKER

*West Indies Central Sugar Cane Breeding Station, Barbados*

Sugarcane is one of the major crops in the tropics and subtropics, being of importance in over 60 countries. With its strong industrial base, and at times good profitability, it has supported a great deal of research, of which plant breeding has been an important sector.

Three centuries of dependence on a few clones, distributed worldwide by European traders, was broken when breeding met with success in Barbados and Java in 1887–1910. As with most other plant breeding, it was prompted by serious disease outbreaks in old varieties and by acute economic pressure—in this case competition from sugar beet which had received serious breeding attention since the 1850s. The later (1910–1920) introduction of interspecific hybridisation by Java marked the major breeding breakthrough in productivity, in extending sugarcane into subtropics and in conferring durable resistance to some of the major diseases.

The main thrust of all breeding programmes is directed towards the production of superior sugar-producing clones. These can be called commercial breeding programmes, and concern parents two or more generations away from the wild species. Certain industries also support research on, and a crossing programme with, the wild species directed towards broadening the genetic base of the commercial breeding programme. Procedures and problems are rather different in these two aspects and will therefore be treated separately.

## 1.1 GENETIC RESOURCES AND GENETIC BASE BROADENING

### 1.1.1 Base collections

Two designated World Collections are under the care of the Governments of India and the USA at centres in Kerala and Florida, respectively. They comprise collections of the primitive chewing canes (the so-called noble canes, *S. officinarum*), wild species clones of the genus *Saccharum* and some clones of related grasses. They also contain up to 100 man-made hybrids that have occupied a place in commercial production in one or more industries. These collections have been heavily augmented in the past 25 years. Fears were expressed about the progressive loss of noble canes (they are not known in a truly wild state) in particular, since their culture for chewing was fast disappearing in their centre of diversity, New Guinea, as the rural population converted to a cash economy and were able to buy imported crystal sugar. Accordingly, collections of *S. officinarum, S. robustum, S. spontaneum, Erianthus (Ripidium) arundinaceus, Miscanthus floridulus* and *M. sinensis* were made in Papua New Guinea in 1957 and 1977, and more extensively in the Indonesian islands in 1976 and 1984 (Warner and Grassl, 1958; Berding and Koike, 1980; Krishnamurthi and Koike, 1982; Tew *et al.*, 1986). Over 400 clones of the diverse *S. spontaneum* from the Indian subcontinent were collected in the Spontaneum Expedition Scheme (Panje, 1956), and other seed and clone collections have been made in the Philippines (Medina *et al.*, 1986) and northern Thailand (Srinivasan and Sadakorn 1983). In Taiwan, workers have also made collections of *S. spontaneum* and *Miscanthus* (Lo *et al.*, 1977).

To maintain these collections is in itself quite problematic. In Florida, the discovery of leaf scald disease *(Xanthomonas albilineans)* in the noble collection in the early 1960s necessitated the entire noble and *robustum* collections be quarantined afresh. The collection is now housed on a shallow coralline soil which is not ideal for these species. The Indian collection of nobles, however, is in excellent condition. *Miscanthus* clones have proved difficult to grow at low elevations and propagation of clones without adventitious buds is difficult. Some intergeneric hybrids *(Saccharum* with *Narenga, Sorghum, Sclerostachya* and *Erianthus)*, made with considerable difficulty, are in danger of being lost through lack of current interest. Despite the steady attrition, these World Collections represent a formidable range of wild materials at the disposal of sugarcane breeders (Table 1.1).

Descriptive work in the collections is incomplete, but sections of them have been screened by breeders for a range of economic characters (reviewed by Roach, 1986). Chromosome numbers have been counted in many, and there has been some taxonomic work using isozyme patterns (Waldron and Glasziou, 1971) and leaf flavonoid patterns (Paton *et al.*, 1977). Some numeric descriptor lists have been formulated (Daniels, 1971).

Trends in Sugarcane Breeding

**Table 1.1.** World collections (no. of clones) held in India and USA, 1980.

| Group | $2n^a$ | India | USA |
|---|---|---|---|
| S. officinarum (nobles) | 80 | 628 | 551 |
| S. spontaneum | 40, 64, 80, 112 | 450 | 261 |
| S. robustum (inc. S. sanguineum) | 80 (60) | 37 | 64 |
| S. sinense (inc. S. barberi) | 82, 92, 116, 124 | 48 | 90 |
| S. edule | 80 | — | 18 |
| Related grasses |  | 23 | 130 |
| Man-made hybrids |  | 1599 [b] | 156 |

Source: Proceedings of the 17th Congress International Society of Sugar Cane Technologists.
[a] Nodal values; aneuploids close to these numbers also exist.
[b] Includes Indian-bred varieties. Neither collection yet contained the clones collected in Indonesia 1976 and 1977, which numbered over 250.

#### 1.1.2 Utilisation

An outline of present-day exploitation of the genetic base is given in Fig. 1.1.

1.1.2.1 Aims

The need for a broader genetic base has been appreciated particularly clearly since Arceneaux (1965) pointed out that the pedigrees of nearly all the world's commercial cane cultivars rest on only two forms of *S. spontaneum* and no more than six noble canes.

Subsections of the World Collection are being used most actively in Louisiana, the West Indies and Australia. In several other breeding programmes there is some more casual use of a few *S. spontaneum* parents, while India is interested in some intergeneric hybrids. The use of *S. robustum*, *Erianthus* and some other genera has hardly begun.

The conventional wisdom in early interspecific programmes was that sugar storage capacity would come from the noble contribution, and vegetative yield, particularly its components stalk number, internode length, good ratooning capacity and disease resistance, would come from the *spontaneum* contribution. More recent work has made it clear that there is scope for the selection of higher sugar storage capacity among *spontaneum* forms (Brown et al., 1969), while the very wide diversity of form and ecological adaptation found in that species indicates equally wide yield potential. However, since today's commercial hybrids have

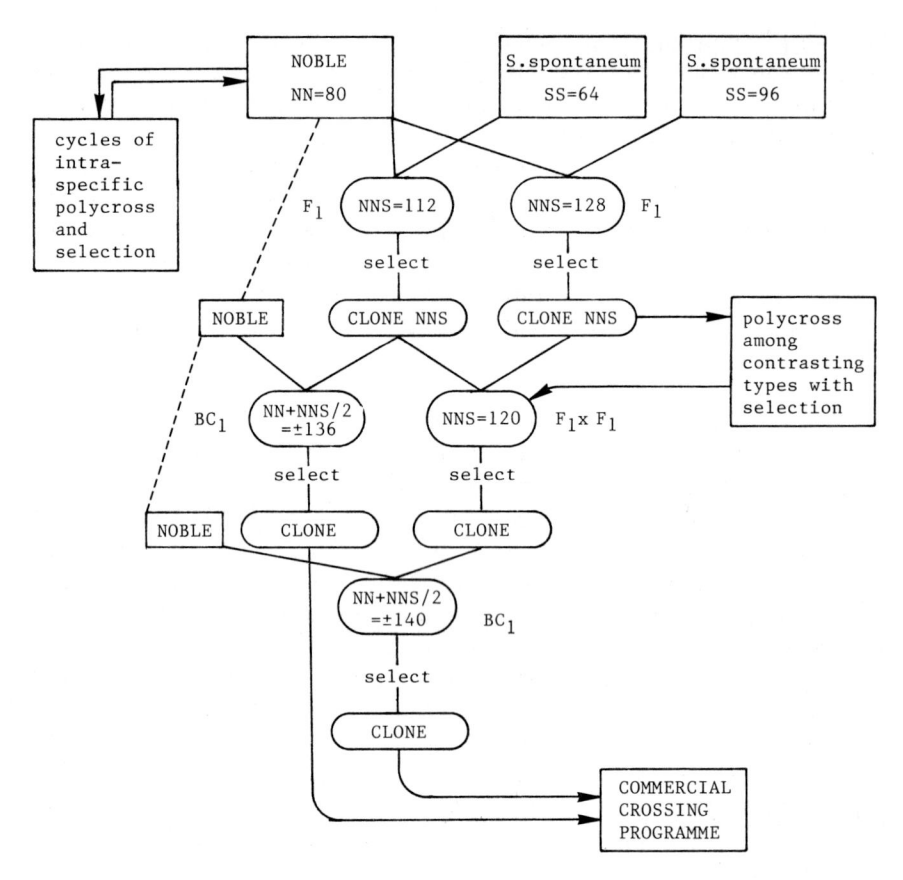

**Figure 1.1.** Outline of the use of nobles and *S. spontaneum* from the genetic base up to introduction into a commercial programme, showing the predominant chromosome inheritance pattern and including hybridisation between contrasting lines through their $F_1$ generations.

85–95% noble chromosomes in their constitution, the noble contribution must also be important.

Three active programmes illustrate these points.

(a) *Louisiana* formulated perhaps the clearest objectives in their genetic base broadening programme. The subtropical climate is marginal for sugarcane, and virulent strains of mosaic disease had not been brought under control by resistance from within the existing commercial germplasm. By screening most of the *S. spontaneum* clones in the World Collection, Dunckelman and Breaux (1969) identified some 70, of diverse origins, that were highly resistant, whereas the two clones common to the old genetic base were very susceptible. Some 30 of these were later subjected to natural and artificial cold tolerance tests (Dunckelman and Breaux, 1971; Irvine, 1977) and five proved outstanding. Significantly, four were from northerly sites in India (>24°N) and one was from Afghanistan. Several forms that combined these traits with better stalk size and stalk solidity were crossed with major Louisiana high-sugar parents, and in this way significant improvements have been directly introduced into the commercial breeding programme. There has been no new noble input, partly because nobles did not contain the variation required and partly because they did not flower.

(b) *The West Indies* had the broader objective of exploiting both nobles and *S. spontaneum*, seeking better adaptations to various environments and more reliable ratooning capacity (Walker, 1971). To date, three cycles of polycross have been achieved from a base of about 90 noble parents. A sample of about 50 *spontaneums* spanning much of the eco-climatic range of the Spontaneum Expedition Scheme collection, plus Indonesian and other forms, have each been crossed with several nobles and selections backcrossed again. Such $BC_1$s now enter up to 30% of the biparental crosses in the commercial programme. The findings suggest that (i) new nobles superior in growth, but not in sugar storage, to the old, can easily be selected (ii) the phenotype of the *spontaneum* and its latitude of origin do not play a large part in determining the performance of $F_1$ hybrids in West Indian conditions; (iii) hybrids between unrelated $F_1$s (i.e. combining two different *spontaneums*) can show heterosis for sugar content and vegetative growth.

(c) *Australia*. With crosses between two nobles and four contrasting *spontaneum* forms, and with selections backcrossed twice to each of the nobles, Roach (1977) studied the variation in quantitative characters. He concluded that (i) it was important to use unrelated nobles in successive generations and (ii) the $BC_2$ generation lacked vigour in all cases, so it would be preferable to introduce $BC_1$ selections into the commercial crossing programme. Bhat and Gill (1985) have suggested that the rather rapid loss of hybrid vigour in interspecific hybrids is due to the inbreeding effect of the diploid contribution from the *officinarum* gametes characteristic of this species when used as a female parent in interspecific crosses and backcrosses. Roach is now using further *spontaneum* forms in this manner, seeking general adaptations to Australian conditions, though only few nobles can be used because of shy flowering.

#### 1.1.2.2 Future prospects

According to the expedition leaders (pers. comm.) many forms of *S. robustum* recently collected in Indonesia and New Guinea should have considerable potential. The weakness of this species would appear to be disease susceptibility and, for conventional sugar extraction, high fibre content. Some (including the red-fleshed $2n = 60$ form) that grow almost in the water on riversides are not tolerant of drought in collections, yet the few hybrids so far made are promising. If sugarcane should be grown for fibre, the high-fibre *S. robustum*, *S. spontaneum* and *Erianthus arundinaceus* will certainly become important. Some preliminary work on selection for total biomass is already using some of these or their hybrids with sugarcane.

Roach (1986) has pointed out that large-scale introduction of primitive germplasm into existing programmes could negate some of the advances made and it is fair to point out that the present genetic base has spawned a great diversity of specialised sugarcane cultivars to suit varied conditions. There are breeders who believe that variability is by no means exhausted and that further recurrent recombination will be productive. The base was not as narrow as, say, oil palm, rubber or cocoa, and contained considerable versatility.

#### 1.1.2.3 Constraints

(a) *Flowering*, or rather the lack of it, has already been mentioned as a major constraint to the breeding of noble canes. Nobles have the most exacting requirements of photoperiod regime and temperature, which are apparently met regularly only in an equable island climate between latitudes 12° and 18°. Great advances in the understanding of flowering, and hence of its manipulation, have been made in the past 25 years. In genetic base broadening a major problem used to be the non-synchrony of many longer-day *spontaneums* with the nobles and commercial hybrids. Now, by understanding the natural photoperiod requirements from floral induction, through initiation and development to emergence, flowering time can be adjusted by extending (by supplementary lighting) or shortening (in dark rooms) the daylength for each clone to imitate natural conditions over a suitably displaced period. The techniques are discussed in a recent review by Moore (1985), while Alexander (1973) and Clements (1980) have useful chapters on flowering.

(b) *Sexual fertility* varies widely. One group of nobles have mostly imperfect florets and therefore produce little pollen or seed. Most of the remainder have fertile pollen but it is becoming clear that seed fertility is consistently poor in some, suggesting a genetic component. *S. barberi* and *S. sinense* are characteristically male sterile and have poor seed set, but most of the other species forms, including *Erianthus*, are very fertile, so long as temperatures are above 21°C. In the subtropics, heated facilities are now standard in order to ensure pollen fertility. Male sterility is much more frequent in backcross generations, probably due to the accumulation of cytoplasmic factors.

(c) *Selection criteria* in noble × *spontaneum* $F_1$ families are of necessity rather different from those applied in a commercial selection programme. Experimental design should allow for appropriate replicated standards, but it must be borne in mind that there is at least one further generation to go before commercial performance in terms of sugar per hectare is to be expected. High yields of dry matter, reliable ratooning, and relatively high sugars/lower fibre are the main criteria. Testing and selection under a poor environment (drought) has been useful in detecting the strongest ratooners. In the West Indies the $F_1$ generation has shown very high smut disease infection (Walker *et al.*, 1977) compared to commercial programme progenies. We do not know why this is, and we know that the heritability of smut, though positive, is not high. A fairly lenient attitude towards this character is therefore reasonable.

### 1.1.3  Commercial breeding programmes

Figure 1.2 gives an outline of a typical present-day commercial crossing and selection programme.

### 1.1.4  Working collections

Most of the larger sugar industries run a breeding station. Typically, each holds a field collection of locally bred and selected clones, plus a sample of cultivars with desirable traits from other industries. International exchange and co-operation at this level is very good indeed, almost the only constraint being adequate inter-nation quarantine facilities. While certain of the major commercially grown cultivars of the past may be maintained in museum plots, for the remainder the tendency in breeding programmes worldwide is to adopt "generationwise assortative mating" as a breeding policy; hence most of the parents of today will be replaced in the collection by the better of their progeny within 15 years. This should be true particularly of a new breeding programme, which is forced

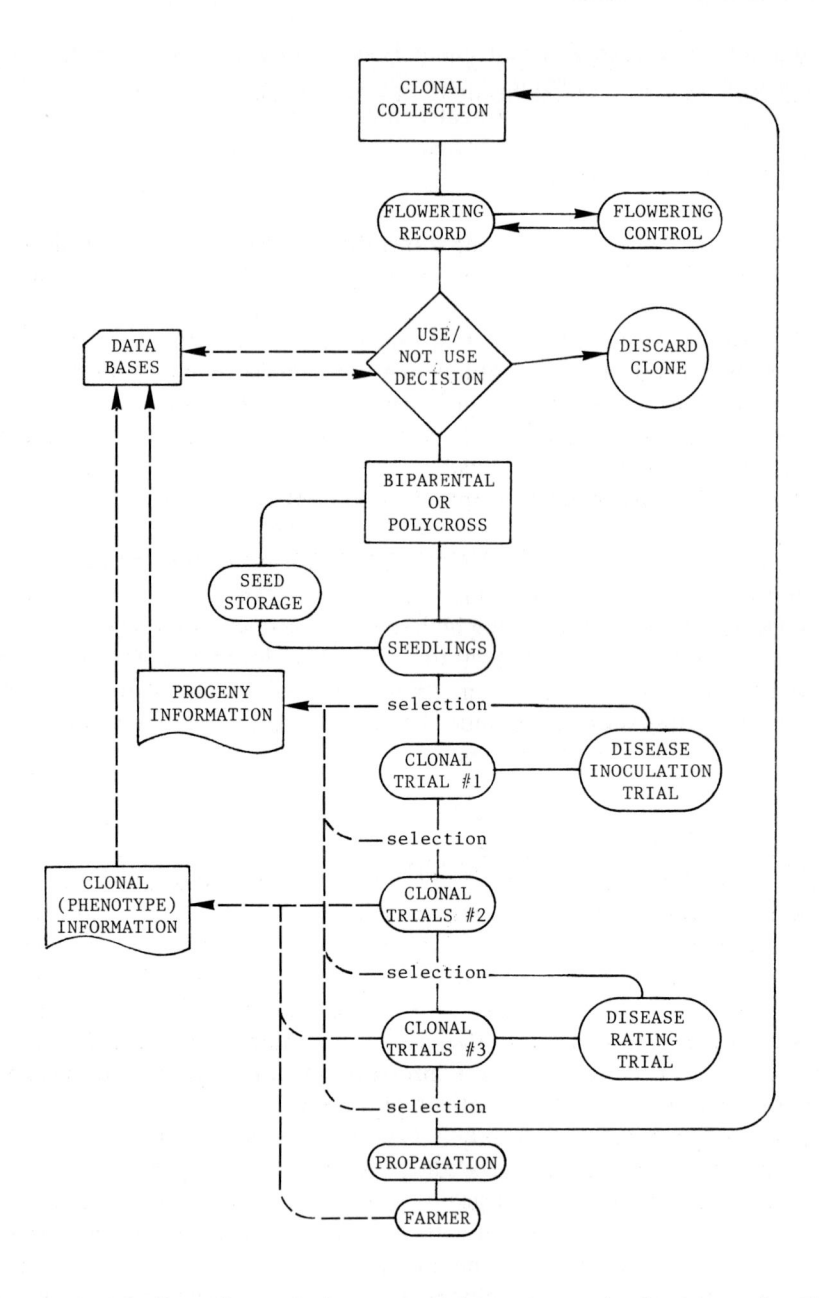

**Figure 1.2.** Outline of a typical commercial crossing and selection cycle. The duration varies from 8 to 14 years, depending on the number of ratoon trials and duration of disease-rating tests.

to start with imported parents but will want to adopt locally selected parents as soon as enough trials data on them have accumulated.

**1.1.5 Size of programmes**

Given a good location, quite simple facilities can be used for hybridisation, and production of viable seed is generally not a limitation to sugarcane breeding.

The tendency is towards larger national programmes, since the likelihood of improving upon existing standards is very much a matter of numbers. However, some industries recognise the limitations of their resources and prefer to keep to a modest scale but endeavour to apply high standards of accuracy in selection and trials.

Overall size has several components, as follows.

*Number of families.* This is mainly a function of the facilities (manpower, crossing facilities) available during the typically short flowering season. In subtropical stations requiring heated glasshouse space, available flowers may be the more severe constraint; in stations such as Barbados and Hawaii with heavy flowering, manpower is the constraint. Clearly, it is desirable to maximise the number of crosses made each year in order to screen as diverse set of families as possible.

*Number of seedlings per family.* This is a subject of some controversy. There is no definite figure for an "optimum" or "sufficient" number of seedlings worth growing. What has emerged empirically in the West Indies programme is that commercial cultivars, with one exception, were selected in the first thousand seedlings of a given family; subsequent thousands gave no further superior cultivars (Walker, 1962). The reasons probably lie in the rising—or changing—standards against which families are tested in successive years, and in the case of certain negatively correlated characters, limitations to the expression of the extremes in progeny performance at acceptable levels of probability.

The compromise now widely followed is to share one year's space among small sample lots (60–200 seedlings) from numerous new or experimental crosses, and larger samples (say 1000) from families that had given higher-than-average selection rates in previous sample evaluations, up to some arbitrary maximum. Short-term seed storage is now an important tool for this approach. It has the added advantage that families can be observed in two or more years, i.e. two or more environments.

Table 1.2 shows the order of magnitude of some major breeding programmes.

### 1.1.6 Selection of parents

1.1.6.1 Heritabilities

Offspring/mid-parent regressions are not high in sugarcane, particularly for vegetative yield and its components. Only for sugar content, flowering and certain disease resistances might they be as high as 0.5.

The variation within a family is often nearly as large as the variation among family means. Hence, close specification of parents on their phenotypic values is not justified. However, experience in the West Indies programme indicates that all the commercial or very-close-to-commercial cultivars of the past 25 years came from families with one if not both parents themselves of commercial performance (Fig. 1.3). While not all commercial cultivars have been good parents, many have had high general combining ability.

1.1.6.2 Data bases

The clonal performance information that accumulates from the series of trials is the first criterion in choosing parent candidates. In most of the larger breeding programmes a fairly liberal selection of new cultivars is evaluated in at least one or few experimental crosses. These will be varieties that came close to the standard in sugar per hectare, perhaps with superiority in one component or with outstanding performance in other desirable features, e.g. erectness and disease resistance, depending on local requirements.

The progeny performance data begins to accrue from the first family raised from a cross. In its simplest form, a progeny record is the percentage selected for advancement to the next stage which reflects the mean "worth" of the family. Among any one set of families grown together in one selection cohort, the relative selection rates (RSR) are in fact used as the measure of a good or poor family mean relative to the overall mean. Occasional families with rather low RSR values do yield good cultivars, and in such cases a bonus point system allows an upward adjustment to the RSR.

As the number of families from a given parent increases, the mean RSR values become more and more a measure of general combining ability (GCA). Part of the performance of individual crosses will also be due to specific combining ability (SCA): in Hawaii, for example, Hogarth

**Table 1.2.** Information on some sugarcane breeding programmes.

| Country | Call sign | Annual breeding effort — Crosses, incl. polycrosses | Annual breeding effort — Seedlings (million) | Sub-stations for stage 1 | Sugar production served (million tonnes) |
|---|---|---|---|---|---|
| *Tropical* | | | | | |
| Caribbean islands including Cuba | B, BJ, BR, BT, C, CR, D, DB, PR, etc. | 1000 | 0.6 | 8 | 8.0 |
| Fiji | LF | 600 | 0.025 | 1 | 0.3 |
| Hawaii | H | 4000 | 1.0 | 4 | 1.0 |
| Mexico | Mex | 700 | 0.2 | 8 | 3.0 |
| Philippines | Phil | 1000 | 0.1 | 2 | 2.6 |
| *Mixed* | | | | | |
| Australia | Q (and CSR) | 700 | 0.075 | 5 | 3.0 |
| Brazil | RB, SP and IAC | 1200 | 4.1 | 9 | 8.0 |
| India | Co | 110 | 0.25 | 8 | 6.0 |
| Mauritius | M | 480 | 0.06 | 2 | 0.7 |
| Reunion | R | 150 | 0.08 | 1 | 0.25 |
| *Subtropical* | | | | | |
| Argentina | NA and TUC | 150 | 0.1 | 2 | 1.4 |
| South Africa | N | 600 | 0.2 | 6 | 2.2 |
| Taiwan | F and ROC | 400 | 0.5 | 9 | 0.8 |
| US mainland | CP, CI and L | 350 | 0.2 | 4 | 1.5 |

Based on scattered literature but relating to the 1970s.

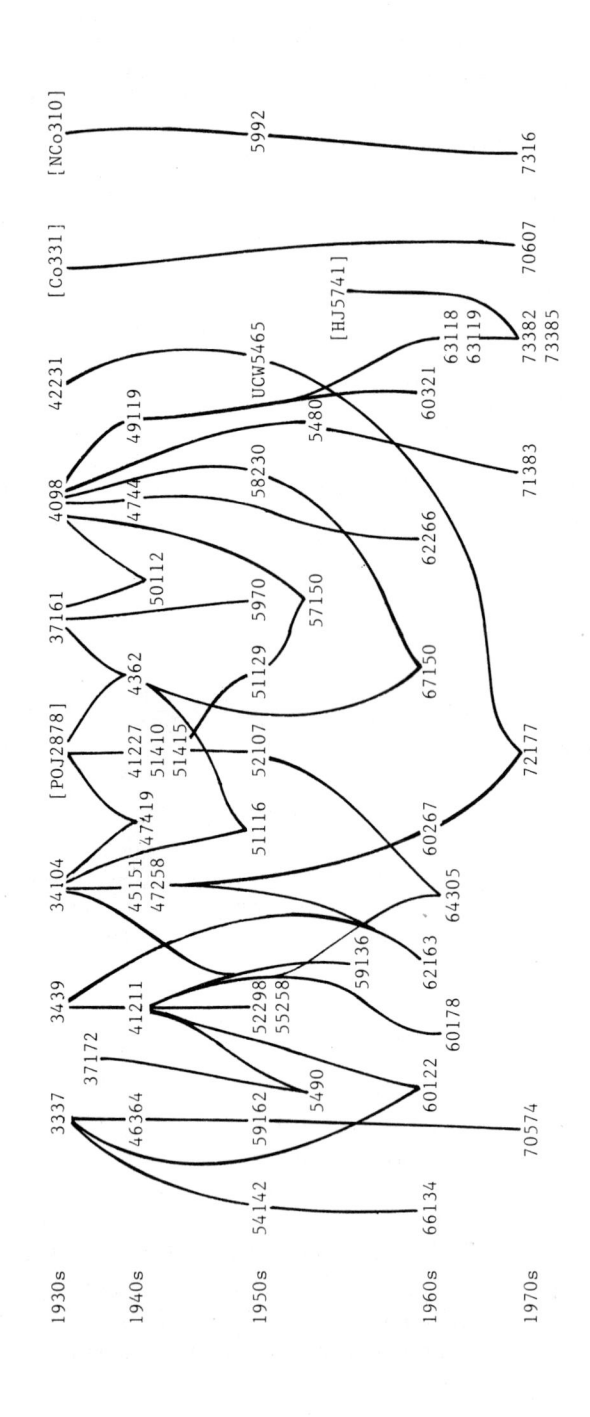

**Figure 1.3.** Pedigree of commercial, or nearly commercial, Barbados cultivars. This diagram contains Barbados cultivars grown in one or more member countries, not only in Barbados, and includes cultivars that were subsequently discarded because of susceptibility to smut or rust, or to mosaic and leaf scald. Where only one parent is given, the other parent was a minor cultivar or unknown polycross pollinator.

*et al.* (1981) found SCA and GCA to be about equally important in vegetative yield component characters. But in sugarcane, as in other clonally propagated heterozygous crops, this information cannot be used in planning the breeding programme, although biparental crossing space may be reserved for crosses between pairs of parents with high GCA.

Parents with high combining ability over many families can be called proven parents. They will remain in this category until superseded by newer cultivars (often their progeny) when their RSR scores begin to fall, or when their total number of progeny pass an arbitrary limit, at which stage the breeder may feel they are redundant.

While it is logical to use accumulated data to guide future breeding, the quality of the data is sometimes questionable. Selection rates, if low and if based on small families, will carry large standard errors. Hence the limitations of any data must be understood and their use adjusted accordingly. Objective use will reduce the element of intuition in designing a breeding programme but will not replace it entirely.

**1.1.7 Crossing systems**

1.1.7.1 Biparental crosses

Biparental crosses made in isolation between male-sterile and male-fertile parents are more costly, but have the advantage that the parentage is completely known. Increasingly, the biparental facilities are being reserved for base-broadening crosses (exploiting a specific *S. spontaneum* parent, for example) and for crosses between proven parents each with high GCA. Polycrosses are cheaper; selection rates have proved broadly similar from polycrosses and biparental crosses in spite of selfs in the former; and polycrosses allow a broader spectrum of cultivars to be used. Most stations use both systems.

1.1.7.2 Polycrosses for experimental evaluation

Parents being used for the first time, i.e. with no selection record, can be screened in suitable polycrosses. One of the best plans is a series of small factorial crosses, whereby a group of new parents (usually 4–6), is placed with three or four parents with known GCA and with strong pollen shedding (Hogarth *et al.*, 1981). They are thus evaluated against a common set of high-GCA parents. A looser system places all new candidate parents together in one polycross, but suffers from the defects that pollen parents are of unknown breeding value and that selfs may be numerous; jointly these could cause an underestimation of breeding values.

### 1.1.7.3 Crossing programmes for particular environments

Australia, Hawaii and India arrange sections of their crossing programmes for different zones of their industries. In Hawaii (Warner, 1953), all cultivars currently performing well in trials in the high rainfall zone, for instance, are placed in a polycross for that zone, and the seedlings assigned to the substation in that zone. In India, the main regions are northern (subtropical) and southern, but there is further subdivision into states within these regions. In Australia the decision making, by a computer program, takes account of parentage (to avoid inbreeding), clonal performance record, family performance record (particularly of proven parents) zone by zone, and disease ratings relevant to the zone, for the four main regions of the industry (Hogarth and Skinner, 1986). In the West Indies our approach to a decision-making programme is similar but we have not yet taken the step to separate programmes by, say, Northern Caribbean and Southern Caribbean.

Table 1.3 gives some typical values for selection rate data and illustrates the use of GCAs to arrive at a recommended cross as being done in Barbados.

### 1.1.7.4 Repeated crosses

If we accept an arbitrary limit to seedling number in a family, over which probability of finding further superior cultivars decreases, and with seed storage now a reality (10-year storage of properly dried and packed seed just below 0°C is easily secured) the need to expend effort on repeating crosses is much reduced.

### 1.1.7.5 Special place of entries from genetic base broadening

Pedigree information other than to guard against close inbreeding is not important in a commercial crossing programme. However, new $BC_1$ entries are expected, on average, to be lower in sugar content and perhaps less desirable in other characteristics than current hybrid parents. If special allowance is not made for this, their intended role, to broaden the genetic base, could be lost. In Barbados an empirical "advancement index" has been awarded to every variety. Its derivation is given in Table 1.4. The GCA threshold (Table 1.3) can be adjusted by such an index in the interests of maintaining a proportion of new genetic variation in the programme but discarding genuinely poor parents that have a higher advancement index.

1.1.7.6  Technical improvements in crossing

*Subtropical problems.* These have already been mentioned in the earlier section. Temperature and photoperiod controls have reached a high degree of sophistication in South Africa, Louisiana and Taiwan, all of them large programmes. A useful summary of these special problems is given by Dunckelman and Legendre (1982).

*Flowering propensity.* Shy flowering of large segments of their collections still plagues many breeding stations, owing to their poor location; they are not so acute as to prevent crossing, but clearly they must bias the crossing programme. In the tropics proper there are also a few shy flowering cultivars. They have been found more likely to provide sufficient flowers for crossing when planted in a larger plot on the windward edge of a high elevation field (Anon., 1970, 1973).

*Flowering time.* As with species crosses, a good understanding of flowering has enabled breeders to alter flowering times. In Barbados most commercial hybrids each have less than 10 days span of flowering out of the seven-week season. A particularly simple photoperiod adjustment has been devised for our latitude (13°N) to delay flowering by 3–4 weeks. It depends on the fact that most of the shortening of daylength occurs in the evening—about 4 min per week during the August–October period. Approximately 12 min of extra light given each morning before sunrise from early August by 100 W lights suspended over the field plots, achieves a three-week delay, and has enabled early and late proven parents to be crossed for the first time.

*Male sterility.* This occurs naturally in about 40% of hybrid parents under high temperature conditions. Not all this male sterility is absolute, but it is now apparent that when 10% or less of the pollen is filled and stainable, anthesis is infrequent and for practical purposes the probability of selfing is small (P. S. Rao, unpubl.). Under cooler conditions, varying with the year, male sterility becomes much higher and needs to be checked at the time crosses are being made. Such male sterility is both a useful tool and a limitation to making desired crosses. Recent experiments on inducing more male sterility could be important to the tropical stations. Divinagracia (1980) reported the use of a brief steam treatment of the inflorescence to kill pollen, while MacColl (1977) reported a photoperiod treatment whereby, on exposure to short days during the lag phase of flower development, the phase was shortened and pollen failed to form. This

**Table 1.3.** Barbados data base usage.
(a) Calculation of relative selection rates (RSR) and combining ability ratings (GCA).

|  | Numbers in stages |  |  |  |  |  | RSR1 | RSR2 |
|---|---|---|---|---|---|---|---|---|
|  | 1 | 2 | % | 3 | % |  |  |  |
| Whole programme, one year | $10^4$ | $10^3$ | 10 | $10^2$ | 1.0 |  |  |  |
| Family A × B, that year | 500 | 100 | 20 | 5 | 1.0 |  | 2.0 | 1.0 |
| A × C | 1000 | 100 | 10 | 20 | 2.0 |  | 1.0 | 2.0 |
| A × D | 400 | 52 | 13 | 24 | 6.0 |  | 1.3 | 6.0 |
| E × B | 1000 | 40 | 4 | 2 | 0.2 |  | 0.4 | 0.2 |
|  |  |  |  |  |  |  | GCA1 | GCA2 |
| For parent A, that year | 1900 |  |  |  |  |  | 1.44 | 3.00 |
| For parent B, that year | 1500 |  |  |  |  |  | 1.20 | 0.60 |

*Note:* GCA values are simple averages of individual RSR values for all crosses made with a parent over several years, except that:
(a) families with fewer than 40 seedlings will not contribute information to the GCA values;
(b) a constant will be added if any selected cultivar was competitive with the standard cultivar at the final stages of trial.

**Table 1.3.** (Cont'd.)
(b) Examples of the use of GCA values in deciding on the use of a parent.

| Parent[a] | Advancement index | Number of Families | Number of Seedlings | Observed GCA1 | Observed GCA2 | Threshold[b] GCA1 | Threshold[b] GCA2 |
|---|---|---|---|---|---|---|---|
| I | 3.0 | 1 | 200 | 1.1 | 1.0 | 0.8 | 0.8 |
| J | 3.0 | 5 | 1 200 | 1.1 | 1.0 | 0.8 | 1.02 |
| K | 0.5 | 5 | 4 000 | 1.1 | 1.0 | 0.7 | 0.85 |
| L | 3.0 | 20 | 4 000 | 1.1 | 2.0 | 1.0 | 1.2 |
| M | 3.0 | 20 | 4 000 | 1.1 | 1.0 | 1.0 | 1.2 |
| N | 3.0 | 100 | 16 000 | 1.1 | 1.3 | 1.17 | 1.4 |

[a] Decision I: only one family—candidate for further crosses
Decision J: just fails to achieve lower threshold in spite of fewer seedlings—suspend
Decision K: advancement index low, allowing low GCA—further crosses
Decision L: passes GCA2 test—expand with further crosses
Decision M: fails GCA2 test—discontinue use
Decision N: fails higher threshold, raised by large number of seedlings already examined—suspend or discontinue

[b] Thresholds are calculated as

$$T_1 = B_1 \log_{10}(\text{seedlings}) \quad \text{and} \quad T_2 = B_2 \log_{10}(\text{seedlings})$$

where $B_1 = K_1/\log_{10}(4000)$ and $B_2 = K_2/\log_{10}(4000)$. To allow for low advancement indices (ADV),

$$K_1 = 1.0, \quad K_2 = 1.2 \quad \text{when ADV} > 1.0$$

$$K_1 = 0.8 + 0.2\text{ADV}, \quad K_2 = 1.0 + 0.2\text{ADV} \quad \text{when AVD} < 1.0$$

To allow for paucity of information, when the number of families < 5, $K_1 = K_2 = 0.8$.

**Table 1.4.** Values of an empirical advancement index representing generations in sugarcane breeding.

| Species or generation | | Assigned or calculated value |
|---|---|---|
| *S. spontaneum* | | −7.00 |
| *S. officinarum* (noble) | | −1.00 |
| *S. robustum* | | −1.00 |
| Noble × *S. spontaneum* | Unselected F$_1$ family | −4.00 = $\frac{1}{2}$(−7 − 1) |
| | Selected F$_1$ clone | −3.00 = $\frac{1}{2}$(−7 − 1) + 1 |
| Noble × F$_1$ | Unselected BC$_1$ family | −2.00 = $\frac{1}{2}$(−1 − 3) |
| | Selected BC$_2$ clone | −1.00 = $\frac{1}{2}$(−1 − 3) + 1 |
| Noble × BC$_1$ | Selected BC$_2$ clone | 0.00 = $\frac{1}{2}$(−1 − 1) + 1 |
| Commercial examples | Selected BC$_2$ × BC$_1$ clone | +0.50 = $\frac{1}{2}$(−1 + 0) + 1 |
| | Advanced commercial family | +2.85 = $\frac{1}{2}$(+2.5 + 3.2) |
| | A selection from this family | +3.85 = $\frac{1}{2}$(+2.5 + 3.2) + 1 |
| | BC$_1$ × advanced commercial clone | +1.43 = $\frac{1}{2}$(−1 + 3.85) |

*Note*: An arbitrary value of 1.00 is added to a family mean for the operation of selecting a clone.

*Trends in Sugarcane Breeding* 21

technique has been used to make successful crosses in Barbados (Midmore, 1980).

*Fertility and possible incompatability* have not been studied sufficiently. The rather low level of success in biparental crosses against polycross (60% vs. 95%) indicates that such problems exist.

**1.1.8 Selection systems**

1.1.8.1 Problems

Sugarcane selection from the seedling stage through a series of trials to the release of a cultivar takes some 10–12 years in all industries. The first problem is one of cost. Experiments are increasingly expensive to record, harvest and analyse, and partial alternatives and solutions are being developed—including mechanised harvesting of plots, and an objective grading substituting for weighing, both in Australia (Skinner, 1965).

The second problem is the question of accuracy. The clonal repeatabilities from stage to stage, over sites or over years, are not high for several important characters. This implies the need for more care and more replication, not less.

1.1.8.2 Decentralisation and environment interaction

Many programmes (see Table 1.2) now decentralise their testing programmes from stage 1, the seedling stage. It seems a sound concept to capture cultivars adapted to each location which might be lost if all are grown at one location, but in some industries it has been shown that year differences make larger relative difference to cultivar performance than sites, apart from perhaps extreme sites such as those with high salinity. The allocation of resources over years rather than sites is a question that deserves investigation in each industry, especially in the later stages of trial which are the more expensive. Sometimes it has been demonstrated that there is a sufficiently good relationship across environments that not every one be included in the testing programme.

1.1.8.3 Selection schedules and selection rates

The unreplicated single plant of each genotype (stage 1) is, as in all plant breeding, very susceptible to error of judgement. Simmonds (1985) has discussed the theory of two-stage selection and has shown that severe selection in stage 1 can actually increase the probability of discarding the best cultivars in the population. Selection has therefore to be lenient.

When clonal repeatabilities are low, a case can be made for deliberately optimising the conditions under which the seedling (unreplicated) and small plot (few replicates) stages are grown and selected (George, 1962). This argument has been adopted in Fiji where the first three stages are intensive care trials, each plant being limited in tillering to reduce competition, grown under a high-moisture, high-fertiliser management regime, with lenient selection being based on single-stalk traits, particularly fibre and sugar contents. All information on tillering, a major component of yield, is foregone in these stages since it is argued that the low repeatability does not justify its use as a criterion in small plots (Stevenson and Daniels, 1971).

At almost the opposite extreme is the Hawaiian selection programme. Again they recognise that repeatabilities are not high. For practical reasons, selection in the first four stages has to be made in one-year-old cane, though the industry grows cane for 2 or 3 years between harvests. The ability to grow for two seasons is vital but no attempt is made to select for this trait in these early stages. Furthermore, genetic variance for sugar is lower than that for vegetative yield. Seedlings are planted in bunches of five which introduces a strong element of intervarietal competition for early growth (Skinner, 1961). These shortcomings are compensated for by growing some 700 000 seedlings per year, and the Hawaiian programme has had a good rate of success in finding improved cultivars.

Table 1.5 is an approximate summary of the two schemes discussed above. There are many others. What can be said is that most selection personnel are aware of the problems associated with low repeatabilities. Perhaps the most objective study, based on known repeatability values (from experiments) applied to specific characters at specific stages, was made by James and Miller (1975). From this they could calculate optimum selection intensities at each stage, for each character independently and for simultaneous selection over several characters, in terms of cost and land use.

#### 1.1.8.4 The place of disease testing

Varietal change has been very successful in combatting most disease outbreaks, even with the present genetic base. Resistance is paramount for those diseases that can generate an epidemic of economic importance, since other means of control are not practical in sugarcane. For the most serious of these, Fiji disease (virus), mosaic (virus), red rot (fungus), downy mildew (fungus), rust (fungus) and smut (fungus), specific inoculation techniques are steadily becoming standardised. Much of the literature on disease reactions has in the past been confused by the use of various

*Trends in Sugarcane Breeding*

**Table 1.5.** Selection schemes in Hawaii (H) and Fiji (F), and estimated proportions (%) rejected for each character in the first three stages and major criteria[*] in the fourth.

| | Stage: | 1 | | 2 | | 3 | | 4 | |
|---|---|---|---|---|---|---|---|---|---|
| | | H | F | H | F | H | F | H | F |
| Character | Name of stage:[a] | FT1 | ICT1 | FT2 | ICT2 | FT3 | ICT3 | FT4 | MSP |
| Growth (height) | | 50 | 5 | 10 | 3 | 15 | | | |
| Stalk size (thickness) | | 25 | – | 5 | | 3 | | | |
| Stalk number | | – | – | 20 | | 25 | | | |
| Brix | | – | 50 | – | | – | | * | |
| Brix × weight | | – | – | – | 65 | – | 65 | | * |
| Hardness (fibre) | | 1 | 15 | 1 | – | 2 | | | |
| Piping in stalk | | 5 | – | 2 | – | 1 | | | |
| Pithiness of stalk | | 1 | – | 1 | – | 1 | | | |
| Splits in stalk | | 2 | – | 2 | – | 1 | | | |
| Protuberant buds | | 1 | – | 2 | – | 2 | | | |
| Brittleness of stalk | | – | – | 2 | – | 2 | | | |
| Anchorage of stump | | – | – | 2 | – | 5 | | * | |
| Short upper internodes | | – | – | 1 | – | 1 | | | |
| Flowering | | 1 | – | 2 | – | 2 | | | |
| Germination rate | | – | – | – | – | – | 1 | * | * |
| Smut disease | | 2 | – | 35 | – | 10 | | | |
| Other diseases | | 2 | – | 3 | – | 5 | | | * |
| Total rejections (%) | | 90 | 70 | 88 | 70 | 75 | 70 | | |

*Sources*: T, L, Tew and S. Prasad (pers. comm.)
[a] ICT, Intensive Care Trial; MSP, Mass Stool Population trial.

techniques, various criteria of assessment and inadequate recognition of the statistical limitations since resistance is quantitative. The latter consideration generally imposes the need for two-stage assessment—a cautious discard of grossly susceptible cultivars when replication is limited, and a more accurate assessment when susceptibility can be measured properly prior to varietal release—plus the use of several controls to put the results into standard perspective and provide a realistic threshold for decision making.

1.1.8.5 Farm trials

The large number of trials operated in some industries is intended to provide the farmer and factory with comprehensive advice. As pointed out by Doggett (1970) and Simmonds (1979), joint interpretation of such data by scatter diagrams is very instructive. The farmer remains the final arbiter of a good or bad cultivar, perhaps field-by-field, and some breeders argue that trials are not intended to estimate farm performance but merely to eliminate those with unacceptable qualities and poor performance over years. In most industries the farmer has 2–4 cultivars in production, providing some flexibility to optimise his harvesting schedule and to guard against excessive losses should a disease appear.

## 1.2 OTHER BREEDING METHODS

Mutation breeding (mutagenic treatment of buds) and tissue culture segregation *in vitro* with or without mutagens or selective agents in culture, are active programmes of research in several major industries. Both systems have given some interesting variants, but the simple aim of removing one defect from an otherwise good cultivar has not in general been realised without the need to check the performance, for yield, of the subclone in a further series of conventional trials. Though they will continue, these programmes will in no way displace conventional breeding in a crop that has such considerable genetic resources readily to hand by conventional hybridisation.

## 1.3 THE FUTURE

The future trends of breeding are, of course, closely tied to the future of the crop itself. Sugarcane as a source of nutritive sugar may decline in the face of competition from beet sugar, high-fructose corn syrups and various non-calorific sweeteners: several traditional sugar-exporting countries are in financial difficulty. Sugarcane's high biomass potential for 12-month growth cycles in the tropics, however, and its political importance to many of those countries, will surely see it retained as a major crop. Product diversification, e.g. fuel alcohol, animal feed, and paper and fibreboards, are receiving major attention. The genetic resources would seem to allow considerable scope for further advances to be made whether we use just the sucrose, or the whole plant.

## FURTHER READING

Much of the literature on sugarcane breeding is in specialist sugar journals. Stevenson's book (1965) gives a good historical account and covers crossing methods but not selection. There are general accounts in Alexander (1973), Blackburn (1984) and a multi-author reference book by Heinz (1987).

## REFERENCES

Alexander, A. G. (1973). "Sugarcane Physiology". Elsevier, Amsterdam.
Anon. (1970, 1973). *Ann. Rep. West Indian Cent. Sugar Cane Breed. Stn* **37**, 39–40; **40**, 34–39.
Arceneaux, G. (1965). Cultivated sugarcanes of the world and their botanical derivation. *Proc. Int. Soc. Sugar Cane Technol.* **12**, 844–854.
Berding, N., and Koike, H. (1980). Germplasm conservation of the *Saccharum* complex: a collection from the Indonesian archipelago. *Hawaii Plant. Rec.* **59**, 1–176.
Bhat, S. R., and Gill, S. S. (1985). The implications of $2n$ egg gametes in nobilisation and breeding of sugarcane. *Euphytica* **34**, 377–385.
Blackburn, F. H. (1984). "Sugarcane". Longman, London.
Brown, A. H. D., Daniels, J., Latter, B. D. H., and Krishnamurthi, M. (1969). Potential for sucrose selection in *Saccharum spontaneum*. *Theor. appl. Genet.* **39**, 79–87.
Clements, H. F. (1980). "Sugarcane Crop Logging and Crop Control". University of Hawaii Press, Honolulu.
Daniels, J. (1971). Description of sugarcane clones 1: Agricultural description. *Proc. Int. Soc. Sugar Cane Technol.* **14**, 112–123.
Divinagracia, N. S. (1980). Emasculation of sugarcane flowers: steam method. *Proc. Int. Soc. Sugar Cane Technol.* **17**, 1287–1295.
Doggett, H. (1970). "Sorghum". Longman, London.
Dunckelman, P. H., and Breaux, R. (1969). Screening for mosaic resistance in *Saccharum spontaneum* at Houma, Louisiana, 1964–1968. *Sugar Azucar* **64**, 16–18.
Dunckelman, P. H., and Breaux, R. (1971). Breeding sugarcane varieties for Louisiana with new germplasm. *Proc. Int. Soc. Sugar Cane Technol.* **14**, 233–239.
Dunckelman, P. H., and Legendre, B. L. (1982) "Guide to Sugarcane Breeding in the Temperate Zone". *Agric. Res. Serv., Agric. Rev. Man., Southern Series*, No. 22. US Department of Agriculture.
George, E. F. (1962). A further study of *Saccharum* progenies in contrasting environments. *Proc. Int. Soc. Sugar Cane Technol.* **11**, 488–497.
Heinz, D. J. (ed.) (1987). "The improvement of Sugar Cane Through Breeding". Elsevier, Amsterdam.
Hogarth, D. M., and Skinner, J. C. (1986). Computerisation of cane breeding records. *Proc. Int. Soc. Sugar Cane Technol.* **19**, 478–491.
Hogarth, D. M., Wu, K. K., and Heinz, D. J. (1981). Estimating genetic variance in sugarcane using a factorial cross design. *Crop Sci.* **21**, 21–25.
Irvine, J. (1977). Identification of cold tolerance in *Saccharum* and related genera through refrigerated freeze screening. *Proc. Int. Soc. Sugar Cane Technol.* **16**, 147–151.
James, N. I., and Miller, J. D. (1975). Selection in six crops of sugarcane. II. Efficiency and optimum selection intensities. *Crop. Sci.* **15**, 37–40.
Krishnamurthi, M., and Koike, H. (1982). Sugarcane collecting expedition, Papua New Guinea, 1977. *Hawaii Plant. Rec.* **59**, 273–313.
Lo, C. C., Chia, Y. H., Shang, K. C., Shen, I. S., and Shih, S. C. (1977). Collecting *Miscanthus* germ plasm in Taiwan. *Proc. Int. Soc. Sugar Cane Technol.* **16**, 59–69.
MacColl, D. (1977). Some aspects of flowering of sugar cane in Barbados and its control in a breeding programme. *Ann. Bot.* **41**, 191–207.

Medina, R. E., Krishnamurthi, M., Lapastora, E., Dosayla, R., Divinagracia, N., Silverio, G., and Cano, I. B. (1986). Sugarcane germplasm collection in the Republic of the Philippines. *Proc. Int. Soc. Sugar Cane Technol.* **19**, 522–527.

Midmore, D. J. (1980). Effects of photoperiod on flowering and fertility of sugarcane. *Field Crops Res.* **3**, 65–81.

Moore, P. H. (1985) In "Handbook on Flowering" (A. H. Halevy, ed.), Vol. 4, pp. 243–262. CRC Press, Boca Raton, Florida.

Panje, R. R. (1956). Studies in *Saccharum spontaneum*: a note on the objectives and scope of work. *Proc. Int. Soc. Sugar Cane Technol,* **9**, 750–756.

Paton, N. H., Smith, P., and Daniels, J. (1977) The leaf flavonoid patterns of Indonesian wild canes collected by the 1976 ISSCT expedition. *Sugarcane Breeders' Newslett.* **38**, 28–34.

Roach, B. T. (1977). Utilisation of *Saccharum spontaneum* in sugarcane breeding. *Proc. Int. Soc. Sugar Cane Technol.* **16**, 43–57.

Roach, B. T. (1986). Evaluation and breeding use of sugarcane germplasm. *Proc. Int. Soc. Sugar Cane Technol.* **19**, 492–502.

Simmonds, N. W. (1979). "Principles of Crop Improvement". Longman, London.

Simmonds, N. W. (1985). Two-stage selection strategy in plant breeding. *Heredity* **55**, 393–400.

Skinner, J. C. (1961). Sugarcane selection experiments 2: competition between varieties. *Tech. Commun. Bur. Sugar Exptl. Stn Queensland* **1**, 1–38.

Skinner, J. C. (1965). Grading varieties for selection. *Proc. Int. Soc. Sugar Cane Technol.* **12**, 938–949.

Srinivasan, T. V., and Sadakorn, J. (1983). Exploration and collection of *Saccharum* germplasm in Thailand. *IBPGR/S. E. Asia Newslett.* **7**, 7–9.

Stevenson, G. C. (1965). "Genetics and Breeding of Sugarcane". Longman, London.

Stevenson, N. D., and Daniels, J. (1971). Rapid screening methods for sugarcane IV: A pot method of growing and ripening sugarcane. *Proc. Int. Soc. Sugar Cane Technol.*, **14**, 195–205, 386–395.

Tew, T. L., Purdy, L. H., Lamadji, S., and Irawan. (1986). Indonesian sugar cane germplasm collecting expedition—1984. Sugar Cane, 1986, pp. 15–17.

Waldron, J. C., and Glasziou, K. T. (1971). Isoenzymes as a method of varietal identification in sugarcane. *Proc. Int. Soc. Sugar Cane Technol.* **14**, 249–256.

Walker, D. I. T. (1962). Family performance at early selection stages as a guide to the breeding programme. *Proc. Int. Soc. Sugar Cane Technol.* **11**, 469–483.

Walker, D. I. T. (1971). Utilisation of noble and *Saccharum spontaneum* germplasm in the West Indies. *Proc. Int. Soc. Sugar Cane Technol.* **14**, 224–232.

Walker, D. I. T., MacColl, D., and Rao, P. S. (1977). Aspects of the use of *Saccharum spontaneum* in the West Indies programme. *Proc. Int. Soc. Sugar Cane Technol.* **16**, 291–303.

Warner, J. N. (1953). Evolution of a philosophy on sugar cane breeding in Hawaii. *Hawaii Plant. Rec.* **54**, 139–162.

Warner, J. N., and Grassl, C. (1958). The 1957 sugar cane expedition to Melanesia. *Hawaii Plant. Rec.* **55**, 209–236.

# 2 Strategies in Rubber Tree Breeding

H. TAN

*The Rubber Institute of Malaysia, Kuala Lumpur, Malaysia*

The Para rubber tree, *Hevea brasiliensis* Muell. Arg. (family Euphorbiaceae), which is extensively cultivated in South-east Asia, is indigenous to the Amazon Basin of South America; it was first introduced into Asia a little more than 100 years ago. Cytological studies show this species to be diploid (it may be an ancient amphidiploid) with $2n = 36$ (Bouharmont, 1960; Ong, 1979). The tree is monoecious with a strong outbreeding tendency, and it is known to cross freely with trees of all other known species in the same genus. The species is propagated naturally by seeds, commercially (mainly) by bud-grafting onto seedling stocks and experimentally by self-rooted cuttings and other methods of grafting.

The economic product of the tree is latex (a cytoplasmic fluid from the laticiferous cells) which can be continually extracted from the tree by repeatedly slicing off a thin layer of the bark of the trunk by a process called "tapping". Latex yield is governed mainly by the genetic potential of the planting material. However, the expression of this genetic potential can be influenced by several other inherent factors of the tree (e.g. vigour, bark thickness, tree dryness, resistance to wind and major diseases, etc.); by environmental factors (e.g. edaphic, climatic and biotic); and by agromanagement practices (e.g. tapping system, chemical stimulation, planting density, fertiliser input, etc.). Rubber breeders have therefore considered most of these factors, with yield as the most important one in improvement programmes.

The emphasis on breeding objectives varies according to the specific needs of individual countries. In general, there are two main emphases in breeding. One is directed primarily towards the improvement of yield, as has been practised in Asia and Africa. The other is concerned mainly with combining high yield and resistance to major diseases such as South American leaf blight (SALB); an approach which has been emphasised in tropical America and recently, to a limited extent, in Asia.

As rubber planting materials can be in the form of seeds and/or clones produced by vegetative propagation, two distinct breeding schemes have been adopted in rubber breeding:

(a) In seedling improvement, populations are upgraded by using seeds from better-yielding plantations, from reformed seed-gardens, from outstanding mother-trees, from elite clones or from polyclonal isolated seed-gardens.
(b) In clonal improvement, clones are mainly produced by selection from outstanding progeny among the heterogeneous seedling populations produced from crossing desirable parents. The process of controlled pollination and vegetative selection is repeated to produce advanced generations for selection. Clonal materials can also be obtained from selection of outstanding seedlings from the seedling improvement programme. For economic, agricultural and social reasons, clonal material has become more popular and now dominates in the rubber industry. Clonal improvement programmes have thus been emphasised in almost all national rubber research organisations.

The history and progress of breeding achieved by various countries (e.g. Malaysia, Indonesia, Sri Lanka, India, Nigeria and Brazil) have been summarised in various literature (Dijkman, 1951; Baptiste, 1952, 1953, 1961, 1962; Brookson, 1953; Tysdal and Rands, 1953; Ross, 1959; Townsend, 1960; De Silva, 1961; McIndoe, 1961; Bos and McIndoe, 1965; Fernando, 1966, 1969, 1973; Ferwerda, 1969; Shepherd, 1969a; Paardekooper and Napitupulu, 1972; Nair *et al.*, 1976; Ayeke and Onokpise, 1977; Gonçalves *et al.*, 1983; see also Simmonds, 1987).

The improvement of yield in rubber through breeding and selection in the past 50–60 years has been spectacular. However, during recent years, progress seems to have slowed down. Consequently, breeders have attempted to identify and formulate strategies to solve the problems impeding rapid progress in the crop improvement programme.

This chapter highlights the major problems faced by rubber breeders. It also discusses recent developments in attempting to solve these problems. Strategies for solving the problems are discussed mainly with reference to the clonal improvement work carried out in Asia in general and in the Rubber Research Institute of Malaysia (RRIM) in particular.

## 2.1. PROBLEMS ASSOCIATED WITH *HEVEA* BREEDING

Several problems are known to hamper rapid progress in yield improvement in rubber. These problems include: narrow genetic base; long breeding and selection cycle; choice of parents; selection of multiple characters; disease resistance; genotype–environment interaction; and seasonality, non-synchronisation of flowering and low fruit-set.

The importance and relevance of each of these factors to *Hevea* breeding are discussed below.

### 2.1.1 Narrow genetic base

Most of the rubber breeding materials used in Asia originated from seeds collected by Sir Henry Wickham in a very small area of the Boim district bordering the Tapajos River in Brazil in 1876 (Wycherley, 1968; Schultes, 1977). In fact, it is widely believed that the initial breeding base in all the rubber-growing countries in Asia came from about 2000 seedlings from the Wickham collection imported into Asia in 1877 (Sri Lanka received a majority of the material while Malaya and Indonesia received only a few seedlings), and probably only a small proportion of it contributed substantially to subsequent selection (Wycherley, 1969). Although there were other introductions of seeds into Indonesia in later years, the poor yield characteristic of these introductions precluded their use in yield improvement programmes. Hence, the genetic base for rubber in Asia (and Africa) is narrow, a problem which has only recently received attention.

This problem of narrow genetic base is further aggravated by the use of directional selection for high yield in earlier breeding programmes, phenotypic assortative mating and the extensive use of clonal vegetative propagation of the crop. Consequently, some undesirable features associated with the crop have become apparent.

First, there is the evolution of selected clones whose genetic origin is limited to only a few dominant parents. As shown in Fig. 2.1, most of the clones bred to date in Malaysia (and other South-east Asian countries) can be traced back to about seven "primary clones". These dominant primary clones include Tjir 1, Pil A 44, Pil B 84, PB 24, PB 49, PB 56 and PB 86. Clones that are currently recommended for commercial planting and potential new clones not yet released commercially are also mainly derived from the same source. Thus, many outstanding clones have close genetic relationships. Some degree of inbreeding has, therefore, featured in the early breeding programmes. If the primary clones are themselves related (as some may well be), the inbreeding may be even greater.

Secondly, there is a diminishing return obtained from breeding efforts. This is illustrated by the progress of yield improvement in the RRIM breeding programme; there was a substantial yield increase in the early phases of breeding, diminishing in later phases (Fig. 2.2).

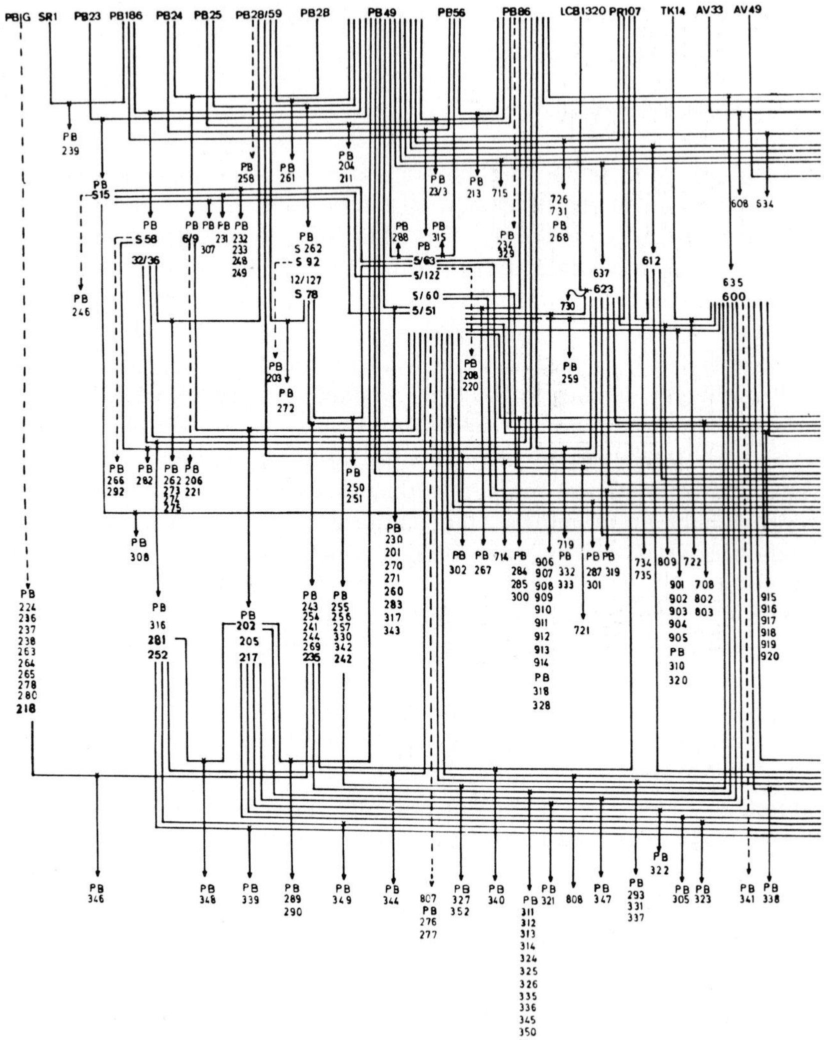

PB|G
SR1
PB23
PB186
PB24
PB25
PB28/59
PB28
PB49
PB56
PB86
LCB1320
PR107
TK14
AV33
AV49

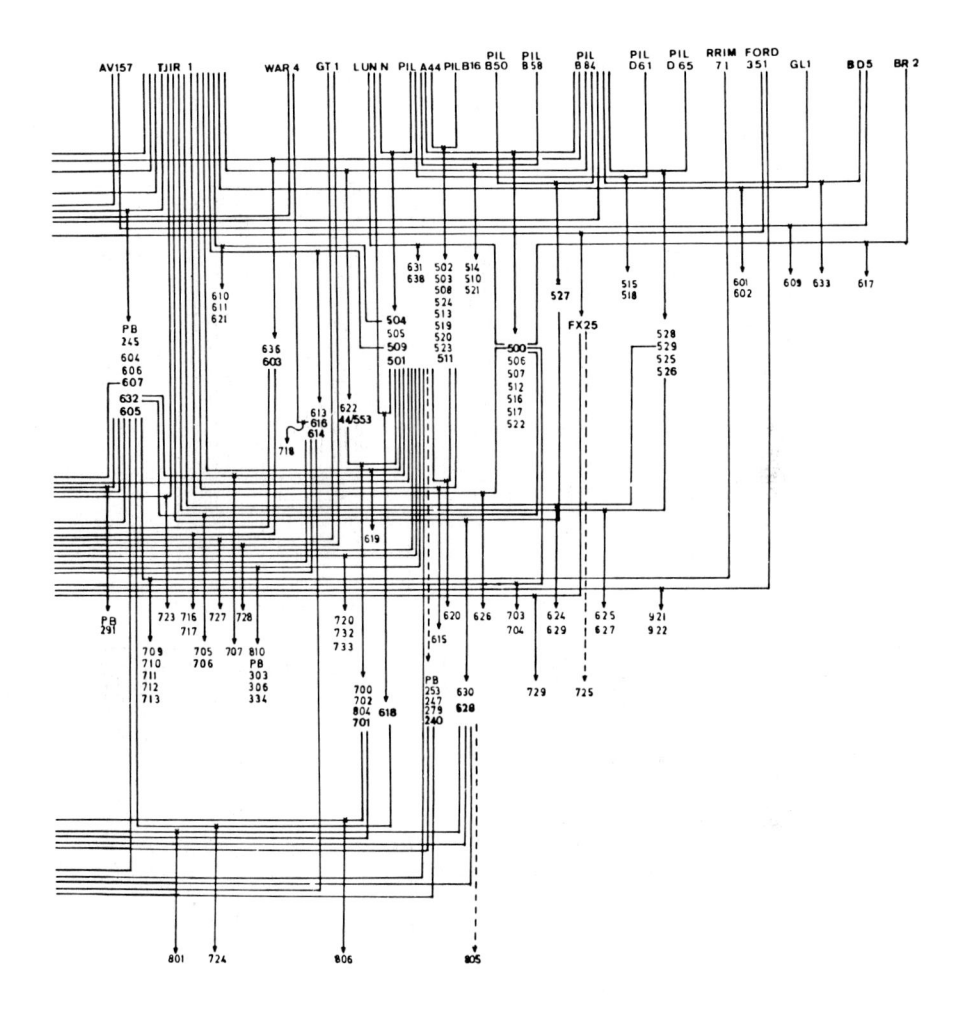

**Figure 2.1.** Pedigrees of *Hevea* clones bred in Malaysia. Clones bred at the Rubber Research Institute of Malaysia have no prefix.

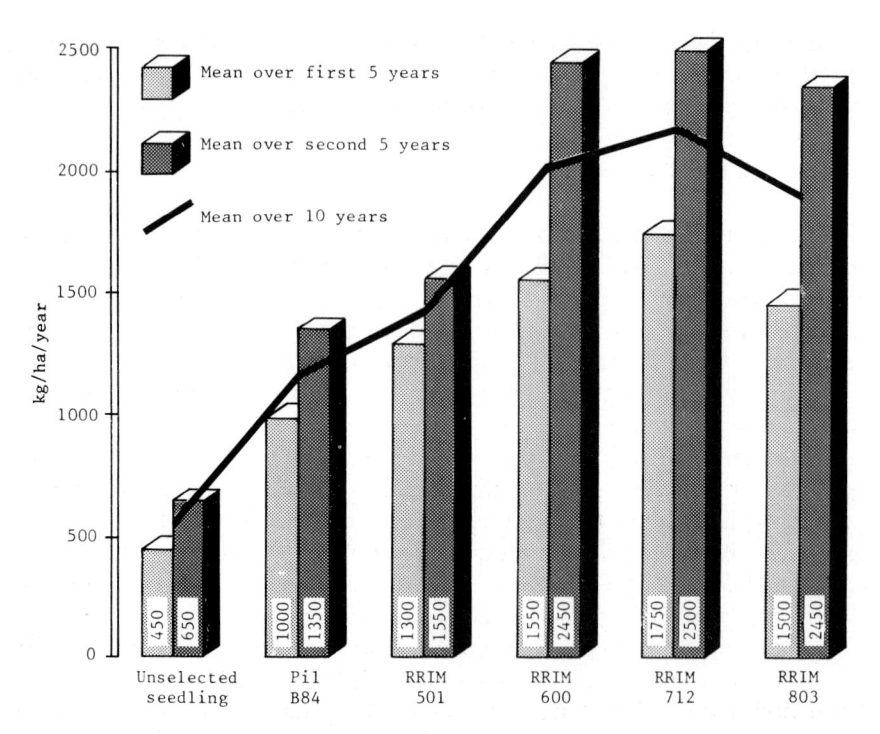

**Figure 2.2.** Progress of yield improvements in the *Hevea* breeding programme at the Rubber Research Institute of Malaysia. Note that "Unselected seedling" represents early planting material. Pil B84 was selected in the 1920s while RRIM 501, 600, 712 and 803 are the best clones bred in Phases I (1928–1931), II (1937–1941), III (1947–1958) and IV (1959–1965) breeding programmes respectively. Data are obtained from trials with 1, 34, 38, 13, 3 and 7 sites for the respective materials. Mean yield over the second 5 years of tapping for unselected seedling, Pil B84 and RRIM 803 are extrapolated. Some degrees of confounding effects with sites, seasons and agronomic practices should be noted.

Thirdly, there is indication of genetic erosion. Materials selected in one country may be found less adaptable in another, due to the different environment in which selection was made. For example, clones bred in Malaysia are found to be more resistant to *Colletotrichum* leaf disease than those selected in Sri Lanka and Indo-China (Wastie, 1967; Simmonds, 1982). Wycherley (1969) suggested that the relative susceptibility of Sri Lankan and Indo-Chinese clones to *Colletotrichum* and of Javanese and Malayan clones to *Oidium* was due to genetic erosion. He remarked that although slight resistance has been reported in PB 86 and RRIM 600, oriental selections are in general susceptible to SALB, whether due to lack of resistance in the Wickham collection or subsequent loss through genetic erosion while conducting selection in the absence of the disease.

### 2.1.2 Long breeding and selection cycle

Rubber generally takes about 4–5 years to produce sufficient flowers for hand pollination, and it has an immaturity period of 5–7 years before economic tapping can begin. Normally, a period of 3–15 years of recording after maturity is thought necessary for reliable assessment of yield and secondary characteristics, depending on the nature of trials conducted. In a seedling progeny trial, a period of at least 5 years of tapping is necessary for effective selection of promising families and desirable parents for inclusion in the next cycle of crossing or for establishment of polyclonal seed-gardens. In small-scale (preliminary proof) trials, a period of 3–5 years of tapping is needed to select outstanding clones for further testing in large-scale (further proof, adaptation or uniformity) trials. In large-scale trials, at least 10–15 years of evaluation is necessary before recommendation for wide-scale planting can be reliably made.

Rubber breeders therefore require about 10–15 years to complete one breeding cycle and about 30 years to complete the selection cycle from hand pollination. This long cycle slows the progress of breeding and is also wasteful of research time and effort.

### 2.1.3 Choice of parents

In the past, the number of clones that could be used as parents for crossing was limited. Most of the parental clones that were used had been selected on the basis of their phenotypic performance. These were crossed in many (but not all) combinations and promising families and progeny were subsequently selected. Today, as a result of the considerable amount of breeding work and clonal exchanges among research institutions, relatively more potential parents have become available. Consequently, a wide range of crosses can now be attempted but this will, in turn, require even more resources for effective exploitation. Clearly, wise choice of parents for crossing is desirable and critical.

### 2.1.4 Selection of multiple characters

In addition to yield and growth vigour, there are many other important secondary characters desired by rubber breeders. According to Simmonds (1969, 1979), as the number of characters sought for in any crop increases, larger population sizes will be required for effective selection. In rubber, because of its perennial nature and low fruit set, the production of large families is often not practicable.

The need for large population sizes is the more acute because rubber has many unfavourable associations between yield and other secondary characters. For example, high yield character is often associated with wind susceptibility, dry tree incidence and poor girthing rate during tapping (Wycherley, 1969; Ho, 1972, 1976). Thus, effective selection for multiple characters is generally difficult.

**2.1.5  Disease resistance**

The rubber trees are attacked by many diseases and pests of economic importance (Hilton, 1955, 1959; Rao, 1974; Chee and Wastie, 1980; Gasparotto et al., 1984; Peries and Liyanage, 1986; Chee and Holliday, 1986). The major leaf diseases include *Microcyclus ulei* (SALB), *Colletotrichum gloeosporioides*, *Oidium heveae*, *Phytophthora* species, *Thanatephorus cucumeris* and the pests include the caterpillar *Erinnyis ello*. (SALB, *Thanatephorus* and *Erinnyis* are reported in South America.) Panel and stem diseases caused by *Phytophthora palmivora* and *Corticium salmonicolor*, respectively, result in bark rot and dieback. Among the diseases, SALB is by far the most damaging. Although SALB is confined to tropical America, it must be regarded as a potential threat to rubber-growing countries in Asia and Africa. The reason is that materials which are widely planted in these regions are of Wickham origin and are known to be susceptible to SALB. In addition, the disease-resistant materials produced through breeding and selection in the past are derived from a few resistance sources and there is clear evidence that some (probably all) of these materials are susceptible to the newly evolved virulent pathotypes (Langdon, 1965; Rao, 1973; Chee, 1977; Simmonds, 1982). Breeding for resistance against SALB is therefore complex and difficult in view of the mutability of the causal organism and the perennial nature of the rubber tree. Breeding for SALB resistance in Asia and Africa is further curtailed by having to conduct disease-screening tests in tropical America where the disease is endemic.

**2.1.6  Genotype–environment interactions**

In large-scale trials and in commercial plantations, there have been instances in which a particular clone performed well in one location but not in another. This phenomenon, which is often known as genotype–environment (GE) interaction, has been confirmed by us and other workers (Paardekooper, 1964a,b; Jayasekera et al., 1977). The occurrence of significant GE interaction suggests that trials should be sited in diverse environments for proper evaluation of the planting materials. This would therefore demand more trial sites which would in turn incur higher costs. Another consequence of GE effects is that planting recommendations of a general nature, as practised in early years, are less effective. Hence, improving planting recommendations are essential to avoid unfavourable interactions.

### 2.1.7 Seasonality, non-synchronisation of flowering and low fruit-set

Rubber breeders face the following three problems associated with flowering and fruit set:

(a) *Seasonality of flowering.* Generally, there is one distinct main flowering season and a less distinct secondary flowering season in a year. Hand pollination work is normally conducted in the main season (with some exceptions) and is restricted to 1–2 months in each season.
(b) *Non-synchronisation of flowering.* Certain parental clones do not synchronise well in flowering time. This means that a number of potentially useful crosses cannot be made.
(c) *Low fruit-set.* The fruit-set success for rubber varies considerably, depending, among other things, on the parents used and environment (e.g. disease and weather). The average success in fruit-set, expressed as number of fruits produced upon number of pollinations made, of the RRIM annual breeding programmes (1928–1980) was about 4.5%, with a range of 0.3–15.4% (Wycherley, 1971). Edgar (1958) reported a range of 2–8% for average fruit-set success with a value of 5% as "fair".

These factors have placed serious constraints on the type of crosses that can be made and also on the number of progeny desired from each cross. This tends to slow down the progress of breeding. Furthermore, low fruit-set increases the cost of hand pollination.

## 2.2 RECENT DEVELOPMENTS

Considering the various problems just discussed, several approaches have been taken by rubber breeders to overcome them, and these are discussed below.

## 2.2.1 Broadening the genetic base

There have been four main approaches to broaden the genetic base, namely: introductions or exchanges of new genetic materials, conservation of germplasm, evaluation and utilisation of new genetic materials, and creating variability through mutation and induction of polyploidy.

### 2.2.1.1 Introductions or exchanges of new genetic materials

Subsequent to Wickham's introduction, there were records of other introductions of seeds from Brazil and Surinam to Indonesia around 1896, 1898 and 1913–1916 for breeding purposes (Dijkman, 1951). The last import included three different *Hevea* species (*H. brasiliensis*, *H. spruceana*, *H. guianensis* and *H. collina* which is now included in *H. guianensis*).

There was a lapse of over 30 years before another introduction of materials was made from Brazil, directly or indirectly, into the Orient (Brookson, 1956; Baptiste, 1958, 1959, 1960, 1961; Wycherley, 1968; Jayasekera and Fernando, 1977; Madjid, 1977). In Malaysia, 1614 seedlings of five *Hevea* species (*H. brasiliensis*, *H. guianensis*, *H. benthamiana*, *H. spruceana* and *H. pauciflora*) and also hybrid seeds from different provenances in Brazil were introduced in 1951–1952. Another import of 25 SALB-resistant clones (which were primary selections of *H. brasiliensis* and *H. benthamiana* and bred materials selected by the Ford Motor Company in Brazil) was made in 1953–1954. Seeds of different *Hevea* species were also imported from the Schultes Museum at Belem, Brazil in 1966. Some of these materials, especially those imported in the 1950s, were also made available to Sri Lanka, Indonesia and Indo-China. Details of these importations and distribution are given by Brookson (1956). In Sri Lanka, new Brazilian materials were introduced through the United States Department of Agriculture (USDA) and Liberia in 1957–1959 (Baptiste, 1958, 1959, 1960, 1961). These were mainly SALB-resistant materials, comprising 11 clones of *H. brasiliensis* and *H. benthamiana* selections of various origins and 105 bred materials selected by the Ford Motor Company and the Institute Agronômico do Norte (IAN), Brazil. Most of these clones were also available in Malaysia (Ong and Tan, 1976).

The most significant recent event has been the introduction of large numbers of wild *Hevea* germplasm in 1981 as a result of an expedition organised by the International Rubber Research and Development Board (IRRDB) and the Brazilian Government (Ong, 1982a; Ong et al., 1983). The germplasm were collected as seed and budwood from Acre, Rondonia and Mato Grosso. These three states in Brazil were thought to have rubber trees of great genetic diversity, possibly of higher yield and/or of good quality rubber (in Acre). In all, 64 734 seeds and budwood from 194 presumed high-yielding native trees were collected from these states. About half of the seeds collected were retained in Brazil while the balance were distributed to Malaysia (37.5%) and the Ivory Coast (12.5%), which serve as *Hevea* germplasm centres for the IRRDB countries.

The budwood material was subsequently brought into Malaysia and the Ivory Coast after a year's quarantine at an intermediate plant quarantine station in Guadeloupe, to prevent contamination by SALB.

There were also exchanges of clones bred in different countries in Asia, notably two major exchanges in 1954 and 1974. Unlike the Brazilian materials, these clones were mainly derived from the Wickham germplasm and hence can not be considered as new germplasm for widening the genetic base.

Further efforts to collect wild *Hevea* germplasm from other states or provenances in the Amazon Basin, and new exchanges of planting materials have also been proposed and emphasised by the IRRDB.

2.2.1.2 Conservation of *Hevea* germplasm

Several workers have discussed the methods of conserving existing *Hevea* germplasm (Subramaniam and Ong, 1973; Ho, 1978; Mohd. Noor *et al.*, 1980). According to these workers, the germplasm in Asia can be grouped under two major categories, namely, Wickham or Wickham-derived materials and non-Wickham materials (including the more recent introductions). These materials have been preserved as living collections.

The emphasis on germplasm conservation varies with countries because of the high cost involved. The RRIM has, on its own, initiated programmes to conserve most of the germplasm available locally as well as the foreign introductions in source-bush (or budwood multiplication) nurseries, museum blocks (clonal or seedling banks) and gene-pool gardens (Mohd. Noor *et al.*, 1980). Among these various approaches for conserving germplasm, the gene-pool gardens deserve further comment. Two special gardens were established in the RRIM to conserve available germplasm and to generate further variability for selection and upgrading of the population. The oriental gene-pool garden consists of obsolete materials of Wickham origin selected in the 1920s and 1930s from Malaysia, Indonesia (mainly from Java and Sumatra) and Sri Lanka. The occidental gene-pool garden consists mainly of Brazilian materials introduced in the 1950s and some hybrid clones bred in the RRIM, meant to reinforce the available Wickham germplasm. As an additional feature, the materials in these gardens are arranged in such a way as to encourage maximum cross-pollination to generate polycross seeds for widening the genetic base (Wycherley, 1969; Subramaniam and Ong, 1973).

In addition to efforts made by individual countries to conserve germplasm introduced before 1981, the IRRDB also supports germplasm centres based in Malaysia and the Ivory Coast to conserve materials of the 1981 introduction from Brazil in seedling bank (museum block) as well as in source-bush nursery.

#### 2.2.1.3 Evaluation and utilisation of new genetic materials

The foregoing genetic materials have been evaluated for yield and secondary characteristics in their respective countries. Until recently, some materials were also tested for resistance to SALB in the RRIM unit in Trinidad. The results have been used to assess the potential of the materials as parents or as planting material *per se*. In Malaysia (see Ong and Tan, 1987), Subramaniam (1969, 1970) reported on the performance of clones derived from seedling materials and also on SALB-resistant clones imported during the 1950s. Ong and Tan (1976) reported on the performance of other imported materials, including the Ford FX and IAN series clones. Ong (1977) reported on the performance of clones derived from various species introduced in 1966. In Sri Lanka, the evaluation of the introduced materials was reported by Jayasekera and Fernando (1977).

In general, most of the Brazilian introductions are relatively low yielding when compared with the good oriental selections. However, a few Brazilian selections which were high yielding and/or have other favourable characteristics (e.g. resistance to SALB and/or *Phytophthora*) have been used as parents. So far, two of the selections, RRIM 725 (selected locally from the imported seed of FX 25 from Brazil) and IAN 873 have been included in RRIM planting recommendations as clonal material (Rubber Research Institute of Malaysia, 1969, 1975a); but RRIM 725 was subsequently removed from the recommendation due to disease problems (Rubber Research Institute of Malaysia, 1983).

Several exploratory (inter- and intra-specific) crosses between introduced and oriental materials have been made with the aim of broadening the genetic base as well as breeding for disease resistance (with special reference to SALB). The results of these crosses have been reported by Malaysian and Sri Lankan workers (Baptiste, 1961; Fernando and Liyanage, 1976; Ong *et al.*, 1977). Although indications are that resistant genes have been incorporated into the oriental materials, yields of the resultant progeny were generally low. This was particularly true of crosses involving low-yielding materials. However, in certain crosses between better-yielding materials of Brazilian and oriental origins, some progenies gave comparable yield to those of the established clones (Fernando and Liyanage, 1976, 1980; Ong *et al.*, 1977). This suggests that it is desirable to upgrade the yield of wild populations before using them in any yield improvement programme. Thus, widening of the genetic base using wild germplasm would necessarily take a long time before any major impact could be realised.

#### 2.2.1.4 Creating genetic variability

Induced mutation and polyploidy have been used, on a very limited scale, in attempts to create genetic variability to widen the narrow genetic base of *Hevea* (Mendes and Mendes, 1963; Shepherd, 1969b; Mendes, 1971; Rubber Research Institute of Malaysia, 1971–1978; Ong and Subramaniam, 1973; Markose *et al.*, 1974, 1977; Anggreani-Hendranata, 1975, 1977; Markose, 1975; Fernando, 1977; Zheng *et al.*, 1980, 1981). Of the two methods, induced polyploidy has shown better progress but it has been plagued by problems associated with chimeras (Shepherd, 1969b; Zheng *et al.*, 1980; Ong, pers. comm.).

To date, some putative polyploid plants have been isolated by workers in Malaysia (Shepherd, 1969b; Ong *et al.*, 1984), India (Markose *et al.*, 1974; Markose, 1975), China (Zheng, *et al.*, 1980, 1981) and Brazil (Mendes and Mendes, 1963; Mendes, 1971; Moraes, 1984). Early yield results of a few putative polyploids were reported to be higher than their original parental clones (Santos *et al.*, 1984), suggesting there may be some scope for creating genetic variability through polyploidy induction technique.

As sufficient emphasis has not been made and progress has been limited, the potential value of artificial induction for widening the genetic base can only be determined if more efforts are given to this area of research. In conjunction with this, *in vitro* (tissue) culture techniques which have shown some success in the development of plantlets from pollen and anthers (Paranjothy and Rohani, 1978; Chen *et al.*, 1979, 1982a, b; Wang *et al.*, 1980; Rubber Research Institute of Malaysia, 1984, 1985; Wang and Wu, 1985; Wan Abdul Rahaman *et al.*, 1985) will probably be a useful aid in overcoming some of the problems associated with this area of work.

### 2.2.2 Shortening the breeding and selection cycle

The long timespan required to produce acceptable cultivars has prompted the rubber breeders to investigate possible methods to shorten the breeding and selection cycle. These investigations include development of early selection techniques and shortening the testing cycle.

#### 2.2.2.1 Early selection techniques

Interest in early selection techniques began in the 1920s with the aim of culling inferior seedlings (or selecting the better ones) at an early stage. Early workers had studied a number of parameters in relation to yield in mature plants (Summers, 1930; Gunnery, 1935). The parameters studied included girth, height, bark thickness, latex vessel number, latex vessel and sieve tube diameters, and rubber hydrocarbon in bark and petiole. The relationships obtained were poor and/or inconsistent. The only parameter which gave a fairly consistent correlation with yield was the latex vessel number. However, measurement of this character was tedious and required skill and facilities which were often unavailable in commercial plantations. Consequently, other characters which showed some association with yield, for example, vigour (expressed as circumference or diameter of trunk), though not as ideal as latex vessel count, were recommended for use in culling inferior seedlings in plantation stands.

Later, direct methods of yield testing were developed. Cramer (1938) adopted a system of grading young rubber plants using a special type of knife to make incisions on 1–2-year-old plants and then made qualitative assessment on the amount of latex exuded. Due to poor associations with yield of mature plants, Cramer's method was considered useful only for culling (Cramer, 1938; Dijkman, 1951). Hamaker (1914), and Morris and Mann (1932–1938) developed another early test (known as the HMM method), consisting of successive tappings of 3–4-year-old trees and weighing the latex produced. Although this quantitative method was found to be more reliable than Cramer's, it too had limitations (Kuneman, 1939). The HMM method was subsequently modified to test-tap 2–3-year-old plants (Tan and Subramaniam, 1976). Mendes (1971) described an early test-tapping method for few-month-old plants using a specially designed knife, while Waidyanatha and Fernando (1972) reported a needle-prick test for very young plants. However, published results are too inconclusive to justify their usefulness in early selection.

Zhou *et al*. (1982) developed two methods to predict the yield potential of one-year-old buddings by the amount of latex oozing out from leaflets or petiolules. The first method involves slicing the leaflet through the lateral veins to obtain a non-quantitative grading of latex yield. The second method is to obtain a quantitative measurement of the latex exuded from the cut end of the petiolule of the leaflet. The authors reported positive correlations between these two juvenile yield characters and mature yields. These methods are adopted in China to identify potential high yielders at their juvenile stage and thus to accelerate selection.

There have also been investigations on indirect methods of yield prediction. These include the use of anatomical characters: latex vessel system in bark (Ho, 1972, 1976, 1979) and leaf (Huang et al., 1981); physiological characters: plugging index (Ho, 1972, 1976) or bursting index (Dintinger et al., 1981) and photosynthetic rates (Samsuddin et al., 1986, 1987a,b); biochemical attributes: rubber hydrocarbon in petiole and leaf (Bolle-Jones, 1954), nucleic acids (Tupy, 1969), cotyledon oil (Fernando and De Silva, 1971), pH (Dintinger et al., 1981); chemical attributes: latex constituents such as N, P, K, etc. (Ho, 1976); and morphological characters: leaf venation (Testam method of Amand, 1962) and stomata number (Senanayake and Samaranayake, 1970). Among these parameters, only plugging index and latex vessel number (or density) have shown consistent significant correlations with yields of mature trees (Ho, 1972, 1976, 1979; Huang et al., 1981).

The main results of recent investigations obtained by Malaysian (RRIM) workers on the relationship between characters of young plants and yield of mature trees are summarised in Table 2.1. In an experiment, Ho (1976) demonstrated a fairly high correlation between the yield of 33-month-old young buddings and that of mature trees. Thus, based on nursery yield alone, one could safely select a certain proportion of the genetically high yielders for the mature phase. Although there was some improvement in correlation when yield and plugging index were considered together, the increase in efficiency of selection was small (Table 2.2). In my own study, which involved a selected sample of a seedling population, a lower correlation between early (nursery) yield and mature yield of derived clones was obtained. Selection using a combination of three nursery characters (viz. yield, plugging index and latex vessel count) did not appear to improve the selection efficiency to any great extent (Table 2.3). One notable feature in my "combined" selection method was that an additional number of progenies in the top-yielding fraction could be identified compared with the use of nursery yield alone as a selection criterion.

Thus, current practice adopted in the RRIM is to carry out early selection on 2–3-year-old seedling progenies based primarily on yield (by the modified HMM method) and, to a lesser extent, on other characters (i.e. vigour, latex vessel number and density, plugging index, disease resistance, tree form, etc.) for further testing in clone trials.

Regarding early selection for disease resistance, some methods have been developed for screening clonal reaction to leaf and panel diseases, for example, *Colletotrichum* in field nursery (Wastie, 1973), *Oidium in vitro* (Lim, 1973), *Phytophthora in vitro* (Chee, 1969) and in field nursery (Satchuthananthavale et al., 1974). In addition, *in vitro* methods for

**Table 2.1.** Simple correlations between mature yields and associated nursery characters of *Hevea* clones and seedlings over 5 years.

| Mature yield versus nursery characters | Age of young plants (months) | Correlation coefficient[a] | Source |
|---|---|---|---|
| *Clones* | | | |
| | 33 | 0.57*** ( 77)[b] | Ho, 1972 |
| | 39 | 0.75*** ( 26) | Ho, 1979 |
| Yield | 24 | 0.77*** ( 26) | Ho, 1979 |
| Yield index[c] | 33 | 0.73*** ( 21) | Ho, 1972, 1976 |
| Latex vessel no. | 33 | 0.25* ( 77) | Ho, 1972 |
| Plugging index | 56 | −0.73*** ( 21) | Ho, 1972, 1976 |
| *Seedlings* | | | |
| Yield | 33 | 0.44* ( 20) | |
| | | 0.52*** ( 37) | |
| | | 0.26** (125) | |
| Yield index[c] | 33 | 0.42* | |
| | | 0.56*** | |
| | | 0.34*** | H. Tan (unpubl.) |
| Plugging index | 36 | −0.44* ( 20) | |
| | | −0.56*** ( 37) | |
| Latex vessel no. | 36 | 0.27 ( 37) | |
| Latex vessel density | 36 | 0.41* ( 37) | |

[a] Significantly different at: * $P = 0.05$; ** $P = 0.001$; and *** $P = 0.001$.
[b] Degrees of freedom in parentheses.
[c] Yield index refers to yield : girth ratio.

screening SALB resistance had been devised (Chee, 1976; Zhang and Chee, 1985) to complement the nursery screening technique developed in the earlier years (Langford, 1945, Holliday, 1970). These techniques could be used as early indications of disease resistance but are not entirely reliable for predicting future field performance.

Investigations into an early selection method for wind-tolerance characteristics (see Nicolas *et al.*, 1979), which is also useful for culling seriously wind-prone materials at their juvenile stage, have not been successful so far.

**Table 2.2.** Number of potential high-yielding *Hevea* trees selected on the basis of yield and plugging index in clonal propagation at the nursery stage.

| Highest yielding fraction | No. of clones in each fraction | No. predicted using Yield index | Yield index and plugging index[a] |
|---|---|---|---|
| Top eighth | 3 | 1 | 1 |
| Top quarter | 6 | 2 | 3 |
| Top half | 12 | 10 | 10 |

*Source*: Ho (1976).
[a] Multiple regression estimate was used.

**Table 2.3.** Number and percentage of common progeny based on selection from nursery and mature phases in selected progeny of *Hevea* PB 5/51 × RRIM 600.

| Proportion selected (%) | No in each fraction | Nursery yield (NY) | Plugging index (PI) | Latex vessel density (LV) | NY LV | NY PI | PI LV | PI LV | NY PI LV |
|---|---|---|---|---|---|---|---|---|---|
| 10 | 4 | 1 | 2 | 2 | 1 | 2 | 3 | 3 | |
| 20 | 8 | 4 | 4 | 5 | 4 | 4 | 6 | 5 | |
| 30 | 12 | 7 | 8 | 5 | 9 | 6 | 9 | 9 | |
| 40 | 16 | 12 | 10 | 10 | 11 | 11 | 12 | 11 | |
| 50 | 20 | 15 | 14 | 12 | 16 | 14 | 13 | 15 | |

Multiple regression estimate was employed for selection based on more than one character. Almost all the clones selected are overlapping; the use of yield index instead of nursery yield does not improve selection efficiency.

From the above, it is obvious that reliable early selection techniques have not yet been developed and there is need for further research on this subject.

2.2.2.2 Shortening the testing cycle

Workers in Malaysia and Sri Lanka have proposed shortening the testing cycle by by-passing one of the testing stages (Senanayake and Wijewantha, 1968; Subramaniam, 1980; Tan *et al.*, 1981), and also by reducing the recording period (Ong, 1980). Some of these approaches have already been implemented in the RRIM; they will now be described.

One approach for shortening the testing cycle is to carefully select a few of the high-yielding progeny based on nursery evaluation results and then to test them directly in various commercial plantations, by-passing the small-scale testing stage. The selected materials are usually tested in two replications with plot size of about 0.2 ha per clone instead of 3–5 replications in the normal large-scale trials. Materials found consistently promising over 5 years of tapping in these trials can then be directly recommended for commercial planting on a moderate scale. This procedure known as "promotion plot" testing is saving some 10 years compared with the conventional method of testing (see Fig. 2.3). This approach was first introduced in 1972. Thus far, promising results have been noted but we must be cautious as the predictability of yield using nursery results is only low to moderate (Tan et al., 1981, 1984; Ong et al., 1986).

Another way of shortening the testing duration is to minimise the yield-recording period during the early stage of clonal testing. Ong (1980)

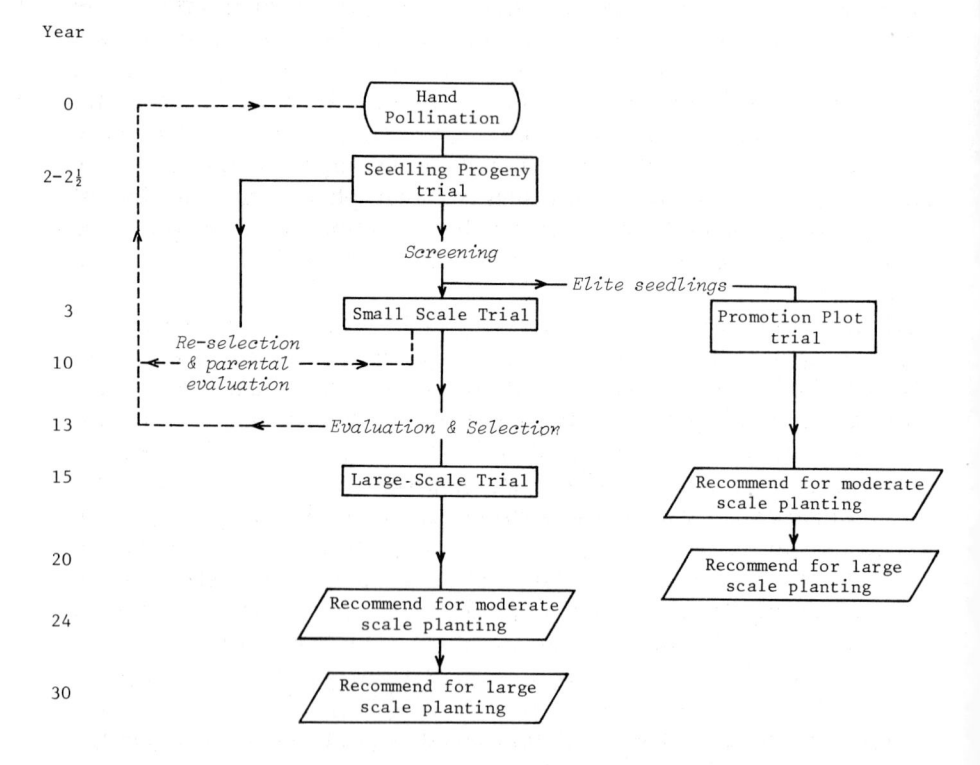

```
Year

 0          ┌─────────►───────┐ Hand
            │                 │ Pollination
            │                 │
            │                 ▼
2–2½        │          ┌─────────────┐
            │          │Seedling Progeny│
            │          │    trial    │
            │          └─────────────┘
            ▲                 │ Screening
            │                 ▼         ────► Elite seedlings ────┐
 3          │          ┌─────────────┐                    ┌─────────────┐
            │          │Small Scale  │                    │Promotion Plot│
            │ Re-selection Trial     │                    │    trial    │
10          │◄── & parental ──►──────┘                    └─────────────┘
            │  evaluation                                        │
            │                                                    │
13          └──────────◄────── Evaluation & Selection            │
                              │                                  │
                              ▼                                  ▼
15                     ┌─────────────┐                   /Recommend for moderate/
                       │Large-Scale  │                   /   scale planting    /
                       │   Trial     │
                       └─────────────┘                          │
20                            │                                 ▼
                              │                         /Recommend for large/
24                   /Recommend for moderate/          /   scale planting  /
                     /   scale planting    /
                              │
                              ▼
30                   /Recommend for large/
                     /  scale planting   /
```

**Figure 2.3** Breeding and selection cycle of *Hevea* as used at the Rubber Research Institute of Malaysia.

reported that it was possible to use 2 years' instead of 4 or more years' yield record of small-scale trials to select almost all the promising clones for large-scale trials. He inferred that such a method would not radically affect the efficiency of selection. Although there is still the possibility of losing materials which might yield better in the later stage of their economic life, this approach is valid because it enables the breeder to economise on time and cost. It also fits into the objective of early selection of precocious high yielders in the breeding programme.

However, it should be noted that, while the techniques for shortening the breeding cycle are useful to ensure an earlier release of new cultivars (clones), the conventional approach should still be regarded as the most reliable method of evaluating clonal performance.

### 2.2.3. Combining ability studies as aid in choice of parents

The study of biometrical genetics of economic characters should, in principle, give some understanding of the pattern of inheritance in the crop. One potentially useful aspect of such a study is that it could assist breeders in the choice of parents in planning their breeding programmes. Several workers (Simmonds, 1969; Gilbert *et al.*, 1973; Nga and Subramaniam, 1974; Tan and Subramaniam, 1976; Tan, 1978a,b, 1981) have concluded that yield and girth variation can be largely accounted for by additive genetic variance. This suggests that phenotypic selection of parents would generally be effective, but selection based on genotypic values (as reflected by general combining ability, GCA) will be more precise and reliable.

Gilbert *et al.* (1973) were the first to estimate GCA values in *Hevea*, using an incomplete diallel model composed of unsystematic crosses (Gilbert, 1967). They obtained GCA constants of individual parents from past data of seedling progeny trials and showed that these could be used for predicting family performance and choice of parents in the breeding programme. In addition to identifying high GCA parents, they were also able to identify some potential good crosses which had been left out of the earlier programme. Tan (1977, 1978a,b) confirmed the usefulness of the method and further reported that it was also possible to assess parental GCA using only 2 years of progeny yield records and that such information would also be useful for improvement of clonal cultivars (Tan, 1978a). In another study, Tan (1978b) reported that some of the high-GCA parents obtained from 2–3-year-old seedlings were found in the high-GCA parental group in the mature stage of different breeding populations. He suggested that it was thus possible to identify at an early stage at least some good parents and accelerate their use.

Although combining ability studies in *Hevea* have some limitations (Tan, 1981), they are found to be useful as an adjunct to the conventional approach of using phenotypic selection, breeder's experience and intuition.

To date, yield and vigour have been considered in GCA studies because of their economic importance. The data available now on wind-damage incidence, disease resistance, bark thickness, and yield determinants such as latex vessel number and plugging index are either very limited to allow for analysis or have not been adequately analysed. However, these secondary characters, among others, are considered with yield and vigour of growth when planning crosses in the RRIM. These criteria have been used by RRIM breeders in their later breeding programmes. Crossing between selected parents whose characteristics are expected to complement each other has been emphasised to reduce unfavourable combinations produced in the progeny. Also, crossing between related parents has been carefully avoided to eliminate inbreeding depression which is a feature of this crop (Sharp, 1940, 1951; Gilbert *et al.*, 1973; Tan and Subramaniam, 1976).

### 2.2.4 Selection of advanced generation polycross progenies

Historically, the base populations for *Hevea* breeding work in Asia were mother-trees or ortets selected from the extensive populations of unselected seedlings in commercial plantings. The concentrated efforts of the 1920s and 1930s resulted in the selection of the primary clones, some of which have been referred to earlier. As indicated earlier, additive inheritance predominates the variation of yield and girth in the breeding population and this clearly suggests that selection among appropriate polycross progenies should be effective.

Commercial plantations in Malaysia have stands, numbered in millions of trees, of polycross seedlings from known advanced generation seed-gardens which have a number of high GCA parents. *Hevea* breeders therefore have the good opportunity to select from very large families with high mean performance. In 1972, the RRIM started an extensive programme to screen these seedlings; early results are encouraging (Ho *et al.*, 1980; Khoo *et al.*, 1981, 1982).

It should be emphasised that this line of work provides material in addition to that from controlled pollination for subsequent screening and selection. It saves time, cost and experimental areas because these seedlings have already been grown in commercial plantations. Problems associated with low fruit-set do not arise, and the relatively large polycross population available allows for a more effective selection for multiple characters.

## 2.2.5 Breeding for disease resistance

Disease-resistance breeding received some attention recently when certain leaf diseases began to cause concern (Ho, 1986; Wastie, 1986). Some efforts have been undertaken by individual countries to breed clones resistant to leaf diseases, as in the case of *Colletotrichum* in Malaysia, *Oidium* in Sri Lanka and *Phytophthora* in India. These attempts have produced some material of various degrees of disease tolerance. Breeding for SALB resistance has also been initiated in Malaysia and Sri Lanka soon after SALB-resistant materials from Brazil were introduced to these countries. However, breeding work (including earlier work in Brazil) has been confined to only a few parents which are not known to be resistant to all the pathotypes of *Microcyclus*. Success in combining two sources of pathotype-specific resistance in a single clone has been reported (Fernando and Liyanage, 1980). However, the long-term stability of this resistance has yet to be tested. In general, breeding for SALB resistance both in South American and in Asia has not been very successful and this has necessitated reconsideration of the approach for breeding resistance against SALB.

Simmonds (1982) pointed out the importance of breeding for "field resistance" (i.e. "horizontal resistance", which is governed by polygenes) as opposed to "vertical resistance" (governed by major genes) to SALB in rubber. He commented that past breeding effort had been concentrated on using immune parents which, although not conclusively studied, may have "vertical resistance", resulting in attack by existing or newly evolved pathotype(s). In Trinidad, Chee (1977) and Chee and Wastie (1980) have noted some materials (e.g. RRIM 600 and PR 107) with restricted lesions, reduced amount of leaf fall and little or no sporulation upon SALB infection. These materials are thought to possess "horizontal resistance" which, if true, should be more stable or durable in resisting SALB. Nursery observation in Trinidad on the "field resistance" of RRIM 600, however, does not seem to hold true on a plantation scale in Brazil. This was evident from the fact that RRIM 600 was severely attacked by SALB in Belem (and other SALB-endemic areas) and required corrective control measures involving crown budding with SALB-resistant clones. If the phenomenon of some degrees of SALB resistance represents "weak horizontal resistance", then perhaps this resistance source could be further enhanced through breeding. Alternatively, it should also be possible to search for potential "field resistance" sources from seedling populations of wild or bred origins in SALB-endemic countries (e.g. Brazil).

However, there is another view which suggested that some of the resistant parents used earlier were governed by polygenes (Wycherley, 1969; Chee, 1977). This was based on evidence of declining resistance in successive generations of recurrent backcrossing for SALB. It is therefore necessary to study the genetic bases of resistance to SALB (and perhaps other diseases) so as to provide a sound base for rubber breeding. Meanwhile, rubber breeders in the RRIM aim towards selection and breeding of materials which exhibit "horizontal resistance". Facilities for disease screening are available in Brazil where SALB is endemic and where different pathotypes of *Microcyclus* exist.

### 2.2.6 Use of crown budding for correction of secondary defects

Many established high-yielding materials are known to have undesirable secondary characters which limit their use in many rubber-growing areas where environmental constraints giving rise to adverse GE interactions exist. In order to overcome this problem and the difficulties of breeding for multiple characters, top-working or crown-budding techniques have been developed and recommended. The defective but high-yielding clone is budded with a desirable crown clone at a required height, so that the resultant crown–trunk combination offsets the inherent defect of the trunk clone.

Yoon (1967, 1971) and Ho and Yoon (1972) have shown that suitable combinations of crown and trunk can result in synergistic effects for higher yield through leaf disease resistance, improved vigour and reduced wind damage losses. Recent observations suggest that crown budding can also modify some physiological, rubber and latex properties of the trunk clone (Leong and Yoon, 1976, 1978; Leong *et al.*, 1986). Several workers have reported that high-yielding but SALB-susceptible clones could be successfully cultivated in tropical America when crown budded with suitable disease-resistant crowns (Rands, 1942; Townsend, 1960; Chee and Wastie, 1980; Sena Gomes *et al.*, 1983).

In using this approach, there are possibilities of correcting undesirable "associations" of high yield and unfavourable secondary characters such as proneness to wind damage and disease susceptibility. In this way, crown budding could widen the choice of clones which are high-yielding but have defective secondary characters.

### 2.2.7 Improving planting recommendations

The final results of large-scale trials culminate in recommendations to the rubber industry. In Malaysia, planting recommendations in the past were of a general nature, for adoption throughout the country. This method of recommendation has a weakness, for it does not take into account interaction of environment with clonal performance. When it became apparent that certain parts of the country showed specific environmental constraints, such as severity of wind damage and incidence of major disease, a revised system known as regional planting recommendation was introduced in 1969 (Ho *et al.*, 1969). It was subsequently modified in 1971 (Ng *et al.*, 1972). This new concept restricted the use of planting materials with known secondary defect(s) to certain localities in order to minimise adverse clone–environment (GE) interactions. Later, in 1974, difficulties were encountered when demarcating boundaries of environmental constraints, and this led to a further refinement in the form of a new concept designed "Enviromax Planting Recommendations" (Ho *et al.*, 1974). The underlying principle is to maximise the yield potential in a particular locality, subject to the inhibitory influence of the environmental factors (including soil type, terrain, drought, etc.), through the choice of planting material.

In order to fulfil the requirements for the above concept, breeders normally conduct their large-scale trials under diverse environments to obtain information about the relative performance of clones and GE interactions. In addition, breeders also establish speculative materials in small blocks of 5–10 ha each (known as block plantings) in various environments on commercial plantations, so as to provide additional information on clonal performance. The information obtained from these block plantings supplements trial results whereby earlier release of cultivars for wider commercial use could be ensured. Feedback from commercial plantations on the performance of recommended cultivars helps to improve subsequent recommendations.

### 2.2.8 Flower induction and improvement of fruit-set

Flower induction is a useful tool for rubber breeders in that it offers the following advantages:

(a) It enables breeders to conduct hand pollinations throughout the year, hence spreading the workload.
(b) It makes hand pollination easier and safer because it can be done near the ground instead of at the top of mature trees. It also reduces the cost of sulphur dusting required to improve fruit-set.
(c) It enables some crosses to be made involving parents which are not normally synchronous in flowering.
(d) It permits parents to be crossed at the relatively early age of 1–2 years, thus helping to reduce the time required for each breeding cycle.

Attempts have been made to induce early flowering in *Hevea* by many workers (Campaignolle and Bouthillon, 1955; De Silva and Chandrasekara, 1959; Tan and Lubis, 1961; Camacho and Jimenez, 1963; Ong, 1972; Combe *et al.*, 1974; Saraswathy Amma, 1975; Nicolas, 1976; Madjid *et al.*, 1977; Najib and Paranjothy, 1978; Rohani and Paranjothy, 1980; Wang and Huang, 1981). These workers experimented on 1–2-year-old buddings, using treatments such as ring barking, bending of branches and application of chemicals such as coumarin, tri-iodobenzoic acid, ethephon, gibberellins, cytokinins, auxins, etc., in various combinations. Generally, ring barking has been the most effective method of flower induction. Flowering can also be intensified by girdling with or without the spraying of coumarin (Ong, 1972; Wang and Huang, 1981). These techniques, however, are found to be ineffective on flower induction for young seedlings (Najib and Paranjothy, 1978). In addition, several practical problems still persist (e.g. variable clonal response to treatment, breakage of fruit-bearing branches or of tree, small number of fruits produced, etc.) and flower-induction techniques have yet to be widely used.

Research on improving fruit-set has not received much attention. Initial studies were made on flower protection and prevention of fruit-drop as a means to improve fruit-set. For flower protection, prophylactic treatments using sulphur and other fungicides (Hilton, 1959; Rubber Research Institute of Malaysia, 1975b) were found to be effective in protecting flowers from fungal attack but only sulphur has been used in practice. For prevention of premature fruit-drop of pollinated flowers, application of various growth hormones have been tried but results have generally been poor (Rubber Research Institute of Malaya, 1962). Another approach reported to increase seed production is to apply extra nitrogen fertiliser (Watson and Narayanan, 1965; Sivanadyan and Ghandimathi, 1985, 1986). Ghandimathi and Yeang (1984) reported a new procedure for hand pollination to improve pollination success and fruit-set but the method is tedious and not practical for normal breeding work.

Clearly, early flowering and improving fruit-set should be further investigated to assist *Hevea* breeding.

## 2.3 BREEDING OBJECTIVES AND STRATEGIES

The current breeding and selection cycle of *Hevea* is undesirably long: materials bred today could only be recommended for wide commercial use in about 30 years' time. Therefore, breeding objectives should be carefully drawn to cater for future needs. However, the latter may change over time, according to technological advances, socio-economic situations and demands of the industry. As such, breeding objectives should also be kept flexible to meet possible changes.

In general, the primary objective is to breed clones with genetically high-yield potential, accompanied by other desirable secondary characters known to modify yield productivity. These secondary characters include:

(a) Good immature vigour to reduce the unproductive period and generate earlier economic gains.
(b) Good girthing on tapping to sustain yield and reduce wind damage losses through trunk-snap.
(c) Thick virgin bark to minimise wounding incidence which is known to affect yield productivity on later panels.
(d) Good bark renewal to enhance economic tapping life.
(e) Resistance to major diseases to ensure better growth and yield and also to minimise the risk of losses, as in severe cases of SALB attack.
(f) Tolerance to wind to minimise losses, thus ensuring a good tapping stand throughout the economic life; this character may be associated with branching habit, tree height, long latex flow or low plugging index (leading to adverse partition and hence imbalance between crown and trunk components), wood property, etc.
(g) Tolerance to dryness to ensure higher yield productivity.

Of these secondary characters, tolerance to wind and major diseases are considered more important. Because of the difficulties of synthesising "perfect" clones, rubber breeders, in practice, have to accept "compromises".

There are additional features or requirements which deserve attention. To meet the short-term requirements, breeders should breed for high-yielding clones which respond well to chemical (yield) stimulation and low frequency tapping, to cut down production costs and overcome situations of labour shortage. To accommodate the projected shortage of land and the increased use of marginal land for rubber cultivation, breeders should perhaps breed for cultivars which can thrive under high density planting, poor soil fertility status and extreme environmental stress (e.g. drought or cold in certain countries). As a long-term requirement, it may be necessary to consider breeding rubber trees not only for rubber yield but also for wood as an additional product to meet the projected shortage and demand of timber. In which case, a compromise between partition for rubber and biomass would have to be reached. It may also be necessary to select or breed for cultivars with special rubber, latex or other technological properties to meet consumers' and/or manufacturers' requirements.

To achieve these objectives, the following strategies can be formulated. In the short term, breeders should aim to produce high-yielding clones with favourable secondary characters in the shortest possible time. The approaches include making best use of potential parents (with high GCA and/or phenotype value) for breeding, selection of advanced polycross progenies, shortening the breeding and testing cycle, flower induction, improvement of fruit-set and further refinements in planting recommendations.

In the medium term, breeders should continue their efforts in widening the genetic base; selection (and breeding, if necessary) of superior crown clones and rootstocks (which influence scion growth and yield performance); establishing new advanced generation polyclonal seed-gardens for selection of improved seedling populations, clones and rootstocks; selection and breeding for "horizontal" disease resistance (particularly SALB) and other special characters not available so far. In the programme for widening the genetic base, rapid evaluation and utilisation of new germplasm should be emphasised. Approaches to shortening the breeding cycle and early flower induction could be attempted to upgrade the wild new germplasm. The other approach is to make use of a special gene-pool garden for gradual upgrading of populations (separately and jointly). The materials which are upgraded at each stage could be incorporated into existing high-yielding materials; material produced, if promising, could also be used directly for commercial planting. It is further hoped that the apparent heterotic effects from wide crosses involving materials from different provenances, species or related genera will be exploited.

In the long term, breeders should aim at using new innovations and special techniques to create new genetic variability and/or to modify certain characters (e.g. genetic dwarf, etc.) on existing high-yielding cultivars. Such teciniques as mutation, polyploidy induction, *in vitro* (tissue) culture (including cell and protoplast cultures) and genetic manipulation may prove to be useful in complementing the conventional long-term *Hevea* improvement programme.

## 2.4 INTERNATIONAL CO-OPERATION

International co-operation is important because rubber breeding is expensive and time-consuming. It is often not wise for a single research organisation or a country to do research on rubber breeding in isolation and, at the same time, attempt to fulfil the requirement of rapid genetic progress. Rubber breeders have recognised this weakness and as a result, there have been some commendable international co-operative efforts in rubber breeding among the research institutions of some natural-rubber producing countries. It is important that the current international co-operation should be further extended and intensified towards the improvement of this crop.

So far, this co-operative effort has gone well in respect of plant material exchange programmes, international exchange clone trials, the collection of wild *Hevea* germplasm, the establishment of germplasm centres and exchange of ideas through regular meetings of plant breeders and scientists of related disciplines (plant pathologists, physiologists and biotechnologists). A joint research programme with particular reference to SALB involving the Malaysian (RRIM) and Brazilian (EMBRAPA and SUDHEVEA) research institutes has been established. This is an extension of the research programme initiated by the RRIM in 1961. Perhaps, the time has now come for a central unit to be formed at international level to formulate, plan and allocate projects to specific institutions. The materials produced and the scientific information obtained should automatically be disseminated to the participating countries. In this way, there will be better utilisation of efforts, manpower and financial resources.

## 2.5 CONCLUSIONS

Rubber breeding has so far been very successful in terms of yield improvement in spite of several problems encountered. These problems have now been identified and steps have been taken to overcome them. Through the concerted co-operative efforts, nationally and internationally, a much higher yield improvement closer to the theoretical maximum potential of about 9500 kg ha$^{-1}$ per annum (Templeton, 1969) could possibly be realised at an earlier date.

Acknowledgements

Acknowledgements are made to Datuk (Dr) Tuan Haji Ani Bin Arope, Director, RRIM, for permission to publish this chapter; and to Professor N. W. Simmonds, Drs P. R. Wycherley, S. H. Ong, C. Y. Ho, P. K. Yoon and E. Pushparajah for helpful comments. I am also grateful to my supporting staff, particularly Mr Chin Kam Heng for his literature searches and information relevant to this review.

## REFERENCES

Amand, H. (1962). Relations entre les productions des descendances génératives et végétatives de *Hevea brasiliensis* et corrélations avec un critère morphologue (Testam). *Ser. Sci. INEAC* **98**, 1–83.

Anggreani-Hendranata (1975). Usaha untuk memperoleh variasi baru pada klon GT 1 dan PR 255 dengan cara iradiasi. *Menara Perkebunan* **43**, 245–249.

Anggreani-Hendranata (1977). Tinjauan mengenai tingkat infeksi secara alam dari pada *Oidium heveae* pada klon GT 1 yang diiradiasi. *Menara Perkebunan* **45**, 59–63.

Ayeke, C. A., and Onokpise, O. U. (1977). "Breeding and Selection of *Hevea brasiliensis*". Workshop on International Cooperation in *Hevea* Breeding and the Collection and Establishment of Materials from the Neo-tropics, Kuala Lumpur, 1977.

Baptiste, E. D. C. (1952). Recent progress in Malaya in the breeding and selection of clones of *Hevea brasiliensis*. *Rep. 13th Int. Hort. Congr., London 1952* **2**, 1100–1121.

Baptiste, E. D. C. (1953). Improvement of yields in *Hevea brasiliensis*. *Wld Crops* **5**, 194–198.

Baptiste, E. D. C. (1958). Director's report. *Rep. Rubb. Res. Inst. Ceylon (1957)* 1–16.

Baptiste, E. D. C. (1959). Director's report. *Rep. Rubb. Res. Inst. Ceylon (1958)*, 1–13.

Baptiste, E. D. C. (1960). Director's report. *Rep. Rubb. Res. Inst. Ceylon (1959)*, 1–16.

Baptiste, E. D. C. (1961). Breeding for high yield and disease resistance in *Hevea*. *Proc. Nat. Rubb. Res. Conf., Kuala Lumpur 1960*, pp. 430–445.

Baptiste, E. D. C. (1962). Les possibilities actuelles de la culture de l'Hévéa (Present possibilities of *Hevea* culture). *Rev. Gen. Caouth.* **39**, 1347–1374.

Bolle-Jones, E. W. (1954). Nutrition of *Hevea brasiliensis* I. Experimental methods. *J. Rubb. Res. Inst. Malaya*, **14**, 183–208.

Bos, H., and McIndoe, K. G. (1965). Breeding of *Hevea* for resistance against *Dothidella ulei* P. Henn. *J. Rubb. Res. Inst. Malaya* **19**, 98–107.

Bouharmont, J. (1960). Recherches taxonomiques et caryologiques chez quelques espéces du genre *Hevea*. *Ser. Sci. INEAC* **85**, 1–64.

Brookson, C. W. (1953). The breeding and selection of *Hevea brasiliensis*. *Arch. Rubbercult.* May 1953 (Extra No. 1), 96–106.

Brookson, E. V. (1956). Importation and development of new strains of *Hevea brasiliensis* by the Rubber Research Institute of Malaya. *J. Rubb. Res. Inst. Malaya* **14**, 423–448.

Camacho, E. V., and Jimenez, E. S. (1963). Resultados preliminares de una prueba de inducción de floración prematura en árboles jóvenes de *Hevea*. *Turrialba* **13**, 186–188.

Campaignolle, J. and Bouthillon, J. (1955). Pollinisation artificielle sur jeunes heveas conduits en espalier. *Rapp. Inst. Rech. Caouth. Indoch. (1954)*, 64–65.

Chee, K. H. (1969). *Phytophthora* leaf disease in Malaysia. *J. Rubb. Res. Inst. Malaya* **21**, 79–87.

Chee, K. H. (1976). Assessing susceptibility of *Hevea* clones to *Microcyclus ulei*. *Ann. appl. Biol.* **84**, 135–145.

Chee, K. H. (1977). Combating South American Leaf Blight of *Hevea* by plant

breeding and other measures. *Planter, Kuala Lumpur* **53**, 287–296.
Chee, K. H., and Holliday, P. (1986). "South American Leaf Blight of *Hevea* Rubber". MRRDB Monograph No. 13, 50 pp. Malaysian Rubber Research and Development Board.
Chee, K. H., and Wastie, R. L. (1980). The status and future prospects of rubber diseases in Tropical America. *Rev. Pl. Path.* **59**, 541–548.
Chen, C., Chen, F., Chien, C., Wang, C., Chang, S., Hsu, H., Ou, S., Ho, Y., and Lu, T. (1979). A process of obtaining pollen plants of *Hevea brasiliensis* Muell. Arg. *Scientia Sinica* **22**, 81–90.
Chen, Z., Quian, L., Xu, X., and Deng, Z. (1982a). Anther culture techniques of rubber tree and sugar cane. (*Proc. Vth Int. Congr. Plant Tissue and Cell Culture, Tokyo 1982.*) *In* "Plant Tissue Culture 1982" (Akio Fujiwara, ed.), pp. 533–534. The Japanese Association for Plant Tissue Culture, Tokyo, Japan.
Chen, Z., Quian, L., Qin, M., Xu, X., and Xiao, Y. (1982b). Recent advances in anther culture of *Hevea brasiliensis* (Muell.–Arg.). *Theor. appl. Genet.* **62**, 103–108.
Combe, J. C., Nicolas, D., and Du Plessix, C. J. (1974). Improvement in *Hevea* breeding: A new approach. *Int. Rubb. Res. Development Board Symp.* (Part 1), Cochin 1974.
Cramer, P. J. S. (1938). Grading young rubber plants with the 'Testatex' knife. *Proc. Rubb. Technol. Conf., London 1938*, 10–16.
De Silva, C. A. (1961). The performance of RRIC clones under Ceylon conditions. *Proc. Nat. Rubb. Res. Conf., Kuala Lumpur 1960*, 378–391.
De Silva, C. A., and Chandrasekara, L. B. (1959). A method of inducing floral stimulus for early flowering of *Hevea brasiliensis*. *Q. Jl Rubb. Res. Inst. Ceylon* **35**, 50–55.
Dijkman, M. J. (1951). *In* "*Hevea*, Thirty years of research in the Far East". University of Miami Press, Florida.
Dintinger, J., Nicolas, D., and Nouy, B. (1981). New early *Hevea* selection criteria: description and first results. *Rev. Gen. Caouth.* No. 609, 85–91.
Edgar, A. T. (1958). *In* "Manual of Rubber Planting (Malaya)". Incorporated Society of Planters, Kuala Lumpur.
Fernando, D. M. (1966). An outline of the breeding, selection and propagation of rubber. *Q. Jl Rubb. Res. Inst. Ceylon* **42**, 9–12.
Fernando, D. M. (1969). Breeding for multiple characters of economic importance in *Hevea*. Preliminary assessment of recent selection. *J. Rubb. Res. Inst. Malaya* **21**, 27–37.
Fernando, D. M. (1973). Trends in the improvement of rubber planting material with particular reference to Sri Lanka. *Int. Rubb. Res. Development Board Symp.*, Bogor 1973.
Fernando, D. M. (1977). Some aspects of *Hevea* breeding and selection. *J. Rubb. Res. Inst. Sri Lanka* **54**, 17–32.
Fernando, D. M., and De Silva, M. S. C. (1971). A new basis for the selection of *Hevea* seedlings. *Q. Jl Rubb. Res. Inst. Ceylon* **48**, 19–30.
Fernando, D. M., and Liyanage, A. De S. (1976). *Hevea* breeding for leaf and panel disease resistance in Sri Lanka. *Proc. Int. Rubb. Conf., Kuala Lumpur 1975* **2**, 236–246.
Fernando, D. M., and Liyanage, A. De S. (1980). South American Leaf Blight resistance studies on *Hevea brasiliensis* selections in Sri Lanka. *J. Rubb. Res. Inst. Sri Lanka* **57**, 41–47.
Ferwerda, F. P. (1969). Rubber, *Hevea brasiliensis* (wild) Muell. Arg. *In* "Outlines

of Perennial Crops Breeding in the Tropics" (F. P. Ferwerda and F. Wit, eds), pp. 427–458. H. Veenman and N. V. Zonen, Wageningen, The Netherlands.

Gasparotto, L. Trindade, D. R., and Silva, H. M. (1984). "Doenças Da Seringueira". EMBRAPA–CNPSD, Technical Circular No. 4, 71 pp.

Ghandimathi, H., and Yeang, H. Y. (1984). The low fruit set that follows conventional hand pollination in *Hevea brasiliensis*: insufficiency of pollen as a cause. *J. Rubb. Res. Inst. Malaysia* **32**, 20–29.

Gilbert, N. (1967). Additive combining abilities fitted to plant breeding data. *Biometrics* **23**, 45–49.

Gilbert, N. E., Dodds, K. S., and Subramaniam, S. (1973). Progress of breeding investigations with *Hevea brasiliensis* V. Analysis of data from earlier crosses. *J. Rubb. Res. Inst. Malaya* **23**, 365–380.

Gonçalves, P. de S., Paiva, J. R. de, and Souza, R. A. de (1983). "Retrospectiva E Atualidade Do Melhoramento Genético Da Seringueira (Hevea spp.) No Brazil E Em Paises Asiáticos". EMBRAPA–CNPSD, Série Documentos No. 2, 69 pp.

Gunnery, H. (1935). Yield prediction in *Hevea*: a study of sieve-tube structure in relation to latex yield. *J. Rubb. Res. Inst. Malaya* **6**, 8–20.

Hamaker, C. M. (1914). Plantwijdte en uitdunning bij *Hevea*. Prae-advies verslagen van het. Int. Rubb. Congress, Batavia.

Hilton, R. N. (1955). South American Leaf Blight: a review of the literature relating to its depredations in South America, its threat to the Far East, and the methods available for control, with appendix of South American leaf disease of rubber by R. A. Altson. *J. Rubb. Res. Inst. Malaya* **14**, 287–354.

Hilton, R. N. (1959). "Maladies of *Hevea* in Malaya". Rubber Research Institute of Malaya, Kuala Lumpur.

Ho, C. Y. (1972). Investigations on shortening the generative cycle for yield improvement in *Hevea brasiliensis*. M.Sc. thesis, Cornell University, USA.

Ho, C. Y. (1976). Clonal characters determining the yield of *Hevea brasiliensis*. *Proc. Int. Rubb. Conf.*, *Kuala Lumpur 1975* **2**, 17–38.

Ho, C. Y. (1978). Conservation and utilization of *Hevea* genetic resources. SABRAO workshop on Genetic Resources of Plants, Animals and Microorganisms, Kuala Lumpur.

Ho, C. Y. (1979). Contributions to improve the effectiveness of breeding, selection and planting recommendations of *Hevea brasiliensis* Muell. Arg. D. Agric. Sci. thesis, University of Ghent, Belgium.

Ho, C. Y. (1986). Rubber, *Hevea brasiliensis. In* "Breeding for Durable Resistance in Perennial Crops". FAO Plant Prod. and Prot. Paper **70**, 85–114.

Ho, C. Y., and Yoon, P. K. (1972). A synergistic approach to *Hevea* breeding. Symp. on International Cooperation in *Hevea* breeding, Kuala Lumpur.

Ho, C. Y., Chan, H. Y., and Lim, T. M. (1974). Enviromax planting recommendation—a new concept in choice of clones. *Proc. Rubb. Res. Inst. Malaysia Plrs' Conf.*, *Kuala Lumpur 1974*, 293–320.

Ho, C. Y., Khoo, S. K., Meignanaratnam, K., and Yoon, P. K. (1980). Potential new clones from mother-tree selection. *Proc. Rubb. Res. Inst. Malaysia Conf.*, *Kuala Lumpur 1979*, 201–216.

Ho, C. Y., Ng, A. P., and Subramaniam, S. (1969). Choice of clones. *Plrs' Bull. Rubb. Res. Inst. Malaya* no. 104, 226–247.

Holliday, P. (1970). South American Leaf Blight (*Microcyclus ulei*) of *Hevea brasiliensis*. *Commonw. mycol. Inst. Phytopathol. Pap.* 12, 1–31.

Huang, X., Wei, L., Zhan, S., Chen, C., Zhou, Z., Yuen, X., Guo, Q., and Lin, J. (1981). A preliminary study of relations between latex vessel system of rubber leaf blade and yield prediction at nursery. *Chinese J. Trop. Crops* **2**, (1), 16–20. (Chinese with English abstract)

Jayasekera, N. E. M., and Fernando, D. M. (1977). Hevea introduction (non-Wickham) into Sri Lanka. Workshop on International Collaboration in *Hevea* Breeding and The Collection and Establishment of Materials from the Neotropics, Kuala Lumpur.

Jayasekera, N. E. M., Samaranayake, P., and Karunasekara, K. B. (1977). Initial studies on the nature of genotype–environment interactions in some *Hevea* cultivars. *J. Rubb. Res. Inst. Sri Lanka* **54**, 33–42.

Khoo, S. K., Tan, H., Ho, C. Y., and Yoon, P. K. (1981). Progress of mother-tree selection in the Rubber Research Institute of Malaysia. Int. Rubb. Res. Development Board Symp., Hat Yai.

Khoo, S. K., Yoon, P. K., Meignanaratnam, K., Gopalan, A., and Ho, C. Y. (1982). Early results of mother-tree (ortet) selection. *Plrs' Bull. Rubb. Res. Inst. Malaysia* no. 171, 33–49.

Kuneman, J. H. (1939). Een beschouwing over het Hamaker–Morris–Mann-systeem. *De Bergcult.* **13**, 1741–1747.

Langdon, K. R. (1965). Relative resistance or susceptibility of several clones of *Hevea brasiliensis* and *Hevea benthamiana* to the races of *Dothidella ulei*. *Pl. Dis. Reptr* **49**, 12–14.

Langford, M. H. (1945). South American Leaf Blight of *Hevea* rubber trees. *U.S. Dept Agric. Tech. Bull.* No. 882, 1–31.

Leong, W., and Yoon, P. K. (1976). RRIM crown budding trials—progress report. *Proc. Rubb. Res. Inst. Malaysia Plrs Conf., Kuala Lumpur*, 1976, 87–115.

Leong, W., and Yoon, P. K. (1978). Some properties of latex and raw rubbers from three-part-trees. Int. Rubb. Res. Development Board Symp., Kuala Lumpur.

Leong, W., Yip, E., Subramaniam, A., Loke, K. M., and Yoon, P. K. (1986). The modification of rubber properties by crown budding. *Proc. Int. Rubb. Conf., Kuala Lumpur 1985* **2**, 3–19.

Lim, T. M. (1973). A rapid laboratory method of assessing susceptibility of *Hevea* clones to *Oidium heveae*. *Exptl Agric.* **9**, 275–279.

Madjid, A. (1977). Introduction and establishment of *Hevea* materials to Indonesia. Workshop on International Collaboration in *Hevea* Breeding and the Collection and Establishment of Materials from Neo-Tropics, Kuala Lumpur.

Madjid, A., Maslichah, Salleh, D., and Bunjamin. (1977). Percubaan menginduksi pembungaan pada *Hevea*. *Menara Perkebunan* **45**, 23–29.

Markose, V. C. (1975). Colchiploidy in *Hevea brasiliensis* (Muell. Arg.) *Rubb. Bd Bull.* **12**, 3–5.

Markose, V. C., Saraswathy Amma, C. K., Sulochanamma, S., and Nair, V. K. B. (1974). Mutation and polyploidy breeding in *Hevea brasiliensis*. Int. Rubb. Res. Development Board Scientific Symp. (Part 1), Cochin.

Markose, V. C., Panikkar, A. O. N., Annamma, Y., and Nair, V. K. B. (1977). Effect of gamma rays on rubber seeds germination, seedling growth and morphology. *J. Rubb. Res. Inst. Sri Lanka* **54**, 50–64.

McIndoe, K. G. (1961). The breeding of *Hevea brasiliensis*. *Rubb. Chem. Technol.* **34**, 413–423.

Mendes, L. O. T. (1971). Poliploidização da seringueira: um nôvo teste para determinação da capacidade de produção de seringueiras jovens. *Polimeros* **1**, (1), 22–30.

Mendes, L. O. T., and Mendes, A. J. T. (1963). Poliploidia artificial em seringueira (*Hevea brasiliensis* Muell. Arg.) *Bragantia* **22**, 383–392.

Mohd. Noor, A. G., Ong, S. H., and Tan, H. (1980). Genetic conservation and utilization of *Hevea* germplasm in the Rubber Research Institute of Malaysia. *Proc. Int. Symp. on Conservation Inputs from Life Sciences, National University of Malaysia, Kuala Lumpur, 1980* 83–85.

Morris, L. E., and Mann, C. E. T. (1932–1938). Report Botanical Division, *Rep. Rubb. Res. Inst. Malaya (1931–1937)*.

Moraes, V. H. de F. (1984). Técnica de obtenção de poliploides de seringueira. *Seminário Nacional Da Seringueira, 4*, Salvador.

Nair, V. K. B., George, P. J., and Saraswathy Amma, C. K. (1976). Breeding improved *Hevea* clones in India. *Proc. Int. Rubb. Conf., Kuala Lumpur 1975* **2**, 45–54.

Najib, L. and Paranjothy, K. (1978). Induction and control of flowering in *Hevea. J. Rubb. Res. Inst. Malaysia* **26**, 123–134.

Ng, A. P., Sultan, M. O, and Yoon, P. K. (1972). Selection of crowns for top working. *Proc. Rubb. Res. Inst. Malaya Plrs' Conf., Kuala Lumpur 1971*, 154–171.

Nga, B. H., and Subramaniam, S. (1974). Variation in *Hevea brasiliensis* I. Yield and girth data of the 1937 hand pollinated seedlings. *J. Rubb. Res. Inst. Malaysia* **24**, 69–74.

Nicolas, D. (1976). Contribution a l'étude de la floration précoce de l'*Hevea. Rev. Gen. Caoutch.* No. 566, 80–82.

Nicolas, D., Magnin, E., and Hofmann, J. P. (1979). "Contribution à l'étude des relations entre certaines caractéristiques du bois et la sensibilité à la casse chez des clones d'*Hevea brasiliensis*". Doc. IRCA, Feb. 1979, 46 pp.

Ong, S. H. (1972). Flower induction in *Hevea*. Symp. on International Cooperation in *Hevea* Breeding, Kuala Lumpur, 1972.

Ong, S. H. (1977). Investigations on clones derived from seedlings of various *Hevea* species introduced in 1966 by the Rubber Research Institute of Malaya. Workshop on International Collaboration in *Hevea* Breeding and the Collection and Establishment of Materials from the Neo-tropics, Kuala Lumpur, 1977.

Ong, S. H. (1979). Cytotaxonomic investigations of the genus *Hevea*. Ph.D. thesis, University of Malaya, Malaysia.

Ong, S. H. (1980). Correlations between yield, girth and bark thickness of RRIM clones trials. *J. Rubb. Res. Inst. Malaysia* **29**, 1–14.

Ong, S. H. (1982a). *Hevea* germplasm collection from South America. *In* "Malaya Peninsular Agriculture Association 1982 Year Book", pp. 73–77. Malay Peninsula Agric. Assn, Penang, Malaysia.

Ong, S. H., and Subramaniam, S. (1973). Mutation breeding in *Hevea brasiliensis* Muell. Arg. *In* "Induced Mutations in Vegetatively Progated Plants", pp. 117–127. IAEA, Vienna 1973.

Ong, S. H., and Tan, A. M. (1976). Performance of Ford, FX and IAN series clones in RRIM trial. Int. Rubb. Res. Development Board Symp., Bogor, 1976.

Ong, S. H., and Tan, H. (1987). Utilization of *Hevea* genetic resources in the RRIM. *Mal. Appl. Biol.* **16** (in press).

Ong, S. H., Tan, A. M., and Chee, K. H. (1977). Breeding for resistance against *Hevea* leaf diseases. Workshop on International Collaboration in *Hevea* Breeding and the Collection and Establishment of Materials from the Neotropics, Kuala Lumpur, 1977.
Ong, S. H., Mohd. Noor, A. G., Tan, A. M., and Tan, H. (1983). New *Hevea* germplasm – Its introduction and potential. *Proc. Rubb. Res. Inst. Malaysia Plrs' Conf., Kuala Lumpur, 1983*, 3–7.
Ong, S. H., Naimah, I., Mohd. Noor, A. G., and Tan, H. (1984). Results of RRIM work on induced mutation and polyploidy of breeding. *In* "Compte Rendu du Colloque Exploitation-Physiologie et Amélioration de l'*Hevea*", pp. 383–399. l'Institute de Recherches sur le Caoutchouc, Paris and Montpellier, 1984.
Ong, S. H., Tan, H., Khoo, S. K., and Sultan, M. O. (1986). Promising clones through accelerated approach of *Hevea* selection. *Proc. Int. Rubb. Conf., Kuala Lumpur, 1985* **3**, 157–174.
Paardekooper, E. C. (1964a). Report on the RRIM 'Exchange Clones' trial (Group B trials) III. *Rubb. Res. Inst. Malaya Res. Arch. Doc. no. 25.*
Paardekooper, E. C. (1964b). Report on the RRIM 600 series distributed block trials (Group A1 trials). *Rubb. Res. Inst. Malaya Res. Arch. Doc. No. 35.*
Paardekooper, E. C., and Napitupulu, L. A. (1972). A brief review of rubber breeding activities in North Sumatra during the period 1950–71. Symp. on International Cooperation in *Hevea* Breeding, Kuala Lumpur, 1972.
Paranjothy, K. and Rohani, O. (1978). Embryoid and plantlet development from cell culture of *Hevea*. 4th Int. Congr. Plant Tissue and Cell Culture, University of Calgary, Canada.
Peries, O. S., and Liyanage, A. De S. (1986). *Hevea* diseases of economic importance and integrated methods of control. *Proc. Int. Rubb. Conf., Kuala Lumpur, 1985* **3**, 255–269.
Rands, R. D. (1942). *Hevea* rubber cultivation in Latin America. Problem and procedures—II. *India Rubber World* **106**, 8–10.
Rao, B. S. (1973). Some observations on South American Leaf Blight in South America. *Planter, Kuala Lumpur* **49**, 2–9.
Rao, B. S. (1974). Current status of diseases and pests of rubber in the South East Asia and Pacific region. *Wld Crops* **26**, 75–77.
Rohani, O., and Paranjothy, K. (1980). Induced flowering in young *Hevea* buddings. *J. Rubb. Res. Inst. Malaysia* **28**, 149–156.
Ross, J. M. (1959). Breeding and selection in *Hevea brasiliensis. Planter, Kuala Lumpur* **35**, 548–554, 558.
Rubber Research Institute of Malaya (1962). *Rep. Rubb. Res. Inst. Malaya (1961)*, p. 40.
Rubber Research Institute of Malaya (1969). Planting recommendations 1969–70. *Plrs' Bull. Rubb. Res. Inst. Malaya* No. 100, 3–23.
Rubber Research Institute of Malaysia (1971–1978). *Rep. Rubb. Res. Inst. Malaysia (1970–77)*.
Rubber Research Institute of Malaysia (1975a). Enviromax planting recommendations, 1975–76. *Plrs' Bull. Rubb. Res. Inst. Malaysia* No. 137, 27–50.
Rubber Research Institute of Malaysia (1975b). *Rep. Rubb. Res. Inst. Malaysia (1974)*, p. 116.
Rubber Research Institute of Malaysia (1983). RRIM planting recommendations 1983–1985. *Plrs' Bull. Rubb. Res. Inst. Malaysia* No. 175, 37–55.

Rubber Research Institute of Malaysia (1984). *Rep. Rubb. Res. Inst. Malaysia (1983)*, 31–32.
Rubber Research Institute of Malaysia (1985). *Rep. Rubb. Res. Inst. Malaysia, (1984)*, 17–18.
Samsuddin, Z., Tan, H., and Yoon, P. K. (1986). Variations, heritabilities and correlations of photosynthetic rates, yield and vigour in young *Hevea* seedling progenies. *Proc. Int. Rubb. Conf., Kuala Lumpur 1985* **3**, 137–153.
Samsuddin, S., Tan, H., and Yoon, P. K. (1987a). Correlation studies on photosynthetic rates, girth and yield in *Hevea brasiliensis*. *J. Nat. Rubb. Res.* **2**, 46–54.
Samsuddin, Z., Tan, H., and Yoon, P. K. (1987b). Studies on photosynthetic rates and its early implications in *Hevea* breeding. *Mal. Appl. Biol.* **16** (in press).
Santos, P. M., Sena Gomes, A. R., Murques, J. R. B., and Virgens Filho, A. C. (1984). Desempenho de clones diplōides e poliplōides de seringueira (*Hevea* sp.) no sul da Bahia. *Seminário Nacional Da Seringueira*, **4**, Salvador.
Saraswathy Amma, C. K. (1975). Induction of early flowering in *Hevea*: a preliminary study. *Rubb. Bd Bull. (India)* **12**, 6.
Satchuthananthavale, Vimala Devi, Satchuthananthavale, R., and Dantanarayanan, D. M. (1974). The evaluation of inherent resistance/susceptibility of *Hevea* cultivars to black stripe disease. Int. Rubb. Res. Development Board Symp. (Part 1), Cochin.
Schultes, R. E. (1977). Wild *Hevea*: an untapped source of germplasm. *J. Rubb. Res. Inst. Sri Lanka* **54**, 227–257.
Sena Gomes, A. R., Virgens Filho, A. C., Marques, J. R. B., and Melo, J. R. V. (1983). Performance de algumas combinações (clones-copa × painel) em seringueira (*Hevea* sp.). *CEPLAC Bolm Técnico* **107**, 15 pp.
Senanayake, Y. D. A., and Samaranayake, P. (1970). Intraspecific variation of stomatal density in *Hevea brasiliensis* Muell. Arg. *Q. Jl Rubb. Res. Inst. Ceylon* **46**, 61–68.
Senanayake, Y. D. A., and Wijewantha, R. T. (1968). Synthesis of *Hevea* cultivars: a new approach. *Q. J. Rubb. Res. Inst. Ceylon* **44**, 16–26.
Sharp, C. C. T. (1940). Progress of breeding investigations wih *Hevea brasiliensis*. The Pilmoor crosses 1928–1931 series. *J. Rubb. Res. Inst. Malaya* **10**, 34–66.
Sharp, C. C. T. (1951). Progress of breeding investigaions with *Hevea brasiliensis* II. The crosses made in the years 1937–1941. *J. Rubb. Res. Inst. Malaya* **13**, 73–99.
Shepherd, R. (1969a). Aspects of *Hevea* breeding and selection investigation undertaken on Prang Besar Estate. *Plrs' Bull. Rubb. Res. Inst. Malaya* No. 104, 207–219.
Shepherd, R. (1969b). Induction of polyploidy in *Hevea brasiliensis*—preliminary observations on trials conducted at Prang Besar Estate. *Plrs' Bull. Rubb. Res. Inst. Malaya* No. 104, 248–256.
Simmonds, N. W. (1969). Genetical bases of plant breeding. *J. Rubb. Res. Inst. Malaya* **21**, 1–10.
Simmonds, N. W. (1979). "Principles of Crop Improvement", Longman, London and New York.
Simmonds, N. W. (1982). Some ideas on botanical research on rubber. *Trop. Agric.* **59**, 1–8.
Simmonds, N. W. (1987). Rubber breeding. *In* "Rubber" (C. C. Webster and

W. J. Baulkwill, eds). Longman, London and New York.
Sivanadyan, K., and Ghandimathi, H. (1985). The interrelationship between nitrogen manuring and vegetative growth, flowering and fruiting in *Hevea brasiliensis*. *In* "Proc. Int. Conf. Soil and Nutrition of Perennial Crops" (A. T. Bachik and E. Pushparajah, eds), pp. 31–40. Malaysian Soc. Soil Sci., Kuala Lumpur.
Sivanadyan, K., and Ghandimathi, H. (1986). Mineral nutrition and reproduction in *Hevea*; effects of nitrogen fertilization on flowering and fruiting in immature trees of some clones. *J. Nat. Rubb. Res.* **1**, 155–166.
Subramaniam, S. (1969). Performance of recent introductions of *Hevea* in Malaya. *J. Rubb. Res. Inst. Malaya* 21, 11–18.
Subramaniam, S. (1970). Performance of *Dothidella* resistant *Hevea* clones in Malaya. *J. Rubb. Res. Inst. Malaya* **23**, 39–46.
Subramaniam, S. (1980). Developments in *Hevea* breeding research and their future. Seminário Nacional Da Seringueira, 3. Manāus.
Subramaniam, S. and Ong, S. H. (1973). Conservation of gene pool in *Hevea*. 2nd Gen. Congr. SABRAO, New Delhi.
Summers, F. (1930). The early diagnosis of high yielding plants. *In* "The Improvement of Yield in *Hevea brasiliensis*", pp. 177–198. Kelly and Welsh Ltd, Singapore, Shanghai and Hong Kong.
Tan, H. (1977). Estimates of general combining ability in *Hevea* breeding at the Rubber Research Institute of Malaysia I. Phases II and IIIA. *Theoret. appl. Genet.* **50**, 29–34.
Tan, H. (1978a). Assessment of parental performance for yield in *Hevea* breeding. *Euphytica* **27**, 521–528.
Tan, H. (1978b). Estimates of parental combining abilities in rubber (*Hevea brasiliensis*) based on young seedling progeny. *Euphytica* **27**, 817–823.
Tan, H. (1981). Estimates of genetic parameters and their implications in *Hevea* breeding. *In* "Crop Improvement Research" (Proc. 4th Int. Congr. SABRAO, Kuala Lumpur 1981) (T. C. Yap, K. M. Graham and Jalani Sukami, eds), pp. 439–446. The Society for Advancement of Breeding Researches in Asia and Oceania.
Tan, H., and Subramaniam, S. (1976). A five-parent diallel cross analysis of certain characters of young *Hevea* seedlings, *Proc. Int. Rubb. Conf., Kuala Lumpur 1975* **2**, 13–26.
Tan, H., Ong, S. H., Sultan, M. O., and Khoo, S. K. (1981). Potential of promotion plot clone trials in shortening clonal evaluation period in *Hevea* breeding. Int. Rubb. Res. Development Board Symp., Hat Yai.
Tan, H., Ong, S. H., Sultan, M. O., and Khoo, S. K. (1984). Accelerated approach to *Hevea* selection. Seminário Nacional Da Seringueira, 4. Salvador.
Tan, H. T., and Lubis, P. (1961). Pertjobaan menpradinikan (vervroegen) pembungaan pada *Hevea*. Nota Bal. Penjelid. Gappersu, No. 3/61, Ser. Karet, No. 6.
Templeton, J. K. (1969). Where lies the yield summit for *Hevea*? *Plrs' Bull. Rubb. Res. Inst. Malaya* No. 104, 220–225.
Townsend, C. H. T. (1960). Progress in developing superior *Hevea* clones in Brazil. *Econ. Bot.* **14**, 189–196.
Tupy, J. (1969). Nucleic acids in latex and the production of rubber in *Hevea brasiliensis*. *J. Rubb. Res. Inst. Malaya* **21**, 468–476.
Tysdal, H. M., and Rands, R. D. (1953). Breeding for disease resistance and

higher rubber yield in *Hevea*, Guayule and Kok-saghyz. *Agron. J.* **45**, 234–243.

Waidyanatha, U. P. De S., and Fernando, D. M. (1972). Studies on a technique of micro-tapping for the estimation of yields in nursery seedlings of *Hevea brasiliensis*. *Q. Jl Rubb. Inst. Ceylon* **49**, 6–12.

Wan Abdul Rahaman, W. Y., Mohd. Ghouse, W., and Cheong, K. F. (1985). *Hevea* tissue culture—prospects and retrospect. The First International Rubber Tissue Culture Workshop of the International Rubber Research and Development Board, Kuala Lumpur.

Wang, Z., and Huang, S. (1981). Induction of dwarfing and early flowering rubber trees. *Chinese J. Trop. Crops* **2**, (1), 10–15. (Chinese with English abstract)

Wang, Z., and Wu, H. (1985). Inheritable characters of anther somatogenic plants of *Hevea brasiliensis*. The First International Rubber Tissue Culture Workshop of the International Rubber Research and Development Board, Kuala Lumpur.

Wang, Z., Zeng, X., Chen, C., Wu, H., Li, Q., Fan, G., and Lu, W. (1980). Induction of rubber plantlets from anther of *Hevea brasiliensis* Muell. Arg. *in vitro*. *Chinese J. Trop. Crops* **1**, (1), 16–26. (Chinese with English abstract)

Wastie, R. L. (1967). *Gloeosporium* leaf disease of rubber in West Malaysia. *Planter, Kuala Lumpur* **43**, 553–565.

Wastie, R. L. ((1973). Nursery screening of *Hevea* for resistance to *Gloeosporium* leaf disease. *J. Rubb. Res. Inst. Malaysia* **23**, 339–350.

Wastie, R. L. (1986). Disease resistance in rubber. *FAO Plant Prot. Bull.* **34**, 193–199.

Watson, G. A., and Narayanan, R. (1965). Effect of fertilisers on seed production by *Hevea brasiliensis*. *J. Rubb. Res. Inst. Malaya* **19**, 22–31.

Wycherley, P. R. (1968). Introduction of *Hevea* to the Orient. *Planter, Kuala Lumpur* **44**, 127–137.

Wycherley, P. R. (1969). Breeding of *Hevea*. *J. Rubb. Res. Inst. Malaya* **21**, 38–55.

Wycherley, P. R. (1971). *Hevea* seed. Part I. *Planter, Kuala Lumpur* **47**, 291–298.

Yoon, P. K. (1967). RRIM crown budding trials. *Plrs' Bull. Rubb. Res. Inst. Malaya* No. 92, 240–249.

Yoon, P. K. (1971). Further progress in crown budding. *Proc. Rubb. Res. Inst. Malaya Plrs' Conf., Kuala Lumpur 1971*, pp. 143–153.

Zhang, K. M., and Chee, K. H. (1985). Distinguishing *Hevea* clones resistant to races of *Microcyclus ulei* by means of leaf diffusates. *J. Rubb. Res. Inst. Malaysia* **33**, 105–108.

Zheng, X., Zeng, X., Chen, X., and Yang, G. (1980). A further report on induction and cytological studies on polyploid mutants of *Hevea* (I). *Chinese J. Trop. Crops* **1**, (1), 27–31. (Chinese with English abstract)

Zheng, X., Zeng, X., Chen, X., and Yang, G. (1981). A further report on induction and cytological studies on polyploid mutants of *Hevea* (II). *Chinese J. Trop. Crops* **2**, (1), 1–9. (Chinese with English abstract)

Zhou, Z., Yuan, X., Guo, Q., and Huang, X. (1982). Studies on the method for predicting rubber yield at the nursery stage and its theoretical basis. *Chinese J. Trop. Crops* **3**, (2), 1–18. (Chinese with English abstract)

# 3 Breeding and Selecting the Oil Palm

J. J. HARDON[1], R. H. V. CORLEY[2] and C. H. LEE[3]

[1] *Centre for Genetic Resources, Wageningen, The Netherlands*
[2] *Unilever Plc, Blackfriars, London*
[3] *Harrisons and Crossfield Oil Palm Research Station, Selangor, Malaysia*

The oil palm, *Elaeis guineensis*, which belongs to the palm subfamily Cocoideae, is an increasingly important tropical perennial crop: in 1980, palm oil represented 14% of world trade in vegetable oils. The palm has a single growing point, and no natural means of vegetative reproduction. It is a cross-pollinating species, producing separate male and female inflorescences in cycles of varying duration, with one inflorescence per leaf axil and usually no overlap between anthesis of successive inflorescences. A mature palm produces 20–26 leaves per year, and up to 18 fruit bunches; each bunch contains 500–3000 fruits, and weighs 5–30 kg, depending on palm age, growing conditions and genotype. The fruit is a drupe, with a thin exocarp, a thick mesocarp containing palm oil, and a lignified endocarp or shell surrounding a white endosperm. The endosperm contains palm kernel oil, which has a different fatty acid composition to the mesocarp oil and is produced in smaller quantity. The economic life of a planting is normally between 20 and 25 years, after which the palms become too tall for harvesting, but individual palms over 100 years old exist.

The development of tissue-culture techniques for vegetative propagation of oil palm (Jones, 1974; Rabéchault and Martin, 1976) necessitates a change in emphasis of breeding programmes, which until now have been aimed solely at seed production. Vegetative propagation can be an important tool in breeding, offering the possibility of rapid multiplication of good genotypes. However, in some perennial crops, instead of genetic programmes aimed at population improvement, study of heritability and combining ability, analysis of yield components, and so on, improvement programmes have been reduced to a search for the chance "wonder tree", to be propagated *ad infinitum*. Oil palm breeders have the advantage of knowing a great deal about the genetics of their crop, and the challenge now is to make full use of this in producing base material for selection for vegetative propagation.

## 3.1 BREEDING FOR SEED PRODUCTION

Breeding and selection started around 1920, when the oil palm began to be developed as a plantation crop in Sumatra, Malaysia and Zaire. The industry in Sumatra and Malaysia commenced with seeds originating from four ornamental specimens in the Bogor Botanic Gardens in Java. This population is referred to as "Deli *dura*", *dura* being the thick-shelled fruit type. The programme in Zaire, at Yangambi, started from ten thin-shelled *tenera* palms. It was found that the offspring of the *teneras* included a percentage of female-sterile palms, in which the fruits aborted early in development, and also a proportion of thick-shelled *dura* palms. Single gene inheritance of the shell thickness character was established by Beirnaert and Vanderweyen (1941). The *dura* (D) is homozygous for the presence of shell, the usually female-sterile *pisifera* (P) is homozygous for absence of shell, and *tenera* (T) is the heterozygote with intermediate shell thickness. Within each fruit type there is some variation in shell thickness, which is under polygenic control. The *tenera* is preferred as planting material to the *dura* because it has up to 30% more oil-bearing mesocarp in the fruit. Thus, significant yield improvement in oil palm has been realised by the exploitation of a single gene.

Elucidation of the inheritance of shell thickness changed breeding strategy from mass selection in available commercial plantings, to the production of *duras* and *pisiferas*, which when crossed give high-yielding progenies of *teneras*. From the 1960s onwards almost all planting material consisted of such *teneras*.

Most commercial plantings of the oil palm in South-east Asia and West Africa can be traced back to the four botanical specimens in Bogor, and male parents (*pisiferas*) originating from a few palms in Zaire. More recent breeding programmes in Nigeria, the Ivory Coast and Cameroon have added additional *tenera* stock, but the genetic base of breeding populations remains very narrow. This narrowness has been recognised, and there is increasing interest in the collection of wild material (Meunier, 1969; Hardon, 1974; Arasu and Rajanaidu, 1977; Rajanaidu *et al.*, 1979).

### 3.1.1 Selection progress

Past work in oil palm breeding has stressed selection for oil yield per palm. Yields have improved significantly over the past few decades, but how much of this improvement is due to breeding and how much to better agronomic techniques is not entirely clear. Plantings of breeding material have usually been in different years, in different locations, and often under difficult agronomic regimes. Thus, straightforward yield comparisons are not valid.

Genetic variance analysis provides a theoretical basis for estimating expected selection progress, but although extensive analyses have been done in oil palm breeding experiments (see Hardon, 1976; Ooi, 1975a,b, 1978), the estimates obtained are rather variable and have large error variances. This is not surprising, if one considers the restricted genetic origin of most material in single experiments; the use of different reference populations with various degrees of inbreeding, and consequent non-random distribution of genes; and the large area required for an experiment, often leading to high environmental variance.

The rate of progress from early mass selection of *dura* palms is indicated by an experiment planted with seedlings from various generations of palms derived from the Deli *dura* population. Random seeds were collected from:

(a) A group of direct ($F_1$) descendants of the Bogor Botanic Garden palms, planted in 1878.
(b) A group of ornamental avenue palms at Tanjong Morawa in Sumatra, planted around 1885, and probably unselected second generation descendants of the Bogor palms.
(c) An old commercial planting at Elmina Estate in Malaysia, planted around 1930 and probably derived from first selections within third generation descendants of the Bogor palms.
(d) As a control, modern Deli *duras* in the third to fourth generation of selection were included.

Results are summarised in Table 3.1. Selection progress is evident if, as seems likely, the Bogor and Tanjong Morawa material represents a reliable sample of the original unselected material. A major improvement in oil yield of approximately 23% was realised in the first generation of selection, and subsequent selection progress was 10–15% per generation. (One generation of a breeding programme takes 8–10 years.)

A further major step forward came with the introduction of *tenera* planting material, which, as noted above, has an oil yield at least 30% higher than the *duras* in the same D×T progeny.

In present-day breeding programmes, the *dura* and *pisifera* parents for seed production are usually selected in different ways. *Duras* are commonly selected on phenotypic performance, but *pisiferas*, because they are mostly

**Table 3.1.** Comparison of unselected sources of Deli *dura* with oil palms in the first (Elmina) and fourth (OPRS) generation of selection.

| Parents[a] | No. of palms | Bunch yield (kg per palm) | Bunch yield (t ha$^{-1}$) | Mesocarp per fruit (%) | Oil per bunch (%) | Oil (t ha$^{-1}$) |
|---|---|---|---|---|---|---|
| 1. F$_1$ Bogor (1878) | 23 | 120 | 16.5 | 58.7 | 17.6 | 2.8 |
| 2. Tanjong Morawa (c. 1885) | 125 | 116 | 16.0 | 59.7 | 17.4 | 2.7 |
| 3. Elmina Estate (c. 1930) | 118 | 146 | 20.1 | 58.2 | 17.0 | 3.4 |
| 4. OPRS (1969) | 94 | 180 | 24.8 | 64.1 | 18.3 | 4.5 |

[a] OPRS refers to selected D×D crosses, while the other material derives from open-pollinated, randomly collected fruits.

female-sterile, must be selected either on the performance of *dura* and *tenera* sibs, or, more reliably, by progeny testing.

The benefit obtained from the first generation of *pisifera* progeny testing is shown in Table 3.2: the progenies of the best *pisifera* in each trial yielded 9–15% above the trial mean. In most trials the *pisiferas* tested would have been taken at random from within families (T×T crosses) selected on the performance of the *duras* and *teneras*. Thus, Table 3.2 gives an estimate of the selection progress achieved by one generation of progeny testing, over and above that already achieved by selection on family performance. No estimate of the superiority of *pisiferas* in selected families over randomly chosen pisiferas is available.

Although progress in the early generations of *dura* selection was good (Table 3.1), results from recent trials suggest that phenotypic selection of *duras* for production of *tenera* crosses is less effective in current material, which has already been subjected to several generations of selection. In several trials, groups of selected *duras* have been crossed with the same *pisifera*. Within 14 such groups, the best *tenera* progeny in each group had an oil yield 7% above the mean (Lee, unpubl.), but phenotypic selection within each group of *duras* was not effective in identifying the best parents, and gave an average improvement of only 3% while parent-offspring correlations indicated very low heritabilities. Other studies have shown that there is little additive variation for yield components left in the Deli *dura* breeding population (Thomas *et al.*, 1969; Ooi *et al.*, 1973).

**Table 3.2.** Results of *pisifera* (male parent) progeny testing, presented as average performance of progenies from the best *pisifera* in each trial, as % of trial mean.

| Trial | No. of pisiferas tested | Bunch yield | Oil per bunch | Oil yield |
|---|---|---|---|---|
| 1 | 6 | 113 | 102 | 115 |
| 2 | 7 | 112 | 100 | 112 |
| 3 | 9 | 107 | 104 | 111 |
| 4 | 9 | 105 | 108 | 113 |
| 5 | 5 | 103 | 108 | 111 |
| 6 | 10 | 113 | 101 | 114 |
| 7 | 5 | 109 | 100 | 109 |
| Mean | | 109 | 103 | 112 |

Each *pisifera* is crossed with several *duras*, the mean being taken to indicate the potential of the *pisifera*.

Where additive variation (general combining ability, or GCA) is limited, new variation may be introduced from wild, or at least unrelated, populations, or we must concentrate on specific combining ability (SCA). All oil palm breeding programmes are now looking for new material, and Ooi (1975b) has shown that, as expected, crossing the Deli *dura* with material from Zaire increases genetic variation. The Institut de Recherches pour les Huiles et Oléagineux (IRHO) programme, in particular, also places considerable emphasis on SCA. In this programme, based on reciprocal recurrent selection, the best combinations are selected from within extensive D×T test-cross programmes. The *dura* and *tenera* parents are selfed and it has been shown that if *duras* and *pisifera* within the selfings are crossed the original progeny can be "reproduced" (Gascon *et al.*, 1981; Jacquemard *et al.*, 1981).

An indication of the progress to be achieved by this technique can be obtained by examining the performance of the best individual D×P crosses in a trial, rather than the average performance of all progenies from the best *pisifera*. The best 10% of progenies in each trial (Table 3.2) averaged 18% higher yield than the trial mean. Gascon *et al.* (1981) show an improvement of 11–30%, with an average of 19%, from more extensive trials in the Ivory Coast. Here, 15% of progenies were selected, not a high selection pressure. For comparison, *pisifera* progeny testing and selection of the best 15% in a few Malaysian trials gave an average improvement of 12% (Table 3.2); combined with phenotypic selection of *duras*, progress of 15% in one generation should be obtainable. The improvement achieved by the IRHO appears slightly greater, but it must be borne in mind that two generations are required (D×T test crosses followed by selfings). In the IRHO programme, time is saved by making the selfings simultaneously with the test crosses, but costs are then increased, as selfings of all parents would have to be planted, rather than just offspring of selected parents.

Those figures reflect the progress achieved in the first generation of test crosses; rate of progress is expected to be lower in subsequent generations, though by how much is not yet known.

## 3.2 SELECTION FOR VEGETATIVE PROPAGATION

Vegetative propagation has many advantages, but also some disadvantages. The advantages include:

(a) more rapid selection progress;
(b) easy selection for combinations of characters;

*Breeding and Selecting the Oil Palm*

(c) easy introduction of new genetic material for commercial planting;
(d) more precise testing of selected parent palms, and improved prediction of field performance of planting material; and
(e) more uniform planting material, allowing precise definition of planting density and fertiliser regimes.

The main disadvantage is genetic uniformity within clones, with the consequent risk of epidemic pest or disease outbreak; this risk can be minimised by planting mixtures of clones, rather than monoclonal blocks. Planting material produced by tissue culture will also have a high cost, which may present a problem for small farmers.

**3.2.1 Expected yield improvement**

The major justification for introducing vegetative propagation is the expected improvement in yield over seedling material; where this material is genetically variable, propagation of the best individual plants may give an appreciable yield increase. The size of this increase will depend on the amount of genetic variation present in the seedling population, on the efficiency with which superior individuals can be identified, and on the number of clones which can be tested. These limitations do not differ from those involved in seed production, but the unit of selection will now be single *teneras*, instead of *dura* and *pisifera* parents. To estimate potential yields, therefore, we look first at the variation in yield of individual *teneras*.

There is considerable variation between palms within experiments; for example the coefficient of variation (CV) for oil yield of individual palms in the trial illustrated in Fig. 3.1 was 30%. For the two components, yield of fruit bunches and oil:bunch ratio, the CVs were 25% and 16%, respectively. The data used in Fig. 3.1 were averages of only 2 years; yield fluctuates considerably from year to year, and variability is somewhat reduced if averages over longer periods are taken. In a long-term fertiliser trial the CV for yield of fruit bunches over 10 years was 20%. Oil yields were not available, but are not likely to be less variable.

Given a CV of 20%, the best 5% of the population would have an expected average yield 40% above the mean. The extent to which this superiority would be shown by clones produced from such palms depends on the broad sense heritability of yield, since clonal propagation will fix both additive and nonadditive genetic variation, including epistasis which may well be a major factor. Heritability of oil yield has rarely been estimated; breeders have usually concentrated on yield components, but Jacquemard *et al.* (1981) and Soh (1986) found rather low broad sense

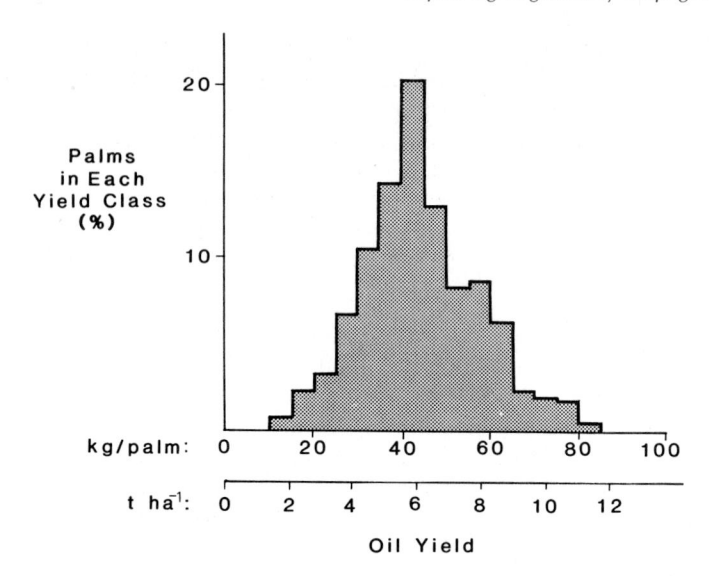

[Histogram showing palms in each yield class (%) vs oil yield, with x-axis in kg/palm (0 to 100) and t ha⁻¹ (0 to 12)]

**Figure 3.1.** Variation in yield of individual palms in a D × P progeny trial (mean of 2 years' data). Potential yield per ha is estimated assuming a standard density of 138 plants per ha.

heritabilities for oil yield in their experiments. Soh (1986) demonstrated that selection for oil yield alone might give an average yield increase of 11–21% in a population of restricted genetic origin.

The work on yield components is more encouraging: estimates of broad sense heritability of yield of fruit in the Deli *dura* population are quite high (Thomas *et al.*, 1969; Ooi *et al.*, 1973), though little of this variation is additive. There is also significant genetic variation in fruit composition and oil : bunch ratio (Ooi, 1975a; Hardon, 1976). Thus, phenotypic selection of palms for clonal propagation can be expected to be more successful than selection for use as seed parents, and it is likely that some outstanding clones could be produced from palms in the best 5% of a population such as that shown in Fig. 3.1. The efficiency of individual palm selection could be significantly improved if the environmental variance component (large in most tree crops) were assessed on a per palm basis. The planting of clonal palms as controls throughout selection trials might make that possible.

Another estimate of the progress possible can be made from performance of the first clones in field trials (Corley *et al.*, 1981). These first clones were propagated from unselected seedlings, rather than from selected elite palms, and show a wide range of variation in yield (Corley, 1982;

see also Jones, this volume). The best clone in one trial had an oil yield 29% higher than seedling controls, over the first three years of production (unpubl.). No firm conclusions can be based on such early yields, but it seems that yield increases of at least 30% should be possible using conventional selection criteria, followed by thorough field testing. This estimate must not be misunderstood: careful ortet selection should give a population of clones among which, after field testing, a few giving yield increases of 30% or more will be identified. Ortet selection alone, without field testing, cannot give such good results.

The upper limit to progress will be set by the physiological potential of the crop. At present, dry matter production rates of 25–30 t ha$^{-1}$ per annum are often obtained under good conditions in South-east Asia. With a harvest index (HI) of 22%, an oil yield of over 6 t ha$^{-1}$ can be obtained (HI is the dry weight of economic product as a proportion of total dry matter production). The best individual palms have an HI of over 35%; a clone combining such an HI with a dry matter production of 30 t ha$^{-1}$ would have an oil yield of over 10 t ha$^{-1}$.

Corley (1973) showed that, at higher than normal planting densities, a total dry matter production of 40 t ha$^{-1}$ per annum is possible, but at such densities the HI is low and oil yield is poor. Selection of palms capable of maintaining a high harvest index at high planting density could give considerable yield increases, and Corley (1983) has estimated a theoretical potential oil yield of 17 t ha$^{-1}$ per annum.

### 3.2.2 Selection objectives

Vegetative propagation does not alter selection objectives, but it should accelerate progress, and allow effective selection for more characteristics simultaneously. The selection criteria discussed below will be applicable both to selection of ortets for clonal propagation, and to selection of parent palms for further breeding.

The primary selection objective remains increased oil yield. The components, yield of fruit bunches per palm and oil:bunch ratio, have been extensively studied, and information on the heritability values for these components is used to establish their relative importance in family and individual palm selection (for a review, see Ooi and Ngah, 1977).

The third yield component, number of palms per unit area, has often been ignored, but breeders are starting to give some consideration to vegetative growth characteristics, both in relation to optimal planting density for selected material, and also in trying to improve selection within populations with varying levels of interpalm competition (Hardon et al.,

1972; Hirsch, 1980). Simple non-destructive methods have been developed for estimating leaf area and rates of dry matter production (Hardon *et al.*, 1969; Corley *et al.*, 1971), and some characters have been shown to be under reasonable levels of genetic control (Hardon *et al.*, 1972; Tan, 1978; Ooi, 1978; Breure and Corley, 1983).

3.2.2.1    Yield of fruit bunches

The yield of fruit per palm has a rather low heritability. The two components, bunch number per year and mean bunch weight have higher heritabilities, but are negatively correlated. Consequently, selection for these components is commonly done in separate lines, bunch weight being emphasised in the *dura* population and bunch number in the *pisifera*. For ortet selection, of course, this approach is not possible, so choice of ortets will have to involve careful consideration of family mean bunch yields and the yield of adjacent palms, with allowances for differences in palm height and leaf area which might give certain palms a competitive advantage.

3.2.2.2    Oil:bunch ratio

The oil content of the bunch is dependent on three components: fruit per bunch, mesocarp per fruit and oil per mesocarp. Standard methods of sampling and analysis have been developed (Blaak *et al.*, 1963; Rao *et al.*, 1983) for routine determination of these components for individual palms. Mesocarp:fruit generally has a high heritability, but Sparnaaij (1969) and Van der Vossen (1974) found poor correlations between mesocarp:fruit and shell:fruit of *duras* and their *tenera* offspring, and suggested that degree of lignification of the endocarp in the *tenera* may be controlled by a polygenic system, which is not expressed in the thick-shelled *dura*. If results from other programmes support this, then good *dura* parents may have to be identified by progeny testing (Hardon, 1976).

Fruit per bunch and oil per mesocarp usually show low heritabilities, but this is probably not due to a lack of genetic variation but rather to a large environmental variance component. Clonal testing may well allow identification of genetically controlled superiority for these characters which may have a very significant effect on yield. The oil:bunch ratio is further affected by the degree of ripeness. Fruit ripening in a bunch proceeds along two axes: from the top to the bottom and from outer to inner fruits. Reducing these gradients may be possible through selection, to increase the overall ripeness of fruits at harvesting.

### 3.2.2.3 Yield per unit area

In the past, oil palm breeders have concentrated on selecting for oil yield per palm, at a fixed planting density. It seems probable, though, that different progenies will have different optimal planting densities (Corley et al., 1971; Breure and Corley, 1983) and this is even more likely to be true of clones. Growth analysis has shown that total dry matter production can be increased appreciably by planting at higher density, and breeders should try to take advantage of this by selecting material capable of maintaining a high HI at high density. Clones can be tested in spacing trials, but selection of individual ortets for potential yield per hectare presents difficulties. Selection for high HI gives material which performs better than average at high density (Breure and Corley, 1983), but Corley (1976) suggested that more specific selection for tolerance of competition might be achieved by using partial defoliation to simulate high density planting, since defoliation has broadly similar effects on palm growth and yield components to dense planting. Table 3.3 shows that some progenies are much less affected than others by defoliation, and those progenies least affected might be expected to tolerate high density planting. For individual palm selection, yield with and without defoliation would have to be compared sequentially.

### 3.2.3 Secondary selection objectives

In addition to the primary objectives, vegetative propagation will allow greater consideration to be given to secondary selection objectives.

**Table 3.3.** Effects of partial defoliation of dry-matter production and distribution in some oil-palm progenies.

|  | Mean of 25 progenies | Progeny 25 | Progeny 7 |
|---|---|---|---|
| *Dry matter production (defoliated as % of control)* |  |  |  |
| Total | 65 | 54 | 69 |
| Yield (fruit) | 34 | 22 | 49 |
| Vegetative dry matter | 96 | 82 | 90 |
| *Per cent of dry matter in fruit* |  |  |  |
| Control | 47 | 47 | 51 |
| Defoliated | 23 | 18 | 36 |

Source: From data of Corley, 1976.

#### 3.2.3.1 Ratio of palm oil to kernel oil

Kernel size has a high degree of genetic control, and clonal selection for this character will be simple, allowing the grower a choice of the desired ratio of oils in clones.

#### 3.2.3.2 Oil composition

Genetically controlled variation in oil composition has been identified (Noiret and Wuidart, 1976). Clonal selection may allow the production of material with a more liquid oil, which would probably assist in expanding palm oil's share of the vegetable oil market. Interspecific hybrids between *E. guineensis* and *E. oleifera* are of interest in this context (Hardon, 1969).

#### 3.2.3.3 Yield stability

Until recently, there was little published evidence for genotype × environment interactions in seedling progenies of the oil palm, but work at several centres in Malaysia shows that significant interactions can occur for most yield components (Rajanaidu *et al.*, 1981; Ong *et al.*, 1985). Clone trials are being planted in a range of environments in Malaysia and elsewhere to investigate this aspect, and early indications are that clones may show larger interactions with environment than genetically mixed seedling progenies.

In general, broad adaptability will be a desirable character in a perennial crop, though specific adaptations, such as drought resistance or cold tolerance, may still be objectives in breeding for more extreme growing conditions.

#### 3.2.3.4 Pest and disease resistance

In South-east Asia pests and diseases occur, but are relatively insignificant. In other parts of the world, disease resistance assumes greater importance. *Fusarium* wilt can be devastating in parts of Africa, and it is possible to breed for resistance to this disease (Renard, 1976). A nursery test for resistance was developed by Prendergast (1963), and de Franqueville (1984) has shown a reasonable relationship between results of nursery tests and disease losses in the field.

There are also major disease problems in South and Central America, where interspecific hybrids between *E. guineensis* and *E. oleifera* may show resistance (Renard, 1976).

3.2.3.5  Efficiency of fertiliser use

Tan and Rajaratnam (1978) showed that heritable genetic differences in leaf nutrient composition occurred between different *tenera* progenies, suggesting that there might be differences between palms in the ability to take up and utilise mineral nutrients. Breure (1982) showed a correlation between the oil yields of D×P progenies and their leaf magnesium content. More recently, he has shown that the leaf magnesium content of *dura* and *pisifera* parents is highly correlated with the yield of their offspring (Breure, 1986). Breure's results may be peculiar to the volcanic soils of West New Britain, but they indicate that attempts to identify palms which are efficient in nutrient uptake can be profitable.

3.2.3.6  Ease of harvesting

Cost of harvesting may, in future, determine the competitive position of palm oil *vis-à-vis* other oil crops. No method of mechanical harvesting has yet been developed, but ease of manual harvesting might be improved by selection for characteristics such as long bunch stalks (making cutting easier), more uniform oil synthesis in the bunch, delayed fruit abscission, high bunch weight, low bunch number and low height increment. The extent and nature of available variability for some of these characters, and methods of selection, require study.

## 3.2.4  Selection populations

Most breeding programmes have large numbers of yield-recorded *tenera* palms from controlled crosses. The problem at present is not the amount of material available, but rather the definition of selection objectives and intensity. Nevertheless, thought must be given to the type of cross and genetic background of future populations.

The alternative types of cross are D×P, D×T and T×T, all of which produce some *teneras*. D×P crosses have the advantage that all of the offspring are *teneras*, whereas in both D×T and T×T crosses only 50% are *teneras*. However, the *pisifera* parent cannot be selected on its own phenotypic performance, as it usually does not produce ripe bunches, while phenotypic selection of *duras* may not be very effective. In D×T crosses, both parents can be selected on phenotypic performance, though perhaps less reliably for *duras*. T×T crosses produce D, T and P in the offspring, so can be used for further breeding as well as for ortet selection. Furthermore, both parents can be chosen on the basis of the agreed selection objectives. A disadvantage is that sterile *pisiferas* (25% of the population) are more vigorous, and thus may bias results by being excessively competitive in a mixed population. If *pisifera*s could be identified at the nursery stage, they could be planted separately, but no method has yet been developed, though some work has been reported using electrophoretic techniques.

In the above discussion we have implicitly assumed that elite individuals for cloning will be selected from populations of mature trees in the field. Soh (1986) and Rajanaidu (1986a) have suggested recreating outstanding D×P crosses and cloning large numbers of seedlings of these crosses, as an alternative strategy. Selection of clones within each cross would only be done after field testing. Their argument is partly based on the premise that mature palms are more difficult to propagate from than seedlings, but Noiret *et al.* (1985) have succeeded in propagating 75% of the mature palms sampled, and our results are similar.

The other point made by Soh (1986) concerned population size: often in an oil palm breeding trial there may be no more than 60 trees of each cross. From his estimates of the heritability of oil yield, Soh calculated that the chance of such a small population including an individual with a yield 30% above the population mean was probably less than 1%. If the best crosses were recreated, and larger numbers of seedlings were cloned, the probability of finding a clone with a given degree of superiority over the mean would be increased.

The final decision as to which approach to adopt should be based on the cost per clone finally selected for commercial propagation. As Rajanaidu (1986b) points out, the major cost in ortet selection and clone development is the field testing of clones, and the larger the proportion of inferior clones, the lower the overall cost effectiveness. Selection of ortets based on family performance alone, which is essentially what Soh (1986) and Rajanaidu (1986a) propose, must be less efficient than selection on family plus individual palm performance, and is therefore likely to give a greater cost per clone finally selected.

As far as the genetic make-up of populations is concerned, variation can be observed at the population level and between individuals within populations or progenies. Comstock (1977), reviewing quantitative genetics and the design of breeding programmes, concluded that overdominance is not a major feature in the genetics of single quantitative traits. Hence, the general observed increase in fitness due to heterozygosity (heterosis) is primarily caused by masking of deleterious recessive genes (referred to as "mutational load"). The degree of mutational load is a population characteristic and depends largely on the population structure. In natural populations of the oil palm, often consisting of small isolated groups of palms, consanguineous mating is likely to occur frequently, probably resulting in elimination of many deleterious mutant genes. The same occurs in breeding populations with a relative narrow genetic origin, as with the oil palm. Hence, in such a species both natural and artificial selection may have resulted in a certain tolerance of low levels of inbreeding or an asymptotic relationship between fitness and heterozygosity (Hardon, 1970). A major effort should be made to increase genetic variability in selection populations of the oil palm, but maximum heterozygosity may not necessarily be the primary objective; this is often linked with higher vigour which, through increased interpalm competition, may in fact depress yield on a per hectare basis (Hardon *et al.*, 1972).

In contrast to most annual crops, space is usually a major limiting factor in tree crop breeding programmes. Hence, decisions are required on the relative emphasis on intra- and inter-population crosses, on the number of genetically different populations to be maintained, and on the balance between number of progenies and number of palms per progeny. Most breeding programmes have ample data available as a basis for such decisions. As an example, some preliminary data in Table 3.4 suggest that, provided further inbreeding is avoided, mean fruit yield of the Deli *dura* may not be increased by outcrossing, but variability is much greater in the outcrossed material. The best 5% of *duras* from the D×T crosses in Table 3.4 have a yield some 8% higher than the best D×D material, and given equal population sizes an even greater difference would be expected.

Further increase in variability will come from incorporation of collections of wild material into breeding programmes. This is relatively easy where vegetative propagation is possible, as individual palms with desirable combinations of characters can be utilised directly, without the lengthy

**Table 3.4.** Variation in yield of fruit of individual *dura* oil palms from three different types of crosses.

|  | Deli *dura* selfed | Deli *dura* × Deli *dura* | Deli *dura* × African *tenera* |
| --- | --- | --- | --- |
| Number of crosses | 2 | 5 | 3 |
| Number of palms | 110 | 270 | 75 |
| Mean yield per palm (kg per annum) | 140 | 173 | 164 |
| Standard deviation (kg) | 34 | 40 | 59 |
| Coefficient of variation (%) | 24 | 23 | 36 |
| Yield of best 5% of palms: |  |  |  |
| (a) Observed | 204 | 251 | 271 |
| (b) Expected from normal distribution | 210 | 255 | 286 |

backcrossing programme which would be needed if seed production were envisaged.

Interspecific hybridisation between *E. guineensis* and its wild American relative, *E. oleifera* (Hardon, 1969, 1974), has interesting possibilities. $F_1$ hybrids are fertile and, although oil yield is usually depressed, show potentially useful secondary characters: low height increment, higher unsaturation of the oil and resistance to some diseases. $F_2$ and backcross generations are extremely variable, but include individual palms that may be of considerable interest for vegetative reproduction. Extensive collections are being established in Malaysia, the Ivory Coast and certain centres in South and Central America.

## 3.3 CLONE TRIALS

In addition to identification of superior clones, the early clone trials have two objectives. First, various trial designs are being examined, in the hope of simplifying future clone testing. For example, it is recognised that the planting density for each individual clone must be defined, and trials are in progress to ascertain whether it will be necessary to test every clone in an elaborate spacing trial, or whether, perhaps, the optimum density could be predicted from performance at only two densities (Bleasdale, 1967), or from vegetative measurements such as leaf area. The second objective is to improve selection methods and examine the value of different selection criteria. Clones selected by different methods will be compared; for example, the performance at high density of one clone from an ortet selected for high HI and another selected on the ortet's response to defoliation is being examined.

At the same time, ortet × clonal offspring regressions, and comparison of variances within clones and between seedlings, will indicate the extent of genetic variation present for the various characteristics, or in other words, the likelihood of transmitting characteristics from ortet to clone. Early results, with unselected clones, are in general agreement with heritability studies (Corley *et al.*, 1981).

Results of current trials may lead to the definition of characters for oil palm "ideotypes". Once these are defined, routine measurements can be adopted and appropriate screening methods designed. Considering the cost of clone testing for oil palm, even small improvements in the efficiency of individual palm selection will justify the required research effort.

The use of vegetative propagation for rapid multiplication of elite genotypes will have considerable consequences for oil palm cultivation. In addition to the mass propagation of clonal planting material, the *in vitro* culture techniques have other potential uses, and will serve as an invaluable tool for the oil palm breeder in his efforts to ensure the future of this important tropical tree crop.

## REFERENCES

Arasu, N. T., and Rajanaidu, N. (1977). Oil palm genetic resources. *In* "International Development in Oil Palm" (D. A. Earp and W. Newall, eds), pp. 16–25. Incorporated Society of Planters, Kuala Lumpur.

Beirnaert, A., and Vanderweyen, R. (1941). Contribution a l'étude genetique et biometrique des variétés d'*Elaeis guineensis* Jacquin. *Publ. Inst. Nat. Etude Agron. Congo Belge, Ser. Sci.* 27, pp. 101.

Blaak, G., Sparnaaij, L. D., and Menendez, T. (1963). Breeding and inheritance in the oil palm (*Elaeis guineensis* Jacq.) II. Methods of bunch quality analysis. *J. W. Afr. Inst. Oil Palm Res.* 4, 146–155.

Bleasdale, J. K. A. (1967). The relationship between the weight of a plant part and total weight as affected by plant density. *J. hort. Sci.* 42, 51–58.

Breure, C. J. (1982). Factors affecting yield and growth of oil palm *teneras* in West New Britain. *In* "The Oil Palm in Agriculture in the Eighties" (E. Pushparajah and P. S. Chew, eds), pp. 109–130. Incorporated Society of Planters, Kuala Lumpur.

Breure, C. J. (1986). Parent selection for yield and bunch index in the oil palm in West New Britain. *Euphytica* 35, 65–72.

Breure, C. J., and Corley, R. H. V. (1983). Selection of oil palms for high density planting. *Euphytica* 32, 177–186.

Comstock, R. E. (1977). *In* "Proceedings of the International Conference on Quantitative Genetics" (E. Pollack, O. Kempthorne and T. B. Bailey, eds), pp. 705–718, Iowa State University Press, Ames.

Corley, R. H. V. (1973). Effects of plant density on growth and yield of oil palm. *Exptl Agric.* 9, 169–180.

Corley, R. H. V. (1976). Effects of severe leaf pruning on oil palm, and its possible use for selection purposes. *Malay. Agric. Res. Devel. Inst. Res. Bull.* 4, 23–28.

Corley, R. H. V. (1982). Clonal planting material for the oil palm industry. *J. Perak Plrs' Assoc. 1981*, 35–49.

Corley, R. H. V. (1983). Potential productivity of tropical perennial crops. *Exptl Agric.* 19, 217–237.

Corley, R. H. V., Hardon, J. J., and Tan, G. Y. (1971). Analysis of growth of the oil palm (*Elaeis guineensis* Jacq.) 1. Estimation of growth parameters and application in breeding. *Euphytica* 20, 307–315.

Corley, R. H. V., Wong, C. Y., Wooi, K. C., and Jones, L. H. (1981). Early results of the first oil palm clone trials. *In* "The Oil Palm in Agriculture in the Eighties" (E. Pushparajah and P. S. Chew, eds), pp. 173–196. Incorporated Society of Planters, Kuala Lumpur.

de Franqueville, H. (1984). La fusariose vasculaire du palmier à huile: relation entre la résistance en pépinière et la résistance en champ. *Oléagineux* 39, 513–518.

Gascon, J. P., Jacquemard, J.-C., Hansson, M., Boutin, D., Chaillard, H., and

Kanga Fondjo, F. (1981). La production des semences sélectionées de palmier à huile *Elaeis guineensis*. *Oléagineux* **36**, 476–481.

Hardon, J. J. (1986). Interspecific hybrids in the genus *Elaeis*. II. Vegetative growth and yield of $F_1$ hybrids *E. guineensis* × *E. oleifera*. *Euphytica* **18**, 380–388.

Hardon, J. J. (1970). Inbreeding in populations of the oil palm (*Elaeis guineensis* Jacq.) and its effect on selection. *Oléagineux* **25**, 449–456.

Hardon, J. J. (1974). Oil palm. In "Handbook of Plant Introduction in Tropical Crops" (J. Leon, ed.), pp. 75–89. FAO Agricultural Studies No. 93, Rome.

Hardon, J. J. (1976). Oil palm breeding—introduction. In "Oil Palm Research" (R. H. V. Corley, J. J. Hardon, and B. J. Wood, eds), pp. 89–108. Elsevier, Amsterdam.

Hardon, J. J., Corley, R. H. V., and Ooi, S. C. (1972). Analysis of growth in oil palm. II. Estimation of genetic variances of growth parameters and yield of fruit bunches. *Euphytica* **21**, 257–264.

Hardon, J. J., Williams, C. N., and Watson, I. (1969). Leaf area and yield in the oil palm in Malaya. *Exptl Agric.* **5**, 25–32.

Hirsch, P. J. (1980). Relations entre l'appareil végétatif et la production chez le palmier à huile en Côte-d'Ivoire. *Oléagineux* **35**, 233–237.

Jacquemard, J. C., Meunier, J., and Bonnot, F. (1981). Etude génétique de la reproduction d'un croisement chez le palmier à huile, *Elaeis guineensis*. *Oléagineux* **36**, 343–352.

Jones, L. H. (1974). Propagation of clonal oil palms by tissue culture. *Oil Palm News* **17**, 1–8.

Meunier, J. (1969). Etude des populations naturelles *d'Elaeis guineensis* en Côte d'Ivoire. *Oléagineux* **24**, 195–201.

Noiret, J. M., and Wuidart, W. (1976). Possibilities d'amelioration de la composition en acides gras de l'huile de palms. Résultats et perspective. *Oléagineux* **31**, 465–474.

Noiret, J. M., Gascon, J. P., and Pannetier, C. (1985). La production de palmier à huile par culture *in vitro*. *Oléagineux* **40**, 365–372.

Ong, E. C., Lee, C. H., Law, I. H., and Ling, A. H. (1985). Genotype–environment interaction and stability analysis for bunch yield and its components, vegetative growth and bunch characters in the oil palm (*Elaeis guineensis* Jacq). Paper presented at 5th Int. Congr. SABRAO, Bangkok, 25–29 November.

Ooi, S. C. (1975a). Variability in the Deli *dura* breeding population of the oil palm (*Elaeis guineensis* Jacq.) II. Within bunch components of oil yield. *Malay. Agric. J.* **50**, 20–30.

Ooi, S. C. (1975b). Variability in the Deli *dura* breeding population of the oil palm (*Elaeis guineensis* Jacq.) III. An outcrossed population. *Malay. Agric. J.* **50**, 147–153.

Ooi, S. C. (1978). Variability in the Deli *dura* breeding population of the oil palm (*Elaeis guineensis* Jacq.) IV. Growth and physiological parameters. *Malay. Agric. J.* **51**, 359–365.

Ooi, S. C., and Ngah, A. W. (1977). Oil palm breeding—some aspects of selection. In "International Developments in Oil Palm" (D. A. Earp and W. Newall, eds), pp. 58–67. Incorporated Society of Planters, Kuala Lumpur.

Ooi, S. C., Hardon, J. J., and Phang, S. (1973). Variability in the Deli *dura* breeding population of the oil palm (*Elaeis guineensis* Jacq.) 1. Components of bunch yield. *Malay. Agric. J.* **49**, 112–121.

Prendergast, A. G. (1963). A method of testing oil palm progenies at the nursery stage for resistance to vascular wilt disease, caused by *Fusarium oxysporum* Schl. *J. W. Afr. Inst. Oil Palm Res.* **4**, 156–175.

Rabéchault, H., and Martin, J. P. (1976). Multiplication végétative du palmier à huile (*Elaeis guineensis* Jacq.) à l'aide de cultures de tissue foliaires. *C. R. Acad. Sci. Paris, Ser. D.* **283**, 1735–1737.

Rajanaidu, N. (1986a). A rapid method of developing oil palm clones. Paper presented at Int. Soc. Oil Palm Breeders Colloquium on "Breeding and selection for clonal oil palms", 21 March, Kuala Lumpur.

Rajanaidu, N. (1986b). Selection criteria for ortets. Paper presented at Int. Soc. Oil Palm Breeders Colloquium on "Breeding and Selection for Clonal Oil Palms", 21 March, Kuala Lumpur.

Rajanaidu, N., Arasu, N. T., and Obasola, C. O. (1979). Collection of oil palm (*Elaeis guineensis* Jacq.) genetic material in Nigeria. II Phenotypic variation of natural population. *Malay. Agric. Res. Devel. Inst. Res. Bull.* **7**(1), 1–27.

Rajanaidu, N., Ooi, S. C., and Lawrence, M. J. (1981). Conservation of oil palm genetic resources and genotype × environment interaction. Paper presented at 4th International SABRAO Conference, Kuala Lumpur.

Rao, V., Soh, A. C., Corley, R. H. V., Lee, C. H., Rajanaidu, N., Tan, Y. P., Chin, C. W., Lim, K. C., Tan, S. T., Lee, T. P., and Ngui, M. (1983). Methods of bunch analysis. *Palm Oil Res. Inst. Malaysia Occ. Paper* **9**, 28 pp.

Renard, J. L. (1976). Diseases in Africa and South America. *In* "Oil Palm Research" (R. H. V. Corley, J. J. Hardon, and B. J. Wood, eds), pp. 447–466. Elsevier, Amsterdam.

Soh, A. C. (1986). Expected yield increase with selected oil palm clones from current D×P seedling materials and its implications on clonal propagation, breeding and ortet selection. *Oléagineux* **41**, 51–56.

Sparnaaij, L. D. (1969). Oil palm. *In* "Outlines of Perennial Crop Breeding in the Tropics" (F. P. Ferwerda and F. Wit, eds), pp. 339–387. University of Wageningen, Misc. Papers no. 4.

Tan, G. Y. (1978). Genetic studies of some morphophysiological characters associated with yield in oil palm (Elaeis guineensis Jacq.). *Trop. Agric. (Trinidad)* **55**, 9–16.

Tan, G. Y., and Rajaratnam, J. A. (1978). Genetic variability of leaf nutrient concentration in oil palm. *Crop Sci.* **18**, 548–550.

Thomas, R. L., Watson, I., and Hardon, J. J. (1969). Inheritance of some components of yield in the "Deli *dura*" variety of oil palm. *Euphytica* **18**, 92–100.

Van der Vossen, H. A. M. (1974). Towards more efficient selection for oil yield in the oil palm (*Elaeis guineensis* Jacquin). Thesis, University of Wageningen, 107 pp.

# 4 Breeding Citrus—Genetics and Nucellar Embryony

R. K. SOOST

*University of California, Riverside, California, USA*

The genetics and breeding of citrus, including the important role of nucellar embryony was thoroughly covered by Cameron and Frost (1968) and Frost and Soost (1968). This material was updated in 1975 by Soost and Cameron. Breeding programmes have been increasingly active since that time, genetic information has increased and the availability of plant materials has improved. This chapter reports on those new developments and discusses the present status, and possible future directions of research.

## 4.1 POLYPLOIDY

Diploidy is the general rule in *Citrus* and its related genera, the gametic ($n$) chromosome number being 9. However, tetraploid individuals have been determined in many cultivars. The Hong Kong wild kumquat, *Fortunella hindsii* (Champ.) Swing. may have been the first reported tetraploid. Although Swingle considered this tetraploid to be representative of the species, it may only be a tetraploid form because diploid sources have subsequently been identified (Swingle and Reece, 1967).

Most tetraploids that have been established were obtained as variant nucellar seedlings in seedling populations grown for other purposes (Soost and Cameron, 1975). A systematic search for tetraploids in a wide range of taxa was reported by Barrett and Hutchison (1978). In general, tetraploid frequency varied from less than 1% to 3% with a few small seed lots having frequencies as high as 25%. Hutchison and Barrett (1982) also concluded that the frequency is genetically controlled, but is modified by environmental and physiological factors. In a study of over 30 000 seedlings of each of two similar cultivars, Troyer and Carrizo [*Citrus sinensis* (L.) Osbeck × *Poncirus trifoliata* (L.) Raf.], they found Carrizo had a lower tetraploid frequency than Troyer in each of three years, but both cultivars had similar year-to-year fluctuations. The position of a fruit on the tree also had a consistent effect in both cultivars, and the percentage was consistently higher in larger seeds. However, the production of tetraploids was not confined to a few fruit or seeds. Therefore, it appears that nucellar tetraploids can be recovered from most cultivars that produce nucellar seedlings although more seedlings will be needed for some than with others. In addition to work reported prior to 1973 (Soost and Cameron, 1975), researchers in Russia (Maisuradze, 1979) continue to report more tetraploid seedlings.

The intentional production of tetraploids by the use of colchicine or other treatments has received little attention. Tachikawa (1971) included the few tetraploids derived from colchicine treatment among his list of chromosome numbers obtained from the literature. Barrett (1974) recovered tetraploids from several cultivars as well as periclinal ploidy chimeras.

Although tetraploids have been available in some cultivars for many years, their effective utilisation in breeding programmes has been limited. All were derived from cultivars that produced moderate or high percentage of nucellar seedlings. The tetraploids also produced nucellar seedlings at the same rates as the diploids from which they were derived. Production of triploid hybrids was difficult, or at low frequency. Thus, while the first hybrid triploid was reported by Longley in 1926 and others subsequently have been obtained (Soost and Cameron, 1975), the numbers have been low.

The barrier of nucellar embryony led to the use of the tetraploids as pollen parents (Russo and Torrisi, 1953; Tachikawa et al., 1961) although the number of triploids reported was still small. Tachikawa et al. (1961) also reported the occurrence of hybrid tetraploids in their populations. Subsequent evaluation of larger populations from several diploid monoembryonic seed parents by Esen and Soost (1977a) showed high percentages of hybrid tetraploids. These tetraploids result from the regular production of diploid megagametophytes by some cultivars. The high frequency of tetraploids results from their differential survival in relation to the triploids. Esen and Soost (1977a) have shown that the low recovery of triploids is due to poor endosperm development and subsequent failure of the embryo. The success or failure of the embryos is clearly not dependent on the ploidy of either embryo or the endosperm, but does correlate with the ploidy ratio of endosperm to embryo (Esen and Soost, 1977a).

The production of diploid megagametophytes among the cultivars tested (Geraci et al., 1975, 1977; Esen and Soost, 1977a) varied from less than 1% to approximately 25% (Table 4.1). With some cultivars that produce

**Table 4.1.** Production of diploid megagametophytes among some citrus cultivars.[a]

| Cultivars | Diploid megagametophytes (%) |
|---|---|
| Sukega | 24.2 |
| Temple | 6.8 |
| Clementine | 1.0 |
| King | 7.0 |
| Wilking | 14.0 |
| Fortune | 20.0 |
| Lisbon | 1.0 |
| Eureka | 5.0 |
| Poorman | 0 |
| Pummelo (CRC 2240) | 0.5 |
| Pummelo (CRC 2241) | 0 |
| Pummelo | 0.1 |

*Sources:* adapted from data of Esen and Soost (1971) and Geraci et al. (1975).

[a] Based on production of triploids in 2x × 2x or tetraploids in 2x × 4x crosses.

very few diploid megagametophytes, significant numbers of triploids can be recovered (Soost, unpubl.). Oiyama et al. (1982) report variability in the recovery of triploid hybrids from some diploids, depending on the pollen parent used. The removal and culture *in vitro* of the embryos prior to full fruit maturity can greatly improve their recovery (Starrantino and Recupero, 1982).

Triploids can also be recovered from diploid by diploid crosses if a cultivar that produces enough diploid megagametophytes is used as the seed parent. This recovery can be enhanced by carefully retaining the smallest seeds. Although development is normal, the triploid seeds are usually smaller than diploid seeds from such crosses (Esen and Soost, 1977a; Wakana et al., 1982; Oiyama and Okudai, 1983).

Esen et al. (1979) obtained evidence that the diploid megagametophytes in the cultivar Sukega develop after first meiotic division. Thus, they are the products of segregation and recombination. If this same mode occurs in other cultivars, variability is introduced by both parents and the progeny will be more variable than if first division did not occur.

The preferred method of producing triploids is to use tetraploids as seed parents. The establishment of monoembyronic tetraploids is critical for this purpose. A few monoembryonic tetraploids were among those recovered by Barrett (1974) but these have not yet been utilised in breeding programmes. Additional monoembryonic hybrid tetraploids have also been identified among the populations evaluated by Cameron and Soost (1980a). A few small populations have been produced from them and evaluation of the seedlings has begun.

Although the horticulturally favourable characteristics of citrus triploids (Fig. 4.1) and the ability of some of them to yield well in spite of known sterility have long been known (Soost and Cameron, 1969), the introduction of new triploid hybrids has been very limited; Oroblanco and Melogold, triploid pummelo–grapefruit hybrids [*C. grandis* (L.) Osbeck × *C. paradisi* Macf.] were introduced by Soost and Cameron (1980, 1985). Maisuradze *et al.* (1978) report a triploid orange hybrid with favourable characteristics although it is unnamed. It is not clear whether any of the triploids produced in the breeding programmes in Japan (Tachikawa *et al.*, 1961) and Italy (Russo and Torrisi, 1953) have been introduced as named cultivars.

**Figure 4.1.** Triploid hybrid (*centre*) with diploid parent (*left*) and tetraploid parent (*right*).

Aneuploids have been found at very low frequencies in several taxa (see Esen and Soost, 1972). Esen and Soost (1972) reported on additional aneuploids that mostly occurred in progeny of $2x \times 4x$ crosses. Chromosome numbers range from 19 to 41. Few of the aneuploids have been successfully established because of greatly reduced vigour. It seems unlikely that they will be useful for breeding or genetic purposes.

Ploidy levels above $4n$ have been very rare (see Barrett and Hutchison, 1982). The highest reported level of $8x$ is a 4–8–8 periclinal chimera. Growth of these higher ploidy plants is even poorer than $4n$. Thus, they appear to have little value for breeding.

## 4.2 INCOMPATIBILITY

The presence of self-incompatibility presents problems for the breeder as well as the opportunity to produce seedless cultivars. Thus far, the problems have been the more prominent. Self-incompatible cultivars generally have problems with maintaining adequate fruit-set. There is considerable literature that deals with pollination and fruit-set of several of these commercially grown cultivars (for example, Mustard et al., 1956; Soost, 1956; Krezdorn and Robinson, 1958; De Lange et al., 1973; Yamashita, 1980). Except for the previous work of Soost (1969), little published information is available on the inheritance of self- and cross-incompatibility. However, additional self-incompatible cultivars continue to be identified (Hearn et al., 1969; De Lange and Vincent, 1972; Moffett and Rodney, 1979; Iwamasa and Oba, 1980; Li, 1980).

Ueno (1978) reported a case of unilateral cross-incompatibility between Citrus tachibana Tan. and C. hassaku hort. ex Tan. and (Natsu Mikan, C. tardiva hort. ex Shirai). Ueno suggests more than one gene to account for the unilateral control; however, other mechanisms cannot be excluded. The presence of unilateral cross-incompatibility could further complicate the breeders' ability to predict the behaviour of progenies in relation to compatibility relationships.

Assuming that the genetic system proposed by Soost (1969) is operating generally in the genus Citrus, then the breeder can predict the occurrence of incompatibility in his progenies if the compatibility relationships of the parents are known. Unfortunately, if the ancestry of the parents is unknown and the parents are effectively self-compatible, some individuals in the progenies of such parents may be self-incompatible.

If the breeder is utilising self-incompatible parents, he needs to be aware that lack of fruiting in some individuals in the progeny may result. However, self-incompatibility may be obscured by adequate fruiting in mixed hybrid plantings. Therefore, hybrids with potential for release as cultivars need to be evaluated for fruitfulness in the absence of other pollen sources. The breeder will need to identify suitable pollinating cultivars if the intended release is self-incompatible and unfruitful in the absence of cross-pollination.

## 4.3 MUTATION BREEDING

Reports of the use of mutagenic agents, particularly radiation, continue to appear in the literature, but it is difficult to substantiate the claims of useful changes. The most active researchers, as judged by the literature, are in Russia. Kerkadze and Kutateladze (1979) describe the recovery of dwarfing mutants with early ripening and good fruit quality after treatment of Kuwano Wase with 30 or 50 Gy γ-ray dosage. Useful mutants from Meyer lemon (*C. limon* (L.) Burm. f.) and oranges are also mentioned. Kerkadze (1979) reports other changes, including increased photosynthetic activity. Kapanadze (1979b) discusses a change in dominance of the inheritance of elaioplast-free juice vesicles after irradiating flower buds that were used as pollen sources for the production of hybrids between elaioplast-free taxa and elaioplast-containing taxa. Inheritance of elaioplast-free juice vesicles shifted to dominant from recessive.

Hearn (1984, 1986) recovered seedless or nearly seedless mutants from several seedy cultivars after irradiating seeds and budwood with gamma radiation. Thus far, these mutants appear to retain all of the characteristics of the source cultivar, including yielding ability. Russo *et al.* (1982) also report decreased seediness in Clementine Monreal (*C. reticulata* Blanco) after radiation of budwood with 40 and 60 Gy from a $^{60}$Co source. If these mutants maintain their performance, this technique could be very valuable for general use. Many hybrids with excellent characteristics fail to be accepted because they have too many seeds.

The isolation of naturally occurring mutants remains an important source of new clones. More of these mutants appear to be reported from those citrus areas that practice regular pruning and have smaller units of production. At the International Citrus Congress in 1981 several mutants were reported from Italy (Russo, 1982), Spain (Bono *et al.*, 1982), Japan (Iwamasa and Nishiura, 1982; Iwamasa *et al.*, 1982), and the USA (Hensz, 1982). Most of these occurred in cultivars for which the total acreage is comparatively small, with the exception of mutants in Satsuma (*C. reticulata*) (Iwamasa *et al.*, 1982). The only new mutants reported from the USA were confined to mutants of Ruby grapefruit (*C. paradisi*).

Perhaps pruning results in the growth of buds that would otherwise not grow. The closer inspection that occurs because of regular pruning and smaller acreage may also result in the discovery of more mutants. One cannot exclude the possibility that mutants are more frequent in some cultivars such as Satsuma than in others or that certain mutants are more readily recognised, such as changes in pigmentation.

## 4.4 CHIMERAS

Chimeras are of interest and importance to the breeder. Useful characteristics may not be transmitted to either the zygotic or nucellar progeny if a periclinal chimera is involved. Cameron et al. (1964) and Olson et al. (1966) clearly showed the presence of periclinal chimeras in some pigmented grapefruit cultivars that make it impossible to establish nucellar bud lines with the characteristics of the parents. A number of chimeras have been reported in Satsuma, including the red-rinded Dobashibeni (Nishiura and Iwamasa, 1970). The authors suggest that it is a sectorial chimera because both red-rinded and orange–yellow rinded nucellar seedlings have been established. Fruits on some parental clone trees also show both colours as rind sectors. However, it may have originated as a periclinal chimera, but clonal propagation has resulted in the loss of the orange–yellow rinded tissue in some sources. Spiegel-Roy (1979) reported that some trees of Shamouti orange (*C. sinensis*) produce nucellar seedlings that have Shamouti characters but others produce nucellar seedlings with characters of the Beledi orange. The Shamouti has long been known to produce limbs with Beledi characteristics and was assumed to be a chimera. Apparently the Beledi tissue has been lost in some sources of Shamouti, but not in others.

The non-recovery of low-acid individuals in the progenies of an acidless orange when crossed with other acidless cultivars, and the occurrence of high-acid individuals in bud progeny from acidless oranges suggests a chimeral nature for acidless orange cultivars (Cameron and Soost, 1979). Alternatively, the inheritance of acidity may be different in the acidless pummelo, where it is semi-dominant with low-acid individuals recovered in the $F_1$ generation (Cameron and Soost, 1977).

The determination of the presence or absence of a periclinal chimera can, therefore, be critical to the breeder, not only as a barrier to transfer of characters but to an understanding of the inheritance of specific characters.

**Figure 4.2.** Training system for earlier fruiting of hybrids (Japan).

### 4.5 JUVENILITY

The long period from seed to fruiting remains a major barrier to rapid progress in citrus breeding. Many different horticultural techniques have been used to shorten this period, but with limited success only. Japanese breeders are currently employing a labour-intensive method that includes budding into older seedlings, training the scions to upright stakes or frames for 2–3 years, and then bending these shoots down (Fig. 4.2). This is reported to hasten fruiting by several years, but the total time from seed to fruit is still about 6 years. Fruiting appears to be heavier.

Rather frequently, citrus propagators note the very precocious flowering of some grapefruit (*C. paradisi*) seedlings. However, flowering does not recur until seedlings are 6–7 years old or more. Iwamasa and Oba (1975) obtained flowering in seedlings of grapefruit, Tengu (*C. tengu* Tan.), Banokan (*C. grandis* var. Banokan Tan), three tangelos [Allspice,

Wekiwa, and Thorton (*C. paradisi* × *C. reticulata*)] and a hybrid between Satsuma and pummelo (*C. grandis*). Precocious flowering was only obtained in seedlings that had been grown at maximum temperatures that did not exceed 20°C and minimum temperatures below 10°C, from November to March. No flowering occurred in 17 other taxa. Precocious flowering appears to occur mainly in grapefruit and grapefruit hybrids or pummelo hybrids. Soost (unpubl.) obtained flowers and fruit during the first year of pummelo (*C. grandis*) × grapefruit hybrids (Fig. 4.3). The occurrence of precocious flowering has stimulated interest in the possibility of breeding for a much shorter juvenile period.

Many flowers of precocious seedlings are abnormal, incomplete and sterile. However, some have fertile pollen and ovules. Iwamasa and Oba

**Figure 4.3.** One-year-old pummelo × grapefruit seedling with fruit.

(1975) describe fruit but have not reported on the progeny. Soost (unpubl.) obtained 10 seedlings from one fruit. None of these seedlings bloomed precociously, nor before a period of 7 years. Yadav et al. (1980) report flowering of a *Poncirus trifoliata* seedling at the age of 9 months. Flowers were perfect and fruits were set in the subsequent season. Seedlings (probably nucellar, but not stated) flowered at 14 months. Unless the seedlings were zygotic, the ability to effectively transfer this character has not yet been demonstrated.

Physiological investigations of juvenility have not determined control mechanisms, and treatments to shorten the period or induce flowering have not been generally effective. Yamashita (1979) obtained no effect from splice-grafting lemon (*C. limon*) and natsudaidai (*C. natsudaidai* Hayata) seedlings to grapefruit seedlings, even when the grapefruit seedlings flowered. Soaking the roots of lemon and natsudaidai seedlings for 48 hours in solutions of 5-FDU, thymidine, uridine, uridilic acid, uracil, xanthine, RNA, and kinetin had no effect on flowering.

Thus, neither treatment nor incorporation by genetic transfer has yet been effective in overcoming juvenility.

## 4.6 NUCELLAR EMBRYONY

Nucellar embryony (Fig. 4.4) is a major barrier to the easy transfer of genetic material. The ability to cross within species where few or no monoembryonic taxa are available is particularly restricted. Within the genus *Citrus*, monoembryony is simply inherited and monoembryonic individuals are easily recovered (Cameron and Soost, 1980a). Therefore, the list of cultivars known to be monoembryonic has gradually been expanded over the last several decades. However, there are still notable gaps in the availability of monoembryonic cultivars in some major species, for example *C. sinensis*, but a monoembryonic orange, Grushevidnyi Korolek, has been listed by Gurtskaya (1981). From time to time, other *C. sinensis* cultivars have produced more zygotic seedlings than the usual very low numbers. However, none of these has been monoembryonic, and the recovery of zygotic seedlings has been variable and unpredictable (Hearn, 1977).

The presence of nucellar embryony has led to two lines of research in efforts to increase the recovery of zygotic individuals. In recent years, several techniques have been utilised to identify nucellar versus zygotic seedlings where both may occur. The second line of research has attempted to reduce the numbers of nucellar embryos and increase the percentage survival and recovery of zygotic seedlings.

Techniques to identify zygotic seedlings have included thin-layer chromatography of leaf flavonoids and coumarins (Tatum et al., 1974), root and leaf isozymes (Button et al., 1976; Spiegel-Roy et al., 1977; Soost et al., 1980), gas chromatography of leaf emissions (Weinbaum et al., 1982) and browning of shoot extracts (Geraci and Tusa, 1976; Esen and Soost, 1977b). Each of these techniques has its disadvantages and advantages. All are more accurate than previous techniques, more widely applicable and usually more rapid. Isozyme techniques have the advantage of specific co-dominate alleles of known inheritance (Table 4.2). The methods and equipment are relatively simple and inexpensive. In citrus, the main disadvantage has been the lack of genetic diversity at the isozyme loci within the major species that are used in breeding programmes. This disadvantage is partially overcome by the identification of several heterozygous loci. The probability of a zygotic seedling having the same isozyme profile as the seed parent becomes very small (Torres et al., 1982). However, it increases the analytical time and costs several times more than running one locus.

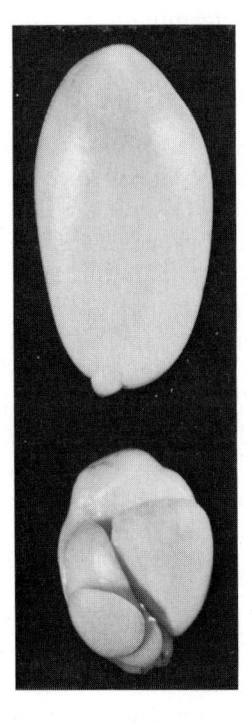

**Figure 4.4.** Monoembryonic seed of *Citrus grandis* (top). Polyembryonic seed of *C. reticulata* (bottom).

Although expensive, the gas chromatographic method described by Weinbaum *et al.* (1982) should be useful for rapid analysis, particularly if other methods are unable to distinguish reliably between zygotic and nucellar seedlings. The thin-layer chromatography technique used by Tatum *et al.* (1974) needs comparatively simple equipment and materials, yet may be useful where other methods fail. The browning of leaf extracts (Esen and Soost, 1977b) is one of the simplest and cheapest methods, but is restricted in application by the lack of variability within major species.

Altogether, breeders now have at their disposal much improved techniques for the identification of zygotic seedlings. This still leaves nucellar embryony as a direct barrier to the production of zygotic embryos and seedlings. Most cultivars that are highly polyembryonic, produce few zygotic individuals. Various techniques have been tried to either reduce the number of nucellar embryos and/or increase the number of zygotic seedlings. Nakatani *et al.* (1978, 1982) reported a reduction in the number of embryos per seed in several cultivars grown under high temperatures. However, they made no determination of recovery of zygotic seedlings. Ikeda (1982) reports reduction in the number of embryos per seed by treating flower buds with 10 or 20 Gy (10 Gy = 1 krad) γ-rays for 20 hours, 20–30 days prior to anthesis. No information on the production of zygotic seedlings is provided. In *in vitro* studies, Esan (1973) demonstrated a graft-transmissible repressor(s) of nucellar embryogenesis in citron (*C. medica* L.) ovules. Tisserat and Murashige (1977a,b) reported the repression of nucellar embryogenesis *in vitro* by ethanol, indolylacetic acid (IAA), abscisic acid (ABA), 2,4-dichlorophenoxyacetic acid (2,4-D), gibberellic acid ($GA_3$), kinetin and ethephon. De Lange and Vincent (1977) reported a large increase in the percentage of zygotic seedlings produced when pollinated Minneola (*C. paradisi* × *C. reticulata*) flowers were treated with 15 mg $GA_3 ml^{-1}$ 30 days after anthesis. However, the percentages of zygotic seedlings were less when IAA, 2,4-D or ABA were applied *in vivo* at anthesis. Timing and concentration may be critical when any of these growth regulators are used for the differential suppression of nucellar embryos. Kobayashi *et al.* (1979) have shown that cells that are capable of developing into nucellar embryos can be identified in polyembryonic cultivars at least 4 days prior to anthesis. Therefore, early application of some compounds may be required.

**Table 4.2.** Modal genotypes of citrus species and *Poncirus trifoliata*.[a]

| | Got-1 | Got-2 | Pgi-1 | Pgm | Mdh-1 | Mdh-2 | Lap | Hk | Idh | Me-1 |
|---|---|---|---|---|---|---|---|---|---|---|
| C. aurantifolia (lime) | ff | sm | ss | fm | fs | fs | mm | si | si | si |
| C. aurantium (sour orange) | ss | mm | ws | fs | ff | ff | ff | ii | ii | ii |
| C. grandis (pummelo) | ff | mm | ss | ss | ff | ff | mm | ii | ii | ii |
| C. jambhiri (rough lemon) | fs | fs | fs | fi | fs | ff | ff | ii | mi | ii |
| C. limon (lemon) | fs | sm | ws | fs | fs | ff | ff | ii | si | ii |
| C. medica (citron) | ff | ss | ss | ff | ss | ff | —[b] | si | mm | if |
| C. paradisi (grapefruit) | fs | mm | ss | ss | ff | ff | fs | ii | ii | ii |
| C. reticulata (mandarin) | ss | fm | ff | ff | ff | ff | ff | ii | ii | ii |
| C. sinensis (sweet orange) | ss | mm | fs | fs | ff | ff | fs | mm | mi | ii |
| Poncirus trifoliata | mp | sm | fs | pm | ss | fs | ff | ff | ff | rf |

*Source*: from Torres et al. (1982).

[a] Got, glutamate–oxaloacetate transaminase; Pgi, phosphoglucose isomerase; Pgm, phosphoglucomutase; Mdh, malate dehydrogenase; Lap, Leucine aminopeptidase; Hk, hexokinase; Idh, isocitrate dehydrogenase; Me, malic enzyme.
[b] No Lap isozymes appear for citron.

## 4.7 SPECIFIC CHARACTERS

### 4.7.1 Cold-hardiness

Because considerable citrus acreage is subject to damage by low winter temperatures, breeding programmes have tried to locate and incorporate cold-hardiness. The earliest breeding work by the United States Department of Agriculture (USDA) had cold tolerance as a major objective. *Poncirus trifoliata* was used as the source of cold tolerance in crosses with numerous scion cultivars in the genus Citrus. Although valuable rootstocks originated from this programme, the $F_1$ hybrids were not suitable as scion cultivars and subsequent generations were not developed. However, a cold-hardy scion breeding programme was restarted by the USDA in 1973 (Barrett, 1982). In addition to using *P. trifoliata*, *Eremocitrus glauca* (Lindl.) Swing. and *Fortunella margarita* (Lour.) Swing. were also used as sources of cold-hardiness in crosses with several *Citrus* species. $F_1$ hybrids with *Eremocitrus* have lacked the bitter, acrid flavours present in $F_1$ hybrids of *Citrus* with *Poncirus*. However, they have been very acid. Backcross progeny from $F_1$ *Poncirus* hybrids have generally been very acid and some still have unpleasant flavours. A few individuals with edible, dessert-like fruit quality have been selected from the backcross of the $F_1$, *P. trifoliata* × *C. paradisi* to *C. sinensis*. They resemble commercial orange cultivars. More advanced crosses have been made but progenies have not been tested. Progress has been faster than anticipated and is encouraging for future advancement in cold-hardiness.

Cold-hardiness has also been of long-time interest in Japan. In fact, cold-hardiness is a major factor in the success of Satsuma mandarin in Japan. Ikeda *et al.* (1980) report on cold damage to named cultivars and hybrid populations after a severe freeze in 1971. *Poncirus*, Yuzu (*C. junos* Tan.), and Troyer citrange were the most cold-hardy. Kiyomi, a recent hybrid of Satsuma × Trovita orange was as cold-hardy as Satsuma. Some hybrid seedlings from crosses of less cold-hardy cultivars were more cold-hardy than either parent and seem to be as cold-hardy as Satsuma. At least the types of cultivars with cold-hardiness equal to Satsuma have increased in number.

Russian researchers view cold-hardiness as a major requirement for citrus. As with other breeding programmes, their main sources of cold-hardiness have been *Poncirus* and Satsuma. Karaya (1981) lists an early, frost-resistant hybrid (Gul'ripshi) of Satsuma × "pomelo" (*C. grandis*) with good fruit quality. Hybrids of Clementine (*C. reticulata*) with *C. grandis* are reported to be more frost resistant than hybrids of Clementine with *C. paradisi*. Kerkadze (1980) suggests using induced mutations to improve the quality of *Poncirus* because hybrids with *Poncirus* have had inferior fruit quality. However, Goliadze and Tutberidze (1977) report an $F_3$ hybrid of orange × *Poncirus* that withstands $-16°C$ and is high in ascorbic acid. Maisuradze (1979) obtained triploid frost-resistant hybrids by crossing *Poncirus* with a tetraploid sweet orange, number 574. They are reported to have "reasonable" fruit quality with very slight resinous taste.

Kokaya (1981) reports promising hybrids from the standpoint of frost resistance and fruit quality. The best were hybrids of mandarins, oranges, lemons and grapefruits with *C. junos*. Hybrids of citrange with kumquat (*Fortunella*) were also reported to have good fruit quality. Gabuniya (1980) even lists a new cultivar, Adreula, which resulted from a cross of Miyagawa Wase Satsuma by *C. ichangensis* Swing. It is said to be very early and cold resistant. In contrast, Gurtskaya (1981) reports $F_1$ hybrids of sweet orange with *Poncirus*, *C. ichangensis*, and *C. junos* to be highly frost resistant, but inedible. Thus, as in the United States programme, the main constraint when *Poncirus* is used is the difficulty of achieving acceptable fruit quality.

#### 4.7.2 Disease and pest resistance

There are few reports on inheritance of resistance or tolerance to specified diseases or pests in scion cultivars. Soost and Cameron (1975) reported on research until about 1972. Considerable effort has since been directed towards the disease "Mal Secco", caused by the fungus *Deuterophoma tracheiphila*. Hybrids suitable for use as substitutes for existing susceptible lemon cultivars have not been obtained. However, Goliadze and Tutberidze (1977) report a hybrid of Monochello lemon × Meyer lemon that is fairly resistant to "Mal Secco". The characteristics of the hybrid are not reported. Koizumi and Kuhara (1982) indicate inheritance of resistance to bacterial canker, *Xanthomonas campestris* pv. *citri*, in hybrids with at least one resistant parent.

In a specialised case, Aubert and Vogel (1981) report the selection of *Citrus hystrix* DC that is resistant to tristeza virus. *Citrus hystrix* is not edible and is little grown but is of interest because of its essential oil. Most research on resistance to tristeza has been directed toward resistant or tolerant rootstocks for use with commercial cultivars.

Sadana and Joshi (1979) have tested cultivars for resistance to the phytophagous mite, *Brevipalpus californicus*, but no breeding work is reported. Satsuma and Dancy mandarins and Marsh grapefruit are listed as the most resistant of 45 cultivars tested.

Although sources of resistance or tolerance to diseases and pests are clearly present, it is difficult to utilise them in scion breeding programmes. The constraints of the horticultural needs of the cultivars and the long life-cycle of citrus make it difficult to establish resistance or tolerance as primary selection criteria.

#### 4.7.3 Other characteristics

Spiegel-Roy and Teich (1972) studied the occurrence of thornlessness in progenies of "thornless" (Shamouti orange and Eureka lemon) and "thorny" cultivars. Based on the segregation ratios that were obtained, they propose that thorniness is controlled by few genes, but by more than one. Kapanadze (1979a) reports on the inheritance of 15 characters in crosses of *Poncirus* by *Citrus* and *Fortunella* Swing. Trifoliolate leaves, thorniness, pubescence and the presence of oil glands in the juice sacs are listed as dominant. Reduction of nectaries and the scaliness of flower buds are said to be recessive. Scora *et al.* (1976) examined the distribution of essential leaf oils in a small $F_2$ population. They propose two-loci control for the level of citronellal and geranial and single-locus control for β-pinene, β-ocime and linalool. Most components did not fit a simple means of inheritance. Without a better understanding of the biosynthetic pathways involved, knowledge of inheritance is bound to be very uncertain. However, some ability to predict the range to be expected in populations is helpful.

Cameron and Soost (1980b) found that the trifoliolate leaf character of *Poncirus* is dominant to the unifoliolate leaf in *Citrus*, with few genes involved, although the exact genetic basis is still uncertain.

Vardi and Spiegel-Roy (1982) have identified genetically controlled chromosome asynapsis in selfed progeny of Wilkins mandarin. Inheritance is simple and may provide a source of seedless cultivars.

### 4.8 ROOTSTOCK BREEDING

#### 4.8.1 Cold tolerance

As with breeding for cold-hardiness for scion cultivars, *Poncirus* has been the main source of cold-hardiness for rootstock breeding. *Eremocitrus* may also be a source (Yelenosky *et al.*, 1978), but it is susceptible to *Phytophthora*. Tutberidze (1980) reports that a source of *C. ichangensis* from China provided good frost resistance when used as a rootstock for lemon. However, it appears that no hybrid rootstocks have been equal to Poncirus in cold-hardiness. (See Yelenosky, 1985, for a complete review of cold-hardiness.)

## 4.8.2 Salt tolerance

For many years, the USDA had an active programme for the selection and breeding of salt-tolerant rootstocks. This programme has been terminated but Reem and Furr (1976) reported on the results of much of this work. Some hybrids of Rangpur lime (*C. limonia* Osbeck) and Cleopatra mandarin (*C. reticulata*) with other *Citrus* cultivars or *Poncirus* resulted in as little accumulation of Cl$^-$ in the scions as did Cleopatra mandarin. A few even had less uptake of Cl$^-$, but were dwarfing. Some of these hybrids may be useful where salinity is a problem but where Cleopatra or Rangpur are not satisfactory rootstocks. Some may be useful as parents.

Mobayen and Milthorpe (1980) show Iranian mandarin (Bakraie) to be more tolerant of Na$^+$ than Cleopatra mandarin. However, their determinations were judged by damage to the seedlings. The performance as rootstocks could be completely contrary. Tolerance of high salt by a rootstock may result in damaging levels in the scion cultivar.

## 4.8.3 Pests and diseases

Most research has been directed to resistance against nematode species and *Phytophthora* species. On a worldwide basis, most nematode research has been on *Tylenchulus semipenetrans*. The main source of resistance has been *Poncirus*. Kaplan and O'Bannon (1981) reported on the comparative resistance of different sources of *Poncirus* and the high level of resistance of Swingle citrumelo. Campos and Ferraz (1980) also reported a hybrid of *Poncirus* with *C. paradisi* to show strong resistance. Others also have reported transfer of resistance from *Poncirus* (see Soost and Cameron, 1975). Ford (1978) selected seven hybrids, two from Ridge Pineapple × Milam rough lemon and three from *Poncirus* × Milam, that are more resistant to the burrowing nematode *Radopholus similis* than Milam or Ridge Pineapple. These two cultivars were released in 1964 from a massive screening programme and have been used widely in Florida.

Although screening programmes for *Phytophthora* resistance have identified several possible sources of resistance, breeding programmes have mainly used *Poncirus*. Recent screening work (Grimm and Hutchison, 1980) shows some resistance in *C. grandis* and *C. aurantifolia* (Christm.) Swing., in addition to *Poncirus* and Poncirus hybrids with *Citrus*. Alexander and Emmett (1981) report two hybrids of Ellendale mandarin by Carrizo citrange that are promising. Tuzcu *et al.* (1984) identified several additional resistant sources on the basis of bark inoculation of seedlings during the winter.

Tristeza virus is a major problem in most citrus producing areas. Incorporation of tolerance to this virus is very slow because of the long-term test needed to detect tolerance or susceptibility. A sweet-orange scion is grafted on the test rootstock, inoculated, and read for symptoms over about a four-year period. Here again, the main source of resistance has been *Poncirus*. Sweet orange is also tolerant but very susceptible to *Phytophthora*. The best hybrids have been those between *Poncirus* and sweet orange.

Garnsey *et al.* (1981) used ELISA (enzyme-linked immunosorbent assay) to determine the titre of tristeza in hybrids that were inoculated with tristeza virus by budding on infected rootstocks. Hybrids with very low titres were found. These need to be tested as rootstocks with infected scions to see if the test can accurately identify tolerant rootstocks. If this technique is valid, it would greatly increase progress in breeding for tolerance or resistance to tristeza.

### 4.9 IN VITRO CULTURE

Because of the major barriers to exchange of genetic material by conventional breeding methods, there has long been interest in *in vitro* methods. To date, progress has been slow because of the difficulty of handling *Citrus* tissue in culture. Only very juvenile tissue, usually ovular or nucellus tissue, has been successfully maintained and manipulated. Probably the research that has shown the most progress has been done in Israel. Button and Botha (1975) reported the development of complete plants from single cells. Vardi *et al.* (1975) developed plants from protoplasts of Shamouti orange, and later Vardi (1982) extended this to several other cultivars. Research on the production of protoplasts from citrus callus has also been reported from China (Anon., 1976). The present problem is to identify the products of protoplast fusion. Ben-Hayyim *et al.* (1982) obtained promising results from their experiments on the use of isozymes for the identification of hybrid callus.

The efforts to exploit culture *in vitro* for breeding purposes have not been limited to research on protoplast production and fusion. Wang and Chang (1978) report the development of triploid plants from the culture of endosperm. If verified, this could be an important means of producing triploids.

Chen *et al.* (1980) reported the production of haploid plants from pollen of *C. microcarpa* Burge. Plantlets from another culture of *C. aurantium* L. and *C. sinensis* were obtained by Hidaka *et al.* (1982) and Hidaka (1984). The technique could be a very important part of the production of diploids from protoplast fusion *in vitro*.

*Breeding Citrus*

Tissue culture research shows the possibility of making genetic combinations not previously possible. However, until adult, rather than juvenile tissue, can be manipulated in culture, it does not offer a means to shorten the long life-cycle (Fig. 4.5).

**Figure 4.5.** (*Left*) Branch of *in vitro* shoot tip micrograft-derived (adult tissue) plant compared with (*right*), *in vitro* nucellus-derived (juvenile tissue) plant (photograph courtesy of C. N. Roistacher).

Culture of embryos that otherwise might not survive to fruit maturity has been a useful adjunct to citrus breeding. In fact, the use of this technique has become routine in some crosses. Starrantino and Recupero (1982) describe its use for the recovery of triploid embryos. Starrantino and Russo (1985) have obtained an improved nucellar line, Navelina ISA 315, by culturing immature ovules of the seedless cultivar, Navelina.

Cell culture has been used to select *C. sinensis* and *C. aurantium* cell lines tolerant of elevated concentrations of sodium chloride (Kochba *et al.*, 1982). In limited testing, one line of plantlets showed better salt tolerance than unselected lines (Spiegel-Roy and Ben-Hayyim, 1985). Spiegel-Roy *et al.* (1983) selected a cell line of *C. sinensis* that was tolerant of increased concentrations of 2,4-D and retained embryogenic capacity.

Another use of *in vitro* culture is the rapid increase of clonal material for large-scale screening tests in rootstock breeding programmes. Barlass and Skene (1982) produced multiple explants from stem internodes of 5-week-old seedlings of five cultivars. Edriss and Burger (1984) also established explants from epicotyl and root tissue of Troyer citrange.

Various attempts have also been made to identify and isolate the zygote in ovules of polyembryonic cultivars that have been used as seed parents. Some success has been reported (L. Navarro, pers. comm.), but reliable identification on a regular basis still appears to be doubtful.

## 4.10 OTHER TECHNIQUES

Techniques for hybridisation of citrus have been standard and have changed little. However, improvements in pollen and seed storage have been made in the past few years. Kobayashi *et al.* (1978) report pollen storage up to 3 years with little decrease in germination. This was achieved by storing pollen over silica gel at $-20°C$. Pollen stored in liquid nitrogen had up to 90% germination after 2 years. Sahar and Spiegel-Roy (1980) obtained increased storage life when pollen was stored in an oxygen-free atmosphere, but their longest reported period was 57 weeks with greatly reduced germination.

Research by King *et al.* (1981) showed the potential for greatly increasing the storage life of citrus seeds. Seeds, dried by placing them over silica gel at 20°C, were then sealed in laminated aluminium-foil bags and stored at reduced temperature. From the results of their research, they estimated that seeds with 5% moisture and stored at $-20°C$ would take 30 years to reach 50% germination. This research contrasts with the present technique of storing in polyethylene bags at 4°C after only surface drying. Previous research has shown drying to be detrimental to germination (Honjo and Nakagawa, 1978). Such long-term storage could assist in germplasm preservation as well as assist the breeder in extending or replacing populations.

## 4.11 CROSSABILITY

Much of the hybridisation in citrus has been interspecific rather than intraspecific. This continues to be the pattern, although the major breeding programmes are now utilising more second- and third-generation hybrids from original interspecific crosses. Hybridisation for development of rootstock cultivars has involved a high percentage of crosses with the closely related genus *Poncirus*. Past hybridisation work succeeded in obtaining hybrids of *Citrus* with *Fortunella* and *Microcitrus* as well. Natural hybrids of *Citrus* with *Eremocitrus* were described by Swingle (Swingle and Reece, 1967). More recently, Barrett (1977) made extensive crosses of *Eremocitrus* with several *Citrus* species, *Fortunella*, *Microcitrus* and *Poncirus*. Although hybrid seedlings were obtained from all of these crosses, hybrids with *Microcitrus* and *Poncirus* died. However, Barrett (1977) found that some barriers could be bridged by utilising a successful intergeneric hybrid as a parent. For example, the use of the $F_1$ hybrid of *Eremocitrus glauca* by *Citrus sinensis* as the female parent in a cross with *Poncirus trifoliata* results in viable offspring, in contrast to the failure of the direct cross between *Eremocitrus* and *Poncirus*. The use of such bridging crosses should result in increasing the transfer of genetic material by conventional hybridisation among *Citrus* and its related genera.

#### 4.12 NEW CULTIVARS

Release of new cultivars developed in breeding programmes is infrequent, and their establishment in the industry is slow. Increased breeding efforts in several countries may result in more cultivar releases than in the past. Several cultivars have been released since those reported by Soost and Cameron (1975). Sunburst, a hybrid of Robinson and Osceola, was introduced in 1979 (Hearn, 1979). Russo *et al.* (1977) have introduced Palazzelli, a hybrid of Clementine by King, and Mapo, a hybrid of Avana by Duncan grapefruit. Yafit, a hybrid of Clementine × Wilking, and Norit, a hybrid of King × Temple, have been released in Israel (Spiegel-Roy and Vardi, 1982). Kiyomi, a hybrid of Miyagawa Wase Satsuma × Trovita, and Sweet Spring, a hybrid of Ueda Satsuma and Hassaku, were released in Japan (Sato, 1979; Nishiura *et al.*, 1983). Oroblanco and Melogold, triploid hybrids of *C. grandis* × 4*n C. paradisi*, were released in California (Soost and Cameron, 1980, 1985). A new hybrid rootstock, the Benton citrange, was released in Australia (Long *et al.*, 1977). Two citranges, C32 and C35, have been released for rootstock trials in California (Cameron and Soost, 1986). It is difficult to determine if any of the several hybrids noted in the Russian literature have been released as new introductions.

New cultivars of the main commercial groups—lemons, oranges and grapefruit—are notable by their absence. This absence is a reflection of the difficulty of breeding within these groups and maintaining the characteristics peculiar to these groups.

New introductions from the selection of natural mutation have been comparatively numerous in some citrus areas. The reader is referred to Chapter 20 (this volume) on mutation breeding for references on these clones.

## 4.13 GERMPLASM

The availability to citrus breeders of a wide range of breeding material has never been better. Exchange of budwood and seeds has greatly increased over the past 20 years. Most breeding programmes have increased their collections of materials. Nevertheless, there is increased concern about the future of these collections and the simultaneous loss of wild sources (Cameron, 1977). The high cost of maintaining collections has brought increasing pressure to reduce or eliminate collections. Ever-increasing demand for building materials and energy sources endangers wild sources. The Citrus Working Group discussed these problems at the 1981 meeting of the International Board for Plant Genetic Resources in Tokyo and made recommendations for support for collecting and maintaining germplasm. Citrus breeders, taxonomists and other researchers will have to continue their efforts to avoid a reduction in the scope and availability of germplasm resource.

## REFERENCES

Alexander, D., and Emmett, R. H. (1981). Screening citrus hybrids for *P. citrophthora* tolerance. *CSIRO Rep. Div. Hort. Res.*, 48 pp.
Anon. (1976). The preparation of protoplasts from citrus callus. *Genet. Plant Breed.* No. 1, 29. Genet. Inst., Chinese Sci. Acad., Peking. (in Chinese).
Aubert, B., and Vogel, R. (1981). First results on the prevention of tristeza disease in Mauritius papeda. *Fruits* **36**, 351–359 (in French).
Barlass, M., and Skene, K. G. M. (1982). *In-vitro* plantlet formation from citrus species and hybrids. *Sci. Hort.* **17**, 333–342.
Barrett, H. C. (1974). Colchicine-induced polyploidy in *Citrus*. *Bot. Gaz.* **135**, 29–34.
Barrett, H. C. (1977). Intergeneric hybridization of *Citrus* and other genera in citrus improvement. *1977 Proc. Int. Soc. Citric.* **2**, 586–589.
Barrett, H. C. (1982). Breeding cold-hardy citrus scion varieties. *1981 Proc. Int. Soc. Citric.* **1**, 61–68.
Barrett, H. C., and Hutchison, D. J. (1978). Spontaneous tetraploidy in apomictic seedlings of *Citrus*. *Econ. Bot.* **32**, 27–45.
Barrett, H. C., and Hutchison, D. J. (1982). Occurrence of spontaneous octoploidy in apomictic seedlings of a tetraploid *Citrus* hybrid. *1981 Proc. Int. Soc. Citric.* **1**, 29–30.
Ben-Hayyim, G., Shani, A., and Vardi, A. (1982). Evaluation of isozyme systems in citrus to facilitate identification of fusion products. *Theor. appl. Genet.* **64**, 1–5.

Bono, R., Fernandez de Córdova, L., and Soler, J. (1982). Arrufatina, Esbal, and Guillermina, three Clementine mandarin mutations recently appearing in Spain. *1981 Proc. Int. Soc. Citric.* **1**, 94–96.
Button, J., and Botha, C. E. J. (1975). Enzymatic macerations of *Citrus* callus and the regeneration of plants from single cells. *J. exp. Bot.* **26**, 723–729.
Button, J., Vardi, A., and Spiegel-Roy, P. (1976). Root peroxidase isozymes as an aid in *Citrus* breeding and taxonomy. *Theoret. appl. Genet.* **47**, 119–123.
Cameron, J. W. (1977). Clonal repositories for preservation of *Citrus* and *Citrus* relatives in the United States. *1977 Proc. Int. Soc. citric.* **2**, 601–604.
Cameron, J. W., and Frost, H. B. (1968). Genetics, breeding, and nucellar embryony. In "The Citrus Industry" (W. Reuther, L. D. Batchelor and H. J. Webber, eds), Vol. II, pp. 325–370. University of California Press.
Cameron, J. W., and Soost, R. K. (1977). Acidity and total soluble solids in *Citrus* hybrids and advanced crosses involving acidless orange and acidless pummelo. *J. Am. Soc. hort. Sci.* **102**, 198–201.
Cameron, J. W., and Soost, R. K. (1979). Absence of acidless progeny from crosses of acidless × acidless *Citrus* cultivars. *J. Am. Soc. hort. Sci.* **104**, 220–222.
Cameron, J. W., and Soost, R. K. (1980a). Mono- and polyembryony among tetraploid *Citrus* hybrids. *HortScience* **15**, 730–731.
Cameron, J. W., and Soost, R. K. (1980b). Leaf types of $F_1$ hybrids and backcrosses involving unifoliate *Citrus* and trifoliate *Poncirus*. *J. Am. Soc. hort. Sci.* **105**, 517–519.
Cameron, J. W., and Soost, R. K. (1986). C35 and C32: citrange rootstocks for citrus. *HortScience* **21**, 157–158.
Cameron, J. W., Soost, R. K., and Olson, E. O. (1964). Chimeral basis for color in pink and red grapefruit. *J. Hered.* **55**, 23–28.
Campos, D. V., and Ferraz, L. C. B. (1980). Susceptibilidade ne nove portaenxertos cítricos as nematóide *Tylenchulus semipenetrans*. In "Trabalhos apresentados á IV Reuniao Brasileira de Nematologia", São Paulo, 16–20 de julho 1979, pp. 85–96. Sociedade Brasileira de Nematolgia (in Portugese).
Chen, Z. Q., Wang, M. Q., and Huihua, L. (1980). The induction of citrus pollen plants in artificial media. *Acta Genet. Sin.* **7**, 189–191.
De Lange, J. H., and Vincent, A. P. (1972). Pollination requirements of Ortanique tangor. *Agroplantae* **4**, 87–92.
De Lange, J. H., and Vincent, A. P. (1977). Citrus breeding: new techniques in stimulation of hybrid production and identification of zygotic embryos and seedlings. *1977 Proc. Int. Soc. Citric.* **2**, 589–595.
De Lange, J. H., Vincent, A. D., and DeLeeuw, J. H. (1973). Pollination studies on Minneola tangelo. *Agroplantae* **5**, 49–54.
Edriss, M. H., and Burger, D. W. (1984). *In vitro* propagation of 'Troyer' citrange from epicotyl segments. *Sci. Hort.* **23**, 159–162.
Esan, E. (1973). The Detailed Study of Adventive Embryogenesis in the Rutaceae. Ph.D. dissertation, Univ. California, Riverside.
Esen, A., and Soost, R. K. (1972). Aneuploidy in *Citrus*. *Am. J. Bot.* **59**, 473–477.
Esen, A., and Soost, R. K. (1977a). Relation of unexpected polyploids to diploid megagametophytes and embryo : endosperm polidy ratios in *Citrus*. In "I. Congreso Mundial de Citricultura, 1973" (O. Carpena, ed.), Vol. II, pp. 53–63. Murcia, Valencia.
Esen, A., and Soost, R. K. (1977b). Separation of nucellar and zygotic citrus seedlings by use of polyphenol oxidase-catalyzed browning. *1977 Proc. Int.*

Soc. Citric **2**, 616–618.
Esen, A., Soost, R. K., and Geraci, G. (1979). Genetic evidence for the origin of diploid megagemetophytes in *Citrus. J. Hered.* **70**, 5–8.
Ford, H. W. (1978). Burrowing nematode resistant hybrids of *Poncirus trifoliata*, Milam and 'Ridge' pineapple. *Proc. Fla State hort. Soc.* **91**, 20–21.
Frost, H. B., and Soost, R. K. (1968). Seed reproduction: development of gametes and embryos. *In* "The Citrus Industry" (W. Reuther, L. D. Batchelor and H. J. Webber, eds), Vol. II, pp. 290–324. University of California Press.
Gabuniya, L. A. (1980). New promising early mandarin varieties. *Subtrop. Kul't* No. 3/4, 142–144 (in Russian).
Garnsey, S., Barrett, H., and Hutchison, D. (1981). Resistance to citrus tristeza virus in citrus hybrids as determined by enzyme-linked immunosorbent assay. *Phytopathology* **71**, 875.
Geraci, G., and Tusa, N. (1976). Distinzione tra semenzali nucellari e zigotici de arancio amaro per mezzo di tests biochimici. *Div. Ortoflorofruttic. Ital.* **60**, 27–32 (in Italian, with English summary).
Geraci, G., De Pasquale, F., and Tusa, N. (1977). Percentages of spontaneoous triploids in progenies of diploid lemons and mandarins. *1977 Proc. Int. Soc. Citric.* **2**, 596–597.
Geraci, G. Esen, A., and Soost, R. K. (1975). Triploid progenies from $2x \times 3x$ crosses in *Citrus* cultivars. *J. Hered.* **66**, 177–178.
Goliadze, Sh. K., and Tutberidze, B. D. (1977). Breeding citrus and other subtropical crops. *Subtrop. Kul't.* No. 5/6, 30–35 (in Russian).
Grimm, G. R., and Hutchison, D. J. (1980). Evaluation of *Citrus* spp., relatives and hybrids for resistance to *Phytophthora parasitica* Dastur. *Rev. Plant Pathol.* **59**, 863–865.
Gurtskaya, V. G. (1981). Some results of breeding sweet orange. *Subtrop. Kul't* No. 1, 30–33 (in Russian).
Hearn, C. J. (1977). Recognition of zygotic seedlings in certain orange crosses by vegetative characters. *1977 Proc. Int. Soc. Citric.* **2**, 611–614.
Hearn, C. J. (1979). 'Sunburst' citrus hybrid. *HortScience* **14**, 761–762.
Hearn, C. J. (1984). Development of seedless orange, *Citrus sinensis*, cultivar Pineapple and grapefruit, *Citrus paradisi*, cultivars through seed irradiation. *J. Am. Soc. hort. Sci.* **109**, 270–273.
Hearn, C. J. (1986). Development of seedless grapefruit cultivars through budwood irradiation. *J. Am. Soc. hort. Sci.* **111**, 304–306.
Hearn, C. J., Reece, P. C., and Fenton, P. (1969). Self-incompatibility and the effects of different pollen sources upon fruit characteristics of four *Citrus* hybrids. *Proc. 1st Int. Citrus Symposium* (H. Chapman, ed.), Vol. I, pp. 183–187. University of California, Riverside.
Hensz, R. (1982). Bud variation in citrus cultivars in Texas. *1981 Proc. Int. Soc. Citric.* **1**, 89–91.
Hidaka, T. (1984). Induction of plantlets from anthers of cultivar Trovita orange, *Citrus sinensis*. *J. Jpn Soc. hort. Sci.* **53**, 1–5.
Hidaka, T., Yamada, Y., and Shichijo, T. (1982). Plantlet formation from anthers of sour orange, *Citrus aurantium* L. *1981 Proc. Int. Soc. Citric.* **1**, 153–155.
Honjo, H., and Nakagawa, Y. (1978). Suitable temperature and seed moisture content for maintaining the germinability of citrus seed for long-term storage. *In* "Long Term Preservation of Favourable Germ Plasm In Arboreal Crops" (T. Akihama and K. Nakajima, eds), pp. 31–35. Fruit Tree Station, Ministry of Agriculture and Forestry, Japan.

Hutchison, D. J., and Barrett, H. C. (1982). Tetraploid frequency in nucellar seedlings from single trees of Carrizo and Troyer hybrids. *1981 Proc. Int. Soc. Citric.* **1**, 27–29.

Ikeda, F. (1982). Repression of polyembryony by gamma-rays in polyembryonic citrus. *1981 Proc. Int. Soc. Citric.* **1**, 39–44.

Ideda, I., Kobayashi, S., and Nakatani, N. (1980). Differences in cold resistance of various citrus varieties and hybrid seedlings based on the data obtained from the frosts of 1977. *Bull. Fruit Tree Res. Sta., Ser. E, Akitsu, Jpn.* No. E, 49–65 (in Japanese, with English summary).

Iwamasa, M., and Nishiura, M. (1982). Recent citrus mutant selections in Japan. *1981 Proc. Int. Soc. Citric.* **1**, 96–99.

Iwamasa, M., and Oba, Y. (1975). Precocious flowering of citrus seedlings (Part I). *Agric. Bull. Saga Univ.* No. 39, 45–56 (In Japanese, with English summary).

Iwamasa, M., and Oba, Y. (1980). Seedlessness due to self-incompatibility in Egani-buntan, a Japanese pummelo cultivar. *Agric. Bull Saga Univ.* No. 49, 39–45 (in Japanese, with English summary).

Iwamasa, M., Nito, N., Yamaguchi, S., Kuriyama, T., Nakamuta, T., and Ehara, T. (1982). Occurrence of very early mutants from the Wase (early ripening) satsumas. *1981 Proc. Int. Soc. Citric* **1**, 199–101.

Kapanadze, I. S. (1979a). Morphological and cytogenetic study of the genus *Poncirus. Subtrop. Kul't.* No. 2, 73–85 (in Russian).

Kapanadze, I. S. (1979b). Effect of cobalt 60 on inheritance of the elaioplasts. *Subtrop. Kul't* No. 3, 131–132 (in Russian).

Kaplan, D., and O'Bannon, J. H. (1981). Evaluation and nature of citrus nematode resistance in Swingle citrumelo. *Proc. Fla State hort. Soc.* **94**, 33–36.

Karaya, R. K. (1981). Results of introducing and breeding grapefruit and pomelo. *Subtrop. Kul't.* No. 2, 24–28 (in Russian).

Kerkadze, I. G. (1979). Mutagenic effects of $\gamma$ rays on citrus crops. *Subtrop. Kul't.* No. 3, 92–93 (in Russian).

Kerkadze, I. G. (1980). Natural and induced mutation and some genetic aspects of frost resistance in citrus crops. *Subtrop. Kul't.* No. 3/4, 125–127 (in Russian).

Kerkadze, I. G., and Kutateladze, D. Sh. (1979). Radiation mutants of *Citrus* and their breeding value. *Subtrop. Kul't.* No. 3, 65–66 (in Russian).

King, M. W., Soetisna, U., and Roberts, E. H. (1981). The dry storage of *Citrus* seeds. *Ann. Bot.* **48**, 865–872.

Kobayashi, S., Ikeda, I., and Nakatani, N. (1978). Long-term storage of citrus pollen. *In* "Long Term Preservation of Favourable Germ Plasm In Arboreal Crops" (T. Akihama and K. Nakajima, eds), pp. 8–12. Fruit Tree Station, Ministry of Agriculture and Forestry, Japan.

Kobayashi, S., Ikeda, I., and Nakatani, M. (1979). Studies on nucellar embryogenesis in *Citrus* II. *J. Jpn Soc. hort. Sci.* **48**, 179–185.

Kochba, J., Ben-Hayyim, G., Spiegel-Roy, P., Saad, S., and Neumann, H. (1982). Selection of stable salt-tolerant callus cell lines and embryos in *Citrus sinensis* and *Citrus aurantium. Z. Pflanzenphysiol.* **106**, 111–118.

Koizumi, M., and Kuhara, S. (1982). Evaluation of citrus plants for resistance to bacterial canker disease in relation to the lesion extension. *Bull. Fruit Tree Res. Stn, Ser. D* No. 4, 73–92.

Kokaya, Ts. D. (1981). Introduction and breeding of wild relatives of citrus crops. *Subtrop. Kul't* No. 1, 34–37 (in Russian).

Krezdorn, A. H., and Robinson, F. A. (1958). Unfruitfulness in the Orlando

tangelo. *Proc. Fla State hort. Soc.* **7**, 86–91.
Li, S. J. (1980). Self-incompatibility in 'Matou' wentan [*Citrus grandis* (L.) Osb.]. *HortScience* **15**, 298–300.
Long, K., Fraser, L., Bacon, P., and Broadbent, P. (1977). The Benton citrange: A promising Phytophthora-resistant rootstock for citrus trees. *1977 Proc. Int. Soc. Citric.* **2**, 541–544.
Longley, A. E. (1926). Triploid citrus. *J. Wash. Acad. Sci.* **16**, 543–545.
Maisuradze, N. I. (1979). Polyploidy in orange breeding. *Sov. Genet.* (English transl.) **15**, 1478–1483.
Maisuradze, N. I., Kukuladze, E. K., and Gurtskaya, V. G. (1978). Intraspecific triploid hybrid of sweet orange. *Subtrop. Kul't* No. 5, 60–63 (in Russian).
Mobayen, R. G., and Milthorpe, F. L. (1980). Response of three citrus rootstock cultivars to salinity. *Austr. J. Agric. Res.* **3**, 117–124.
Moffett, J. O., and Rodney, D. R. (1979). Pollination requirements of Fremonts, Fairchilds. *Citrograph* **64**, 243, 252.
Mustard, M. J., Lynch, S. J., and Nelson, R. O. (1956). Pollination and floral studies of the Minneola tangelo. *Proc. Fla State hort. Soc.* **69**, 277–281.
Nakatani, M. Ikeda, I., and Kobayashi, S. (1978). Studies on an effective method for getting hybrid seedlings in polyembryonic citrus I. *Bull Fruit Tree Res. Stn, Ser. E, Akitsu, Jpn* No. 2, 25–38 (in Japanese, with English summary).
Nakatani, M. Ikeda, I., and Kobayashi, S. (1982). Studies on an effective method for getting hybrid seedlings in polyembryonic citrus 3. Artificial control of the number of embryos per seed in Minneola tangelo and sweet orange cultivars by high temperature treatment. *Bull Fruit Tree Res. Sta., Ser E,* No. 4, 29–40 (in Japanese, with English summary).
Nishiura, M., and Iwamasa, M. (1970). Reversion of fruit color in nucellar seedlings from the Dobashibeni Unshu, a red color mutant of the Satsuma mandarin. *Bull. Hort. Res. Sta., Ser. B* No. 10, 1–5 (in Japanese, with English summary).
Nishiura, M., Shichijo, T., Ueno, I., Yamada, Y., Yoshida, T., Kihara, T., Iwamasa, M., and Iwasaki, T. (1983). New cultivar Sweet Spring tangelo. *Bull. Fruit Tree Res. Sta., Ser. B* No. 10, 11–23 (in Japanese, with English summary).
Oiyama, I., and Okudai, N. (1983). Studies on the polyploidy breeding in citrus 3. Occurrence of triploids in the progenies of diploid sweet oranges, *Citrus sinensis*, crossed with diploids. *Bull. Fruit Tree Res. Sta., Ser. D* No. 5, 1–8 (in Japanese, with English summary).
Oiyama, I., Okudai, N., and Takahara, T. (1982). Ploidy level investigation of seedlings obtained from $2x \times 4x$ crosses in citrus. *1981 Proc. Int. Soc. Citric.* I, 32–34.
Olson, E. O., Cameron, J. W., and Soost, R. K. (1966). The Burgundy sport: Further evidence of the chimeral nature of pigmented grapefruits. *HortScience* **1**, 57–59.
Reem, C. L., and Furr, J. R. (1976). Salt tolerance of some *Citrus* species, relatives and hybrids tested as rootstocks. *J. Am. Soc. hort. Sci.* **101**, 265–267.
Russo, F. (1982). Bud variations in citrus cultivars in Italy. *1981 Proc. Int. Soc. Citric.* **2**, 597–601.
Russo, F., and Torrisi, M. (1953). Problems and objectives of citrus genetics. 1. Selection of hybrids, nucellar embryos, and triploids and the artificial production of mutations. *Ann. Sper. Agric., Rome* **7**, 883–906 (in Italian).
Russo, F., Donini, B., and Starrantino, A. (1982). Mutagenesis applied for citrus improvement. *1981 Proc. Int. Soc. Citric.* **1**, 91–94.

Russo, F., Starrantino, A., and Recupero, G. (1977). New promising mandarin and mandarin hybrids. *1977 Proc. Int. Soc. Citric.* **2**, 597–601.
Sadana, G. L., and Joshi, R. (1979). Comparative susceptibility of different varieties of citrus to the attack of phytophagous mite, *Brevipalus californicus*. *Sci. Cult.* **45**, 413–414.
Sahar, N., and Spiegel-Roy, P. (1980). Citrus pollen storage. *HortScience* **15**, 81–82.
Sato, H. (1979). Introduction of new varieties of summer crops. Tangor. *Jap. J. Breed.* **29**, 337–338 (in Japanese).
Scora, R. S., Esen, A., and Kumamoto, J. (1976). Distribution of essential oils in leaf tissue of an $F_2$ population of *Citrus*. *Euphytica* **25**, 201–209.
Soost, R. K. (1956). Unfruitfulness in Clementine mandarin. *Proc. Am. Soc. hort. Sci.* **67**, 171–175.
Soost, R. K. (1969). The incompatibility gene system in *Citrus*. *In Proc. 1st Int. Citrus Symposium* (H. Chapman, ed.) Vol. I, pp. 189–190. University of California, Riverside.
Soost, R. K., and Cameron, J. W. (1969). Tree and fruit characters of citrus triploids from tetraploid by diploid crosses. *Hilgardia* **39**, 569–579.
Soost, R. K., and Cameron, J. W. (1975). Citrus. *In* "Advances In Fruit Breeding" (J. Janick and J. N. Moore, eds), pp. 507–540. Purdue University Press, West Lafayette, Indiana.
Soost, R. K., and Cameron, J. W. (1980). 'Oroblanco', a triploid pummelo-grapefruit hybrid. *HortScience* **15**, 667–669.
Soost, R. K., and Cameron, J. W. (1985). 'Melogold', a triploid pummelo-grapefruit hybrid. *HortScience* **20**, 1134–1135.
Soost, R. K., Williams, T. E., and Torres, A. M. (1980). Identification of nucellar and zygotic seedlings with leaf isozymes. *HortScience* **15**, 728–729.
Spiegel-Roy, P. (1979). On the chimeral nature of the Shamouti orange. *Euphytica* **28**, 361–365.
Spiegel-Roy, P., and Ben-Hayyim, G. (1985). Selection and breeding for salt tolerance *in vitro*. *Plant Soil* **89**, 243–252.
Spiegel-Roy, P., Kochba, J., and Saad, S. (1983). Selection for tolerance to 2,4-D in ovular callus of orange. *Z. Pflanzenphysiol.* **109**, 41–48.
Spiegel-Roy, P. and Teich, A. H. (1972). Thorn as a possible genetic marker to distinguish zygotic from nucellar seedlings in citrus. *Euphytica* **21**, 534–537.
Spiegel-Roy, P. and Vardi, A. (1982). Yafit and Norit, two new easy peeling mandarin hybrids. *1981 Proc. Int. Soc. Citric* **1**, 57–59.
Spiegel-Roy, P., Vardi, A. and Shani, A. (1977). Peroxidase isozymes as a tool for early separation of nucellar and zygotic citrus seedlings. *1977 Proc. Int. Soc. Citric* **2**, 619–624.
Starrantino, A. and Recupero, G. (1982). Citrus hybrids obtained 'in vitro' from 2× females by 4× males. *1981 Proc. Int. Soc. Citric.* **1**, 31–32.
Starrantino, A. and Russo, F. (1985). Una nova varieta di arancio precoce: la Navelina ISA 315. *Rivista di Frutticoltura* 9/10, 56–59. (Italian).
Swingle, W. T. and Reece, P. C. (1967). The botany of *Citrus* and its wild relatives. *In* "The Citrus Industry" (W. Reuther, L. D. Batchelor and H. J. Webber, eds) Vol I, pp. 190–430. University of California, Berkeley.
Tachikawa, T. (1971). Investigations on the breeding of citrus trees IV. On the chromosome numbers in citrus. *Bull. Shizuoka Prefect. Citrus Exp. Stn*, No. 9, 11–25. (Japanese, English summary).
Tachikawa, T., Tanaka, Y. and Hara, S. (1961). Investigations on the breeding

of citrus trees I. Study on the breeding of triploid citrus varieties. *Bull. Shizuoka Prefect. Citrus Exp. Stn*, No. 4, 33–44. (Japanese, English summary).
Tatum, J. H., Berry, R. E., and Hearn, C. J. (1974). Characterization of *Citrus* cultivars and separation of nucellar and zygotic seedlings by thin layer chromatography. *Proc. Fla State hort. Soc.* **87**, 75–81.
Tisserat, B., and Murashige, T. (1977a). Probable identity of substances in citrus that repress asexual embryogenesis. *In Vitro* **13**, 785–789.
Tisserat, B., and Murashige, T. (1977b). Repression of asexual embryogenesis *in vitro* by some plant growth regulators. *In Vitro* **13**, 799–805.
Torres, A. M., Soost, R. K., and Mau-Lastovicka (1982). Citrus isozymes: Genetics and distinguishing nucellar from zygotic seedlings. *J. Hered.* **73**, 335–339.
Tutberidze, B. D. (1980). Results of breeding rootstocks for citrus at the All-Union Institute of Tea and Subtropical Crops. *Subtrop. Kul't* No. 3/4, 120–124 (in Russian).
Tuzcu, O., Cinar, A., Goksedef, M., Ozsan, M., and Bicici, M. (1984). Resistance of citrus rootstocks to *Phytophthora citrophthora* during winter dormancy. *Plant Dis.* **68**, 502–505.
Ueno, I. (1978). Studies of cross-incompatibility in *Citrus tachibana* Tanaka. I. Fruit set of *Tachibana* after cross pollination with eight citrus varieties. *Bull Fruit Tree Res. Stn, Ser. B, Akitsu, Jpn* No. 5, 1–7.
Vardi, A. (1982). Protoplasts from different *Citrus* species and cultivars. *1981 Proc. Int. Soc. Citric.* **1**, 149–152.
Vardi, A., and Spiegel-Roy, P. (1982). Gene control in meiosis of *Citrus reticulata*. *1981 Proc. Int. Soc. Citric.* **1**, 26–27.
Vardi, A., Spiegel-Roy, P., and Galun, E. (1975). *Citrus* cell culture: Isolation of protoplasts, plating densities effect of mutagens, and regeneration of embryos. *Plant Sci. Lett.* **4**, 231–236.
Wakana, S., Iwamasa, M., and Uemoto, S. (1982). Seed development in relation to ploidy of zygotic embryo and endosperm in polyembryonic citrus. *1981 Proc. Int. Soc. Citric.* **1**, 35–39.
Wang, T. Y., and Chang C. J. (1978). Triploid citrus plantlets from endosperm culture. *Sci. Sin.* (English edn) **21**, 823–827.
Weinbaum, S., Cohen, E., and Spiegel-Roy, P. (1982). Rapid screening of 'Satsuma' mandarin progeny to distinguish nucellar and zygotic seedlings. *HortScience* **17**, 239–240.
Yadav, I. S., Jalikop, S. H., and Singh, H. P. (1980). Recognition of short juvenility in *Poncirus. Curr. Sci.* **49**, 512–513.
Yamashita, K. (1979). A few trials to induce precocious flowering in citrus. *Bull. Fac. Agric. Miyazaki Univ.* **26**, 237–243 (in Japanese, with English summary).
Yamashita, K. (1980). Studies on self-incompatibility of Hassaku I. *J. Jap. Soc. hort. Sci.* **49**(1), 49–56 (in Japanese, with English summary).
Yelenosky, G. (1985). Cold-hardiness in *Citrus. In* "Horticultural Reviews", (J. Janick, ed.), Vol. 7, pp. 201–238. Avi Publishing Co., Fairfield, Conn., USA.
Yelenosky, G., Barrett, H., and Young, R. (1978). Cold-hardiness of young hybrid trees of *Eremocitrus glauca* (Lindl.) Swing. *HortScience* **13**, 257–258.

# Part II Improving Temperate Crops

# 5 Strategy for Apple and Pear Breeding

F. H. ALSTON

*Institute of Horticultural Research, East Malling, Kent*

In common with other clonally propagated crops, new fruit tree cultivars are derived from segregating $F_1$ generations and multiplied vegetatively. However, tree fruits differ significantly from most clonally propagated crops, in that prior to cropping they pass through a juvenile period which can vary from 2 years for certain *Malus* and *Pyrus* species to over 15 years in some domestic pear hybrids. The juvenile period delays the selection of new cultivars and also the testing and establishment of commercial orchards of new cultivars, since it is closely related to the length of the non-productive phase of adult trees.

Orchards are usually established as production units for 15–20 years and represent considerable investment in land and other resources. Therefore, great care is taken to choose only cultivars which are a major improvement on existing ones. Market and consumer requirements are also important factors and growers are not prepared to commit large areas to new cultivars without assured markets.

The area of land required for breeding plots also restricts the progress of tree fruit breeding. Individual trees must remain in the ground for 4–8 years, and only 1600 apple or pear trees can be managed on 1 ha, if they are grown on their own roots.

Methods developed over the past 20 years have alleviated many of the constraints to tree fruit breeding (Alston and Spiegel-Roy, 1985). For example, early selection techniques have been developed for use in the first 2 years of growth and during the first two cropping seasons. The delineation of genes governing important commercial characters (Alston and Watkins, 1975) has improved the predictive basis of the crossing programmes.

The aim is to produce regularly heavy cropping cultivars with high quality and long storage potential. By careful choice of breeding aims and selection procedures, new cultivars can be released as early as 12 years from germination, a timescale equivalent to most cereal cultivars.

IMPROVING VEGETATIVELY PROPAGATED CROPS
ISBN 0-12-041410-4

*Copyright © 1987 Academic Press Limited*
*All rights of reproduction in any form reserved*

## 5.1 METHODS USED AT EAST MALLING

About 15 000 apple seedlings and 6000 pear seedlings are produced annually at East Malling. In the first 3 months after germination, apple seedlings are screened for resistance to scab (*Venturia inaequalis*), mildew (*Podosphaera leucotricha*) and rosy apple aphid (*Dysaphis plantaginea*). Selection for habit and foliage characters results in only about 2000 seedlings being retained for fruiting after the second growing season. These are budded on to the dwarfing rootstock M 27 and grown as cordons. Apple seedlings grown in this way fruit 4 or 5 years after germination compared with 8 or 9 years for seedlings grown on their own roots; 5000 trees ha$^{-1}$ can be planted with this system as compared with only 1600 trees ha$^{-1}$ if they are grown on their own roots.

Assessment for fruit quality, yield and storage potential, made during the first two cropping seasons (4 and 5 years from germination) results in up to ten selections each year being propagated for orchard trials. Outstanding selections (about 10%) are propagated for grower trials after 4 years (10 years from germination), and a cultivar may be released after 6 years (12 years from germination) if it continues to perform well in all the orchard trials (Smith and Bates, 1982).

Pears are propagated on the dwarfing rootstock Quince C which, like M 27, permits efficient land use, and a similar procedure is followed.

## 5.2 DISEASE AND PEST RESISTANCE

Sources of resistance to all the main diseases and pests of the apple are known (Alston, 1971). After careful consideration of economics, breeding efficiency, spray costs and integrated control possibilities (Alston, 1981a) work at East Malling is concentrated on breeding for resistance to mildew, scab and rosy apple aphid.

### 5.2.1 Apple scab (*Venturia inaequalis*)

Five genes, each of which provides strong resistance to all known races of the pathogen are known (Williams and Kuc, 1969). The most widely used gene, $V_f$, is derived from *Malus floribunda*, and a number of recent scab-resistant cultivars incorporate this gene. Rouselle *et al.* (1975) have shown that the resistance provided by $V_f$ is considerably enhanced in the presence of polygenes for scab resistance. The large differences (Table 5.1) found in the proportion of seedlings with a high level of scab resistance amongst progenies segregating for the gene $V_f$, demonstrate the importance

**Table 5.1.** Percentage segregation for scab resistance in two apple progenies with different scab-resistant parents, each derived from Malus floribunda ($V_f$).

| Progeny | Scab grade[a] | | | | |
|---|---|---|---|---|---|
| | 0 | 1 | 2 | 3 | 4 |
| Cox × A423/2[b] | 17 | 15 | 20 | 30 | 18 |
| Cox × A423/10[b] | 6 | 7 | 20 | 16 | 51 |

Source: Alston (1977).
[a] Scab grades: 0, no symptoms; 1, pin-point pits; 2, chlorotic or necrotic spots and no sporulation; 3, restrictive sparsely sporulating lesions; 4, extensive abundantly sporulating lesions.
[b] Scab resistant ($V_f$).

of using parents which carry $V_f$ combined with polygenes (Alston, 1977).

Seedlings are inoculated at 2 weeks and selection is done in the glasshouse only 6 weeks after germination. Only seedlings showing no sign of sporulation are retained. The proportion of resistant plants varies in the range 10–50% according to the degree of polygenic resistance available in each cross.

### 5.2.2 Mildew (*Podosphaera leucotricha*)

Many apple cultivars carry polygenic resistance to mildew, which provides a reduced level of susceptibility. Very susceptible seedlings are discarded after field selection in the second growing season. Although high levels of resistance can be found occasionally in a few progenies from cultivars, they are less resistant than some *Malus* species (Table 5.2).

**Table 5.2.** Percentage segregation for mildew resistance in apple progenies.

| Parentage | Mildew grade[a] | | | | | | |
|---|---|---|---|---|---|---|---|
| | 0 | 1 | 2 | 3 | 4 | 5 | 6 |
| Res. × Res.[b] | 0 | 0 | 7 | 31 | 42 | 19 | 1 |
| Malus robusta × Sus.[b] | 39 | 3 | 10 | 27 | 15 | 4 | 2 |
| Malus zumi × Sus.[b] | 38 | 3 | 10 | 7 | 14 | 15 | 13 |

[a] Mildew grades: 0, no visible infection; 6, severe attack on leaves, all shoots badly damaged (Alston, 1977).
[b] Res., resistant selection; Sus., susceptible selection.

Two species, *M. robusta* and *M. zumi*, carry a very high degree of resistance, determined in each case largely by two single dominant genes, although supporting polygenes are also important (Alston, 1977). Breeding from these small-fruiting species started in 1963, and by the second backcross had produced precocious, heavy-cropping selections combining good fruit size and appearance with high mildew resistance (Alston, 1983a). Further crosses have been made to introduce all the qualities needed in a commercial dessert apple.

Glasshouse selection of seedlings following spray inoculation is very effective (Alston, 1983a).

Under unsprayed conditions in the field, resistant selections of *M. zumi* show no sign of mildew sporulation and appear immune to the disease, while the undersides of leaves of *M. robusta* resistant selections show amounts of necrosis that are related to the degree of polygenic resistance carried. It is therefore possible to select progenies of *M. robusta*, combining a high degree of major gene and polygenic resistance, by field selection for a low incidence of necrotic patches amongst glasshouse pre-selected seedlings (Table 5.3).

Although mildew is not a serious problem in commercial pear orchards some progenies become infected during the second growing season. To

**Table 5.3.** Percentage segregation for necrotic reaction in mildew-resistant seedlings derived from *Malus robusta*.

|  | Mildew reaction[a] |  |  |  |
|---|---|---|---|---|
|  | Resistant |  |  | Susceptible |
|  | 0 | 0+ | 1 | (2–6) |
| A121/16[b] × A329/73[c] | 13 | 26 | 1 | 60 |
| A121/4[b] × A329/73[c] | 16 | 26 | 5 | 46 |
| Howgate Wonder[d] × A329/73[c] | 0 | 26 | 4 | 71 |
| Cox[d] × A329/73[c] | 0 | 31 | 0 | 69 |
| Edward VII[d] × A329/73[c] | 5 | 47 | 0 | 47 |

Source: Alston (1977).

[a] Mildew reaction: 0, no visible infection; 0+, no sporulation but necrotic flecks on undersurface of leaf; 1, occasional minute sporulation, no necrosis; 2–6, slight to very severe sporulation.

[b] Polygenic resistant (grade 3).

[c] *Malus robusta* resistant derivative (grade 0 with necrosis).

[d] Susceptible (grade 4).

maintain the present satisfactory level of mildew resistance, only seedlings without the disease are retained for fruiting (Alston, 1975).

### 5.2.3 Rosy apple aphid (*Dysaphis plantaginea*)

This is the most serious pest in English apple orchards. Severe infestations prevent shoot and fruit growth. The most effective resistance is derived from the clone of *M. robusta* which provides strong resistance to mildew (Alston and Briggs, 1970). It is determined by a single dominant gene $Sm_h$ for hypersensitivity. Pre-selection for hypersensitivity is done 5 days after inoculation, on 2-month-old seedlings in a glasshouse.

### 5.2.4 Fireblight (*Erwinia amylovora*)

This disease is the only pear disease that merits a significant breeding programme for resistance in England. Early glasshouse tests cannot identify individual field resistant material (Zwet and Bell, 1981), but tests on maiden trees correlate closely with the responses of mature trees (Zwet, 1977). This method is used at East Malling in an isolation glasshouse, cooled to 22°C, to provide conditions for selecting trees with promising levels of resistance (Alston, 1983b). Seedlings are tested for resistance only after being first selected for yield and fruit characters.

### 5.2.5 Resistance breeding policy

Where possible, major genes are combined with polygenes for resistance as a means of safeguarding the crop against a new gene-specific race of a pathogen. At present, resistance is regarded as a valuable bonus in cultivars bred primarily for improved yield and quality. New improved cultivars will be grown whatever their response to pathogens. If resistance breaks down, pathogens would have to be controlled by other methods, but growers and consumers would still benefit from a commercially superior cultivar provided that an adequate number of resistance polygenes have also been included. Nevertheless, the future economic and environmental importance of resistance, achieved by combining major genes and polygenes, cannot be ignored.

## 5.3 YIELD

Screening the plant habit of 2-year-old seedlings is a valuable means of early selection for high yield. Seedlings with leader length : plant height ratios between 0.35 and 0.55 are selected; precision is improved by selecting from plants with height × spread measurements greater than the progeny mean. Such selections yield up to three times more than their remaining sibs (Alston and Bates, 1979). The type of plant selected by this method is very similar to that preferred in maiden trees of commercial cultivars for high-yielding orchards (Shepherd, 1979).

Effective selection at 18 months, prior to propagation on M 27 rootstocks, results in a larger proportion of high-yielding plants in the cordon rows, and allows later selection for quality and storage amongst mainly high-yielding seedlings.

Further selection for heavy cropping is made at 5 years by retaining only those seedlings with an accumulated crop of 5 kg or more. The yield of young apple seedlings grown as cordons on M 27 is highly correlated with their subsequent performance as orchard trees (Alston and Bates, 1979).

These pre-selection techniques have produced new cultivars such as "Malling" Greensleeves and "Malling" Jupiter which crop economically in the early years, yielding up to three times more than Cox's Orange Pippin, the main commercial variety in England.

There are good prospects for producing new apple cultivars with a compact columnar habit which can be grown as natural cordons and planted less than 1 m apart. These tree types are based on a compact mutant of the Canadian cultivar McIntosh (Tobutt, 1985). Experimental orchards of this type have given yields in the second year equal to those normally obtained 8 years after planting in a conventional orchard.

In pear breeding, precocity is a major yield component and the aim is to select early-bearing cultivars with a semi-dwarf habit, wide branch angle and a high fruit : shoot ratio (Alston, 1983c).

## 5.4 QUALITY

Good fruit quality is the principal aim of the breeder, there being little merit in increasing yield and disease resistance if not accompanied by high quality.

Quality can be considered under four main categories:
(a) *Appearance* which concerns size and shape as well as colour and skin finish; (b) *eating* quality which covers all aspects of flavour, flesh colour and texture; (c) *storage* which depends on a number of factors including resistance to bitter pit, core flush, low temperature breakdown, senescent breakdown, scald and diseases such as *Gloeosporium* and *Phytophthora* fruit rots; and (d) *processing* quality which includes canning and cooking

performance, ease of peeling and juice quality. Breeding for quality in apples has been reviewed by Alston (1981b).

Selection for quality characters is usually done at 4 or 5 years after germination using the early crops of fruit on cordon trees on M 27. Fruit is screened for appearance in the field and in the laboratory, prior to grading.

During storage tests, high humidity is maintained to minimise weight losses. Sealed plastic bins, with a regulated airflow system, simulate conditions in commercial stores. Selections are tested at 3.5°C (to mid-January) and 0°C (to mid-February). Fruit is kept to 10°C for 14 days after removal from store, to simulate conditions during marketing. Fruit is then scored for storage disorders. Eating quality is first recorded instrumentally using a pH meter (acidity), a refractometer (sugar content) and a penetrometer (texture). Large numbers of samples are screened and the best selected for assessment by a taste panel. The incidence of many storage disorders is affected by variation in the environment and in the soil of orchard sites. Therefore, it is important to continue large-scale experiments on new selections, on a number of sites and harvest dates, over several years.

Disorders associated with a low calcium concentration in the fruit (bitter pit, water core and senescent breakdown) and a low phosphorus content (low temperature breakdown) present a challenge to the breeder. The very late maturing cultivar, "Malling" Kent, has fruit with a high calcium concentration (Perring, 1982), and is relatively free from bitter pit and breakdown during storage. Another approach is to breed rootstocks which are more efficient in calcium uptake (Kennedy et al., 1980).

## 5.5 BREEDING PRIORITIES

Combining resistance, habit, yield and storage characters with all the important quality characters is hampered by the heterozygosity of apples and pears, their long juvenile phase and the necessity for several crosses to combine all features. Priorities are decided on the basis of feasibility and the prospects of a new cultivar making an early impact on the fruit industry. Optimum values have been defined for apple selection criteria (Alston, 1981b) and these are used at the first fruiting stage. Priority is given to breeding dessert apples, and while there is provision for consumer choice between "Cox-type", green and red apples, emphasis is given to improving colour. The need for longer-storing cultivars with improved texture has prompted changes in the initial specifications (Table 5.4).

In planning the apple breeding programme, seven principal characters are considered; namely, yield, skin finish, fruit colour, fruit size, acidity, storage and texture. None of these characters are determined solely by major genes, although basic genes, subject to the modifying effects of polygenes, have been identified for fruit colour, skin finish and acidity.

An estimate of the number of progeny required and of the most desirable parental combinations is based on previous progenies and variety trials, and estimates of the varying parental contributions to each character (Table 5.4). Thus Cox, a poor parent for skin finish, is given a selection level of 10% for that character while the best parents are given a selection level of 40%. Cultivars storing well at 0°C are given a selection level of 80% for storage. Tests have shown that this storage character is transmitted to a high proportion of the progeny. Account is also taken of parents which store well at 3.5°C. The best progeny, Gloster 69 × "Malling" Fiesta, had a combined selection level of nearly 1%. Cox was a poor parent, giving combined levels of 0.1, 0.1 and 0.03% when crossed with, respectively, Gloster 69, Idared and Starkrimson.

If crosses are likely to have a 1% success rate, then ten selections for orchard trials, will require a minimum progeny size of 1000 at the fruiting stage. Initial progeny sizes should approach 10 000 to allow for those not meeting the habit and some disease resistance requirements in the first 2 years.

While progeny population sizes are assessed numerically, the number of advanced selections sent for trial is based solely on their individual merits. Tree fruit trial progress can be hindered by submitting large numbers of only moderately good selections. Breeding and selection must be on a sufficient scale to justify critical variety trials of only those selections likely to prove better than existing cultivars.

## 5.6 ESTABLISHMENT OF NEW CULTIVARS

The principal English apple, Cox's Orange Pippin, is poor yielding, has poor size and appearance and is difficult to grow. Its special market position, a consequence of its high eating quality, is threatened by cheaper and more attractive imported apples. We need a new high-yielding, high-quality, long-storing cultivar.

We have produced two cultivars which crop well, carry good quality apples and promise to extend the marketing period of "Cox-style" apples in the UK; they are "Malling" Jupiter (Cox × Starking), and "Malling" Fiesta (Cox × Idared). Both produce substantially more good quality fruit than Cox and their eating quality is equal to Cox. Jupiter completed

**Table 5.4.** Notional parental and progeny selection levels (%) and the principal apple-breeding aims.

| Parent | Yield (5 kg by 5 yr from germination) | Skin finish (0–10% russet) | Colour (75–100% red) | Size (70–75 mm) | Storage (6 months at 0°C) | Acidity (ph 3.4–3.6) | Texture (3.0–4.0 kg, 8 mm probe) |
|---|---|---|---|---|---|---|---|
| Cox | 20 | 10 | 30 | 20 | 40 | 30 | 40 |
| Idared | 60 | 40 | 20 | 80 | 60 | 45 | 80 |
| Gloster 69 | 40 | 40 | 35 | 80 | 80 | 45 | 40 |
| Starkrimson | 40 | 10 | 50 | 20 | 80 | 60 | 40 |
| "Malling" Fiesta | 60 | 40 | 35 | 60 | 60 | 45 | 80 |

Progeny selection levels (%) calculated using means of parental levels:

Cox × Idared = 0.1406  Gloster 69 × Idared = 0.8316
Cox × Gloster 69 = 0.1097  Gloster 69 × Fiesta = 0.9261
Cox × Starkrimson = 0.0259  Starkrimson × Fiesta = 0.4686

**Table 5.5.** Eight years' accumulated crop (kg per tree) of "Malling" Jupiter and Cox's Orange Pippin on rootstock M26, at the National Fruit Trials, Brogdale EHS.

|  | Total | ⩾65 mm | Class I Total | ⩾65 mm |
|---|---|---|---|---|
| "Malling" Jupiter | 108.6 | 92.3 | 65.2 | 55.4 |
| Cox | 38.6 | 15.8 | 18.5 | 7.6 |

orchard trials in 1980 (Table 5.5) and was released in 1981; Fiesta, released in 1986, has excellent texture and stores well for marketing in May and June.

A new heavy-cropping cultivar of pear, "Malling" Concorde (Doyenné du Comice × Conference), has also been produced at East Malling and entered in grower trials in winter 1982–1983. It promises to provide earlier, more regular and higher yields than any existing commercial variety including the main English pear cultivar, Conference.

### 5.7 CONCLUSIONS

Greater chances of success are likely to result from breeding programmes that have limited and specific aims and use large progeny populations. The main aim is to breed for increased precocity with regular high yields combined with high quality and storage attributes.

It is essential that long-term breeding programmes should aim to combine good scab and mildew resistance, and cropping and quality characters, with compactness and self-rooting which are advantageous for ultra-intensive orchard systems. The development of pear cultivars resistant to fireblight is now also important.

A close liaison between pomologists, fruit storage workers and breeders at the trial stage helps in the early identification of growing and storage problems and allows an assessment of their likely resolution. Finally, more consumer testing is welcomed since good, well-authenticated results must involve an effective and significant consumer contribution if a new cultivar is to reach profitable commercial acceptance.

### REFERENCES

Alston, F. H. (1971). Integration of major characters in breeding commercial apples. *Proc. Eucarpia Symp. Tree Fruit Breed., 1970, Angers*, pp. 231–248.

Alston, F. H. (1975). Early stages in pear breeding at East Malling. *Proc. Eucarpia Symp. Tree Fruit Breed., 1973, Canterbury*, 1–13.

Alston, F. H. (1977). Practical aspects of breeding for mildew (*Podosphaera leucotricha*) resistance in apples. *Proc. Eucarpia Symp. Tree Fruit Breed., 1976, Wageningen*, pp. 4–13.

Alston, F. H. (1981a). Pest resistance in apple breeding. *Bull. Int. Org. Biol. Control/West Pal. Reg. Sect. New Series* 4(1), 83–88.

Alston, F. H. (1981b). Breeding high quality high yielding apples. *In* "Quality in Stored and Processed Vegetables and Fruit" (P. W. Goodenough and R. K. Atkin, eds), pp. 93–102. Academic Press, London.

Alston, F. H. (1983a). Progress in transferring mildew (*Podosphaera leucotricha*) resistance from *Malus* species to cultivated apple. *Bull. Int. Org. Biol. Control/ West Pal. Reg. Sect. New Series* 6(4), 87–95.

Alston, F. H. (1983b). Fireblight (*Erwinia amylovora*) resistance in the East Malling pear breeding programme. *Bull. Int. Org. Biol. Control/West Pal. Reg. Sect. New Series* 6(4), 165–170.

Alston, F. H. (1983c). Pear breeding, progress and prospects. *Proc. 21st Int. hort. Congr.1982, Hamburg*, pp. 127–137.

Alston, F. H., and Bates, J. W. (1979). Selection for yield in apple progenies, *Proc. Eucarpia Symp. Tree Fruit Breed, 1979, Angers*, pp. 15–27.

Alston, F. H., and Briggs, J. B. (1970). Inheritance of hypersensitivity to rosy apple aphis *Dysaphis plantaginea* in apple. *Can. J. Genet. Cytol.* **12**, 257–258.

Alston, F. H., and Spiegel-Roy, P. (1985). Fruit tree breeding: strategies achievements and constraints. *In* "Attributes of Trees as Crop Plants" (M. G. R. Cannell and J. E. Jackson, eds), pp. 49–67. Institute of Terrestrial Ecology, Huntingdon.

Alston, F. H., and Watkins, R. (1975). Apple breeding at East Malling. *Proc. Eucarpia Symp. Tree Fruit Breed., 1973, Canterbury*, pp. 14–29.

Kennedy, A. J., Rowe, R. W., and Samuelson, T. J. (1980). The effects of apple rootstock genotypes on mineral content of scion leaves. *Euphytica* **29**, 477–482.

Perring, M. A. (1982). Mineral concentrations in the fruit of varieties other than Cox's Orange Pippin and Bramley's Seedling. *Rep. East Malling Res. Stn for 1981*, pp. 193–195.

Rouselle, G. L., Williams, E. B., and Hough, L. F. (1975). Modification in the level of resistance to apple scab from the $V_f$ gene. *Proc. 19th Int. hort. Congr. 1974, Warsaw* **3**, 19–26.

Shepherd, U. M. (1979). Effect of tree quality at planting on orchard performance. *Rep. East Malling Res. Stn for 1978*, p. 40.

Smith, R. A., and Bates, J. W. (1982). The progress to release of new East Malling apple selections. *Exptl. Hort.* **32**, 1–7.

Tobutt, K. R. (1985). Breeding columnar apples at East Malling. *Acta Hort.* **159**, 63–68.

Williams, E. B., and Kuc, J. (1969). Resistance in *Malus* to *Venturia inaequalis*. *Ann. Rev. Phytopath.* **7**, 223–246.

Zwet, T. van der (1977). Comparative sensitivity and response of various *Pyrus* tissues to infection by *Erwinia amylovora*. *In* "Current Topics in Plant Pathology". *Symp. Hung. Acad. Sci., 1976, Budapest*, pp. 236–276.

Zwet, T. van der, and Bell, R. L. (1981). Some factors affecting selection for fireblight resistance in pear. *Acta Hort.* **117**, 55–61.

# 6 Plum Breeding

R. P. JONES[1] and D. WILSON[2]

[1] *Institute of Horticultural Research, East Malling, Kent*
[2] *Long Ashton Research Station, Bristol.*

Plums have been an important fruit crop in England for centuries but there has been a rapid decline in commercial production during the past 40 years; over 19 500 ha of plums were grown in 1947 but only 2900 ha in 1985. During these years there was little sustained research into plums and no breeding at all until the Long Ashton programme (now transferred to IHR, East Malling) was started in 1969 with the aim of producing new cultivars suited to the present and future needs of the plum industry.

Fruit growing has changed much recently with the consequence that increases in the cost of labour, chemicals and equipment used must result in consistent high yields of good-quality fruit. Regular cropping is a particular problem of plums and although much can be achieved using improved pest and disease controls, better rootstocks, more efficient orchard management and more effective marketing of fruit, it is the breeder who must raise the potential for improvement by introducing new cultivars.

## 6.1 IMPORTANT PLUM SPECIES

Bailey (1935) described many plum species but only relatively few have been used to produce commercial cultivars (Table 6.1). *Prunus domestica* ($2n = 48$) is the most important source of present cultivars grown in the major plum areas of the world. It is used almost exclusively in Europe and is the predominant species in the USA. It is an outbreeding hexaploid producing wide variation in its progeny (Crane and Lawrence, 1952). This accounts for its adaptability to differing climatic conditions and the diverse uses for its fruit. *Prunus salicina* ($2n = 16$) is also important, particularly in the warmer conditions of California and South Africa where there is

**Table 6.1.** Important species in plum breeding programmes.

| Species | Chromosome no., 2n | Origin | Use |
|---|---|---|---|
| Prunus domestica L. | 48 | West Asia | Fruit |
| P. cerasifera Ehrh. | 16 | Southwest Asia | Fruit, rootstock |
| P. insititia L. | 48 | West Asia | Fruit, rootstock |
| P. salicina Lindl. | 16 | China | Fruit |
| P. spinosa L. | 32 | Europe and Asia | Rootstock |

little winter chilling required and no danger of frosts at flowering time which prevents its use in Europe and the northern states of the USA. *Prunus cerasifera* (2n = 16) and *Prunus insititia* (2n = 48) are grown locally in parts of Europe for their fruit but are now more important as sources of new rootstocks.

## 6.2 BREEDING PROGRAMMES

Plums are an important world crop with over 2.5 million tonnes being produced annually, mainly in the USA and Europe. Most of the breeding work is situated in these areas. There are several breeding programmes in the USA; Washington State and the New York Agricultural Experiment Station are breeding for improved processing cultivars which are used for prune production. In California, more emphasis is placed on breeding for dessert plums, using *Prunus salicina* which has very good size and is suitable for transporting.

Most of the Eastern European countries have their own breeding programmes using *Prunus domestica* to replace their present cultivars which are used for prune and plum brandy production.

In France, as well as introducing new prune cultivars, they are also breeding new dessert plums of the greengage type. There is a similar programme in Italy with more emphasis on dessert plums.

## 6.3 OBJECTIVES OF THE BREEDING PROGRAMME

It is a feature of perennial, clonally propagated crops that cultivars are introduced over a long period and the establishment of new ones is a slow process. Most of the important English cultivars are well over 100 years old and several arose as chance seedlings (Table 6.2). For reasons

Table 6.2. The main plum cultivars grown commercially in England.

|  | Introduced | Parentage |
| --- | --- | --- |
| Victoria | 1840 | Unknown |
| Pershore | 1827 | Unknown |
| Purple Pershore | 1877 | Diamond × Rivers |
| Czar | 1875 | Prince E. × Rivers |
| Giant Prune | 1893 | D'Agen × Ponds |
| Marjorie's Seedling | 1928 | Unknown |
| Damson | c. 1700 | Unknown |
| Early Rivers | 1834 | Preococe de Tours O.P. |

that will follow, these cultivars are now unsuitable for the present plum industry but, because of the lack of progress in plum breeding, the grower has few alternatives to turn to. Of the main characters and qualities required of any new plum there are three primary objectives that have been regarded as essential for a successful introduction to the industry.

**6.3.1 Regular cropping**

Unpredictable and erratic cropping have always been major problems with plums (Fig. 6.1). A comparison of national yields with winter temperatures has shown that high yields were often associated with lower-than-average temperatures in January and February. It had previously been assumed that winter temperatures had little effect on bud development because they would still be dormant. However, when the dormancy of a range of plum seedlings was studied (Wilson et al., 1975) it was found that 30% of seedlings were capable of growth by November and that this figure had increased to 74% by February. This means that bud development and subsequent flowering date are dictated by temperatures in February, March and April which often fluctuate quite markedly. Figure 6.2 shows the spread of flowering of a seedling population on the years with quite different temperature patterns. In a very mild year (1975), those seedlings with a low chilling requirement were flowering by early February. The flowering season was protracted and the national plum yield that season was only 3.3 t ha$^{-1}$. In 1978, a fairly typical year, most of the flowering occurred in April. In the following year, 1979, the daily maximum temperature for 3 months never rose above 7°C, with the result that flowering did not start until May and was finished within 4 weeks. The yield that year was 10.1 t ha$^{-1}$.

Because the winter was mild in 1975 the prolonged flowering season was primarily due to differences in chilling requirement. However, in

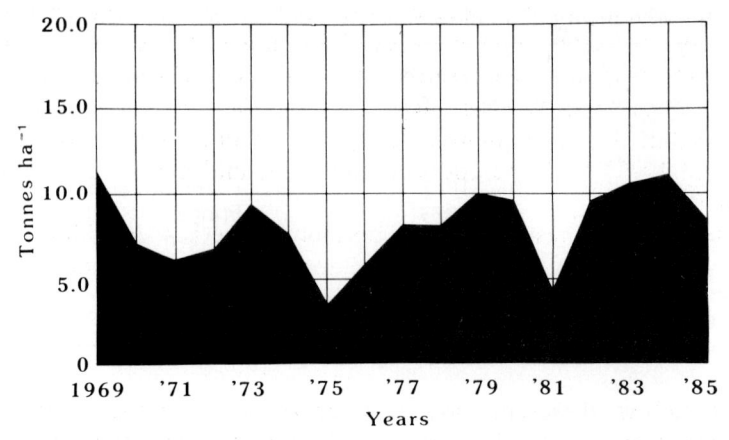

**Figure 6.1.** National plum yields between 1969 and 1985.

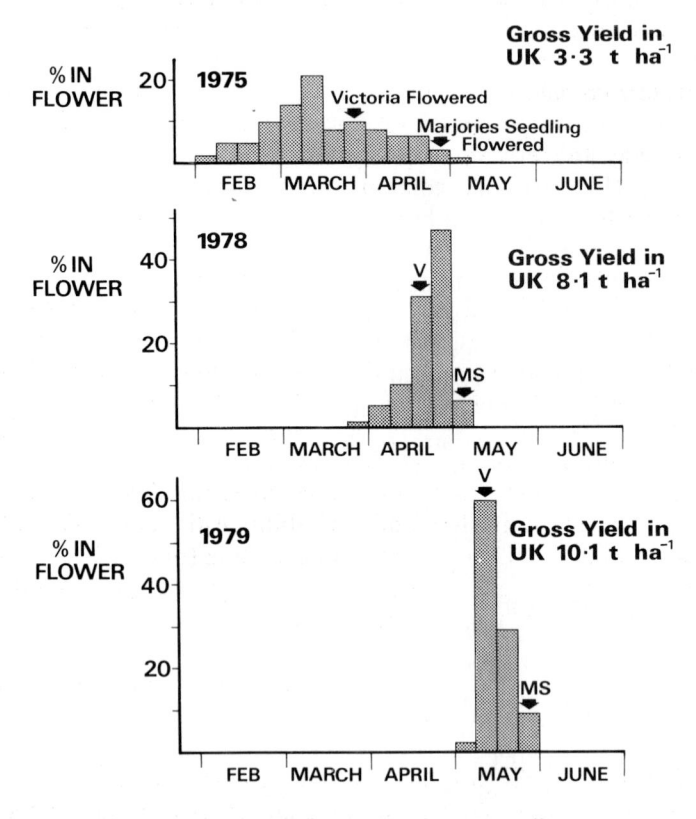

**Figure 6.2.** The variation in time of flowering of plum seedlings in 3 years at Long Ashton Research Station.

1979, the prolonged cold weather would have released the buds from true dormancy but held them in an imposed dormancy. The spread of flowering that season was then due to differences in rates of bud development at the prevailing temperatures. Hence, if long dormancy can be combined with slow bud development in some selections then, hopefully, the flowering season of 1979 can be reproduced each year and more regular cropping obtained.

Of the cultivars tested for long dormancy, the best are Merton, Blue, Pacific, Marjorie's Seedling, Cox's Emperor and Utility, and these have been used extensively in the breeding programme to meet this objective.

### 6.3.2 Dessert-quality fruit

Traditionally, most English plums are regarded as dual-purpose or culinary fruit. The gages, which have the finest flavour, crop poorly and are not grown on a wide scale. Habits and preferences have changed with increasing affluence, and there is much competition from fresh fruits which are available all year round. This has resulted in a much reduced demand for processed plums but there is a large, and as yet unsatisfied demand for large, attractive, juicy, good-flavoured plums with good texture. Parents used to combine these characters are Count D'Althans Gage, Merton Gage, Reeves, Utility and Opal.

### 6.3.3 Fruit size

Because of this past reliance on processing, most of the present-day cultivars produce most of their fruit in the 30–40 mm range. However, with the change in emphasis from culinary to dessert plums, fruit size has become one of the most important characters in the breeding programme; 40 mm is now regarded as the minimum size necessary, with 50 mm being preferable. Parents used for large fruit size are Reeves, Edwards, Valor, President, Pond's Seedling and Thames Cross.

Hansche *et al.* (1975) have shown that flowering date and fruit size are largely genetically controlled, and heritability estimates indicate that they are additive in effect. Consequently, intermating (mass selection) between the best progeny should be effective in increasing the potential of the breeding stock and this is the strategy employed in the breeding programme.

Important, but secondary, characters that are required in combination with the primary ones are as follows.

(a) *Fruit colour*. Reds, purples and blues are selected rather than yellow or green, as the latter show blemishes and bruises, reducing the market quality.

(b) *Free-stone*. Consumers much prefer this to cling-stone, particularly where the fresh market is concerned.

(c) *Tree habit*. Good branch angles, producing a spreading tree, are important to minimise the branch breakage when carrying heavy crops and to reduce the labour input on tree shaping.

(d) *Disease resistance*. The most important diseases of plum are bacterial canker (*Pseudomonas mors-prunorum*), silver leaf (*Chondrostereum purpureum*) and fruit rot (*Monilinia fructigena*). No screening is done at the seedling stage but minimal spraying in the orchard allows identification of the most susceptible progeny.

### 6.4 LIMITATIONS ON PLUM IMPROVEMENT

The two main enemies of any breeding programme are time and space. In tree fruits these can be identified as juvenility and plant size, and anything which minimises these factors will increase the selection response and hence the improvement of the breeding stock.

Hansche (1983) gives examples for cherry and walnut where highly heritable traits of economic importance which one would predict to yield large selective responses in the order of 15–25% per selection cycle are, in fact, reduced to 2–3% because of a juvenile phase lasting 6–8 years. With plums, the juvenile phase can vary between 4 and 6 years for different seedlings. Any techniques that can shorten this time will obviously result in a more rapid turnover of seedlings. In apple, this has been achieved by grafting the seedlings on to a dwarfing rootstock, M 27 (Tydeman and Alston, 1965). This shortens the juvenile phase, with the additional advantage of allowing much closer planting because of the reduced tree size. Unfortunately, no similar dwarfing rootstock exists for plum, so the juvenile phase is shortened in a more conventional way by growing the seedlings continuously under protection throughout the summer. The average height of one-year-old plum seedlings at planting is 2.3 m (Fig. 6.3). This reduces the juvenile phase by 1–2 years and so assessment of the population will be complete within 5–6 years instead of the more normal 7–9 years for conventional nursery-raised plants (Table 6.3).

In order to increase the frequency of these seedlings with the required characters, large populations must be raised which can be more efficiently used if the size is reduced and inferior seedlings identified at the earliest possible stage and eliminated. Hopefully, rootstock breeding will produce

# Plum Breeding

**Table 6.3.** The effect of continuous protection and conventional nursery conditions on the length of the juvenile phase in plum seedlings.

| Family | Seedling treatment | Age of trees (years) | Number of seedlings | % flowering | % fruiting |
|---|---|---|---|---|---|
| Marjorie's Seedling × Pacific | Continued growth | 4 | 263 | 80 | 23 |
| Reeves O.P. | Nursery-raised | 5 | 243 | 51 | 4 |
| Victoria O.P. | Nursery-raised | 5 | 201 | 55 | 2 |

**Figure 6.3.** Plum seedlings, 10 months after germination.

material that will induce dwarfness and precocity in addition to that bred into the scion but neither characters are present in a rootstock yet.

We have tried to identify characters which can be assessed in the juvenile phase and are correlated with important flower and fruit characters in the adult stage. To be of any practical value, the correlations must be high and the character easily measured to allow large numbers of seedlings to be assessed. As mentioned, late flowering is of considerable importance and there is a good correlation between vegetative bud burst in the juvenile phase and date of flower in the mature tree (Table 6.4). Consequently, if any seedling with a bud-break rating of 4 or less had been rejected, then 28% of the population could have been rejected with the loss of only ten seedlings which flowered with a rating of 5 or more. Fruit size is also very important, and more seedlings are rejected because of small fruit than for any other character. There is not a good correlation with any juvenile character but there is a reasonable correlation with leaf size in the adult tree (Table 6.5). As plums usually flower for 1 or 2 years before setting a crop, this could help by allowing us to discard small-leafed seedlings after the first year's flowering. This means that seedlings could initially be grown closer together and so increased populations could be obtained without the use of more land.

## 6.5 PROSPECTS

One of the major problems in many vegetatively propagated crops is the length of time between producing the seed and the eventual naming of a

**Table 6.4.** Correlation between times of bud break in juvenile phase and flowering in adult phase of plum seedlings.

| Budbreak rating in 1974[a] | First flower rating 1977[a] |||||||
|---|---|---|---|---|---|---|---|
| | 1 | 2 | 3 | 4 | 5 | 6 | 7 |
| 1 | 0 | 10 | 23 | 13 | 4 | 2 | 0 |
| 2 | 1 | 10 | 17 | 16 | 3 | 0 | 0 |
| 3 | 0 | 6 | 31 | 17 | 1 | 0 | 0 |
| 4 | 0 | 10 | 36 | 42 | 6 | 1 | 0 |
| 5 | 0 | 15 | 111 | 167 | 36 | 10 | 0 |
| 6 | 0 | 2 | 48 | 103 | 42 | 10 | 0 |
| 7 | 0 | 1 | 4 | 37 | 29 | 9 | 1 |
| 8 | 0 | 0 | 0 | 2 | 2 | 2 | 1 |
| 9 | 0 | 0 | 0 | 1 | 0 | 0 | 0 |

$r = 0.47^{***}$.
[a] Weekly intervals.

**Table 6.5.** Correlation between leaf size at first flowering and fruit size at the time of first cropping in plum seedlings.

| Leaf size | Fruit size |   |   |   |   |
|---|---|---|---|---|---|
|  | 1 | 2 | 3 | 4 | 5 |
| 1 | 3 | 1 | 0 | 0 | 0 |
| 2 | 16 | 35 | 27 | 6 | 1 |
| 3 | 15 | 105 | 100 | 36 | 8 |
| 4 | 3 | 28 | 53 | 22 | 7 |
| 5 | 1 | 0 | 0 | 7 | 0 |

$r = 0.39$***.

new cultivar. With plum it is about 20 years. Anything which can reduce that time is very important and although our breeding programme started only 16 years ago, the benefits gained from similar crops may well help substantially. Identification and elimination of inferior seedlings as early as possible must be of major concern in order to increase the frequency of more promising material.

There is more awareness of the need to conserve genetic resources for all crop plants on a worldwide scale; the availability of this material combined with greater co-operation with other breeders must be of considerable benefit.

**REFERENCES**

Bailey, L. H. (1935). "The Standard Cyclopaedia of Horticulture". Macmillan, New York.
Crane, M. B., and Lawrence, W. J. (1952). "The Genetics of Garden Plants". Macmillan, London.
Hansche, P. E. (1983). Response to selection. *In* "Methods in Fruit Breeding", (J. Janick and J. N. Moore, eds), pp. 154–171. Purdue University Press, West Lafayette, Indiana.
Hansche, P. E., Hesse, C. O., and Beres, V. (1975). Inheritance of fruit size, soluble solids and ripening data in *Prunus domestica* cv. Agen. *J. Am. Soc. hort. Sci.* **100**, 522–524.
Tydeman, H. M., and Alston, F. H. (1965). The influence of dwarfing stocks in shortening the juvenile phase of apple seedlings. *Rep. East Malling Res. Stn for 1964*, pp. 97–98.
Wilson, D., Jones, R. P., and Reeves, J. R. (1975). Selection for prolonged winter dormancy as a possible aid to improving yield stability in European plum (*Prunus domestica* L.) *Euphytica* **24**, 815–819.

# 7 Raspberry and Blackcurrant Breeding

D. L. JENNINGS, M. M. ANDERSON and R. M. BRENNAN

*Scottish Crop Research Institute, Dundee*

## 7.1 RASPBERRIES

Two recent trends in the raspberry industry have influenced our breeding objectives. First, raspberry acreage has increased, especially in England, mainly due to an expansion of "pick-your-own" and fresh fruit sales. The traditional English markets have always preferred late cultivars whose marketing does not clash with that of strawberries, but "pick-your-own" operators now require a range of cultivars that mature over a more extended period. Secondly, machine harvesting of raspberries has become common practice in America and Canada, and although this has not yet happened in the UK, all breeders are concerned with cultivar suitability for machine harvesting.

The ideal plant for machine harvesting has not been defined, but it will certainly need to have firm fruit, which will absciss readily but not too easily. Resistance to fruit mould is also important because the fruit is harvested at a more advanced stage of ripeness and fruits left after early passes of the machine are liable to infect later-ripening fruits. A range of cultivars which mature in sequence are needed. In the UK, but not in America, resistance to cane blight (*Leptosphaeria coniothyrium*) is especially desirable for cultivars grown for machine harvesting, because infection of machine-inflicted wounds by this fungus can cause considerable losses of new canes in our climate.

Fruit quality is more important, and the increased use of frozen fruit has placed a premium on good appearance in terms of colour and brightness. However, high yield remains an important objective in these times of economic difficulty.

### 7.1.1 Yield

A characteristic of the raspberry is that it has biennial stems which are normally vegetative in their first year and fruit-bearing in their second.

Vegetative and fruit-bearing canes compete with each other and the relative vigour of first- and second-year canes consequently has a big effect on yield.

The cultivar Glen Clova, released by the Scottish Crop Research Institute (SCRI) in 1969, shows this competition very well. It grows vigorously, and though it yields some 25–30% more than most other cultivars, the improvement is further increased when the vigour of its first-year canes is controlled. This is done by chemically burning down the new growth when it is about 30 cm high; a treatment which gives yield increases of up to 50% due to the production of stronger fruiting branches (known as fruiting laterals) carrying more and bigger fruit (Dale, 1977). Clearly, reduction of vegetative growth releases a higher proportion of the plants' resources for fruit production. The treatment has become standard practice for this cultivar, and the good results have led to Glen Clova now occupying about 60% of the UK raspberry acreage.

In most other cultivars, this treatment prejudices the numbers and quality of the new canes produced and hence the following year's crop. It does not do so in Glen Clova because the cultivar is vigorous, has high cane numbers and its new canes commence growth early. Most cultivars lack one or more of these attributes and cannot be managed routinely in this way.

Glen Clova has subsequently been outyielded by some 25% by the cultivar Glen Moy, which was released by SCRI in 1981. The higher yields of this cultivar owe much to its larger fruit, but attempts to improve them further by controlling cane vigour have been less successful than with Glen Clova, largely because the cultivar produces inadequate replacement canes. Joy, a cultivar released by East Malling in 1980, shows a similar yield improvement and achieves it by a combination of large fruit size and high fruit numbers per fruiting lateral. Like most genotypes with high fruit numbers per lateral, its laterals are exceptionally long and late-ripening. Further improvement in fruit number per lateral has been achieved at East Malling Research Station by introducing germplasm from *Rubus cockburnianus* and *R. flosculosus*. But it is difficult to obtain selections which combine high fruit numbers per lateral with early ripening, though progress has been made with selections whose laterals are shorter (Knight, 1986) or start to develop earlier (Dale, 1982a).

In these examples, a steady improvement in yield potential has been achieved by improvement in fruit size or fruit numbers, or by a combination of these two yield components. Parents have also been identified which confer improvements in lateral number per cane (Jennings and Dale, 1982), sometimes through a high incidence of multiple laterals (Knight, 1986). Prospects for further advances seem good, but to obtain cultivars which achieve their yield potential, we must select plants which have a high harvest index and have either a good inherent balance between first and second year's growth, or achieve it by reliable response to cane vigour control.

### 7.1.2 Season of ripening and autumn fruiting

Prior to 1974, British raspberry cultivars could conveniently be classed as early, mid-season or late. The release from East Malling of cultivars Leo (in 1974) and Joy (in 1980) has extended the season and added a fourth group, and we can now redesignate our cultivars as early, mid-season, late mid-season or late.

There is also progress in producing earlier cultivars. In North America the cultivar Prestige has inherited earliness from *Rubus pungens oldhami* and in the UK early ripening derivatives of *R. crataegifolius* and *R. spectabilis* are being tested (Keep *et al.*, 1980; Knight, 1986).

Primocane-fruiting cultivars which produce fruit in late summer on the tips of first-year canes have not yet achieved major importance in the UK because the fruit produced is too little and too late. However, the cultivar Heritage, bred at Geneva, New York, gives high yields of good quality fruit and has become important in areas with a longer growing season, and a major advance in the earliness of primocane fruiting has now been obtained by introducing new germplasm from *Rubus arcticus*, the arctic raspberry. This led to the release from East Malling Research Station of Autumn Bliss in 1983. This cultivar crops over a high proportion of its canes' length, even in a short growing season, and gave more than twice the yield of Heritage in a trial in southern England, largely because its fruit started to ripen so much earlier (Helliar and Turner, 1984). Its autumn crop overlaps that of late summer-cropping cultivars like Leo and continues until October (Keep *et al.*, 1984). It thus provides the first opportunity in the UK for continuous production over an extended season. In North America wild raspberries from Wyoming have contributed a short stocky growth habit which is especially desirable for primocane-fruiting cultivars (Lawrence, 1980).

### 7.1.3 Fruit quality

The progress made in improving the firmness of raspberry fruit by introducing genes from the black raspberry (*R. occidentalis*) is illustrated by objective measurements made in North America on fruits of the cultivars Glen Prosen and Glen Isla (Barritt *et al.*, 1980). These show that fruit firmness has been improved to nearly twice the levels present in Glen Clova, Willamette and Meeker, which are the standard cultivars of Britain and North America. Glen Prosen, in particular, travels well and has the potential for supplying distant markets. Fruit colour has also been improved, as illustrated by the Canadian cultivars Chilcotin and Chilliwack, the fruit of which freezes well.

### 7.1.4 Resistance to diseases and pests

In most raspberry breeding programmes, resistance to the important virus diseases is achieved by routine screening for major-gene controlled resistance to either *Amphorophora idaei*, the main aphid vector in Europe, or *A. agathonica*, the main aphid vector of North America. Screening for resistance to an important pollen-borne virus (bushy dwarf) is more laborious because it requires graft inoculation, and though we have a major gene for strong resistance or immunity (Jones *et al.*, 1982) the discovery of a resistance-breaking strain of virus has caused some concern.

In recent years most breeding for resistance has been directed against fruit moulds, especially mould caused by *Botrytis cinerea*. The reasons for this are its importance in relation to machine harvesting and the marketing of fresh fruit, and the limited success of chemical control. Good sources of resistance have been identified; in Canada and the USA the cultivar Cuthbert and its derivatives have been used extensively (Barritt and Torre, 1980; Daubeny, 1980); in the UK we have concentrated on related species, notably *R. occidentalis* and *R. crataegifolius* (Jennings, 1980; Knight, 1980). Russian breeders have also used *R. crataegifolius* (Kichina, 1976).

There are several components of resistance: for example, fruit firmness in cultivars like Glen Prosen, but there are probably unidentified components in soft-fruited cultivars such as Cuthbert and its derivatives, and in related species (Barritt and Torre, 1980; Jennings, 1980; Pepin and MacPherson, 1980). In addition, some cultivars escape the disease because they have erect fruiting laterals and an open plant habit which facilitate rapid drying after rain. We now know that stylar invasion is an important infection route for *B. cinerea*, and that styles of different genotypes differ in the amount of mycelial growth that they allow to occur after germination of the spores on the stigmas. This may prove to be a resistance component to be exploited in breeding (Williamson and McNicol, 1986).

At SCRI we are studying cane resistance to *B. cinerea* because it is an important disease in its own right, and because resistant canes sporulate sparsely and provide less inoculum for fruit infection. Progress has been helped by the development of a mycelial inoculation technique which provides accurate quantitative assessments of resistance; these are based partly upon autumn and spring measurements of the lesions which result from August inoculation, and partly on assessments of spring sporulation on these lesions. We have identified good sources of resistance, and shown that resistance is inherited as a simple additive character with no interactions when resistances from more than one source are combined. Hence, we can predict accurately the mean resistance of progenies from the mean resistance of their parents (Jennings, 1982a; Jennings and Williamson, 1982).

Progress in breeding for resistance to *Leptosphaeria coniothyrium* (cane blight) and *Didymella applanata* (spur blight) has also been helped by the use of mycelial inoculation techniques (Jennings, 1979, 1982b). Many Asian species contribute resistance to several diseases, and there is evidence that most of the resistance to *B. cinerea* and *D. applanata* is conferred by the same genes (Jennings, 1983; Williamson and Jennings, 1986).

The incidence of raspberry yellow rust (*Phragmidium rubi-idaei*) has increased following the widespread planting of susceptible cultivars. Immunity from the disease is determined by the major gene *Yr* and a "slow rusting" type of resistance is determined by minor genes (Anthony et al., 1986). The latter is highly correlated with resistance to cane spot (*Elsinoe veneta*). Hence, we have two instances in the raspberry of a resistance which is common to two diseases. Common resistance to *B. cinerea* and *D. applanata* can possibly be attributed to the sharing by the two pathogens of the same niche at nodes on the raspberry canes, but yellow rust occurs predominantly on leaves and cane spot occurs predominantly on stems, and the only obvious common factor is that both of these diseases can invade only immature tissues (Jennings and McGregor, 1987).

Cane midge (*Resseliella theobaldi*) attacks plants with canes whose rind (epidermis and primary cortex) splits and peels in spring to provide breeding sites for the larvae. Canes of *R. crataegifolius* and its hybrids escape because their splits are small and because the canes rapidly produce suberised wound periderm to "heal" the wounds produced by the splits (McNicol et al., 1983). We do not know whether the character can be transferred to raspberries.

## 7.2 BLACKCURRANTS

Until recently, the blackcurrant industry in the UK relied almost exclusively upon the cultivar Baldwin. This cultivar is preferred by the processors because of its superior flavour, but it is risky for the growers because, if unprotected, frosts at flowering time devastate the crop once or twice every 10 years, and low (non-freezing) temperatures frequently reduce fruit set. The cultivars Ben Lomond, Ben Nevis and Ben Sarek, released by SCRI in the 1970s, were bred from cold-hardy Nordic types and are more frost tolerant and less risky commercially.

When considering these improvements it is useful to remember that centres of diversity in *Ribes* occur in places such as northern Russia, Finland, Sweden and northern Canada, to the north of the UK. Baldwin and the cultivars grown in the south of Britain have been obtained by selection in the south for quality and yield, and they do not represent the full range of the available germplasm. These cultivars are more adapted to England than to Scotland, where they ripen so irregularly that they cannot always be picked on the strig. In addition, they are vulnerable to frost and low temperatures at flowering time in all parts of the UK. At SCRI we have produced better-adapted cultivars by the simple expedient of exploiting gene sources from the northern geographical limits of blackcurrant distribution. Successful cultivars are often those developed by breeders who have used the northern climate to provide natural selection for cold tolerance, thereby exploiting the phenological response most appropriate for the area (Stushnoff, 1974).

The Finnish cultivar, Brödtorp, and the Swedish cultivar, Janslunda, have strigs with remarkably regular berry size and uniformity of ripening. A cross between them made at SCRI gave selections bearing many large fruit, but unfortunately they also inherited an undesirable spreading growth habit. Use of an erect Canadian cultivar as a parent corrected this fault, and gave us the cultivars Ben Lomond and Ben Nevis, both of which are notable for their large, uniformly ripening fruits, good quality juice and improved tolerance of spring frost (Anderson, 1975).

The cultivars Öjebyn and Sunderbyn II, selected from the Swedish wild by Larsson and Trajkovski (Larsson, 1979, pers. comm.) have also been successful parents at SCRI and elsewhere (Larsson, 1959; Oydvin, 1974; Keep *et al.*, 1982; Hiirsalmi, 1985; Trajkovski, 1986). They are tolerant of cold and spring frosts at flowering, as are many of their hybrids, and possess other desirable attributes, including high yield potential and disease resistance combined with a compact but spreading growth habit. Two SCRI Öjebyn hybrids have been named Ben More and Ben Sarek, and a third is under consideration for commercial release.

Ben More can give high yields, but it has not done so consistently, apparently because it is prone to fruit losses from "run-off" (Daugaard, 1981). However, the trial results of more recent selections derived from Ben Lomond, soon to be named, suggest that cultivars which combine frost tolerance and late flowering (frost escape), and are not prone to "run-off", will take much of the risk out of blackcurrant growing and may well double the national average blackcurrant yield.

Natural frosts occur too infrequently to be relied upon for selection of frost tolerance in progenies, and when they do occur the tolerance shown by a given genotype is largely determined by the stage of flower development reached. Instead, selected plants at known stages of flower development are subjected to 4 hours of simulated frosts at −4°C in a controlled frost-testing chamber (Stushnoff, 1972; Levitt, 1978; Quamme, 1978; Weiser et al., 1979; Marshall, 1982). These tests have shown that Baldwin and two complex Ben Lomond × Öjebyn hybrids represent the two extremes for a wide range of frost tolerances, and that Ben Lomond, Ben Nevis, Ben Sarek and Ben More occupy intermediate positions (Dale, 1981). Therefore, much scope still exists for breeding new hybrids with greatly improved frost hardiness.

A similar technique is currently being evolved to screen selected hybrids for tolerance of cold temperatures (above freezing) at critical stages of flowering (Dale, 1981, 1982b).

Late bud break (and hence late flowering) is valuable for frost-avoidance and confers yield stability. The season of bud break is determined by a cultivar's winter chilling requirement, and flowering dates are modified by the mean daily temperatures encountered after its chilling requirement has been met. Season of leafing-out is controlled by two additive genes, *Lf1* and *Lf2* (Keep, 1985a). Effective donors of late flowering include the cultivars Goliath, French Black and the Californian blackcurrant species *R. bracteosum*.

### 7.2.1 Yield

Ben Lomond and Ben Nevis have bigger fruit than older cultivars, and some new selections at SCRI have even larger fruit; others currently in trials have a high harvest index combined with a compact growth habit. These improvements, and control of some of the factors which formerly limited yield, have introduced a new requirement for stronger branches able to carry heavy crops without breaking, and preferably without bending excessively and thereby reducing the effectiveness of harvesting machines. The need is shown by Ben Lomond and Ben Nevis whose growth habits are good except when their branches are borne down by sheer weight of crop. Fortunately, there are several sources of germplasm for greater branch strength, including selections of blackcurrant × redcurrant, blackcurrant × *R. bracteosum* and, in the longer term, blackcurrant × *R. sanguineum*, the ornamental flowering currant, which combines elasticity of branches with high mechanical strength.

As with raspberries, there are indications that the highest-yielding genotypes are those with only moderately vigorous vegetative growth and a high harvest index. This characteristic is yet another valuable feature introduced from northern germplasm. Exceptionally high yields have been attained from closely planted hedgerows of small, compact plants with short internodes, but the commitment to annual machine harvesting makes it likely that this type of bush will not be acceptable for large-scale, high-density growing systems unless the machines available are capable of harvesting the crop.

### 7.2.2 Juice quality

More than half of the UK blackcurrant crop is used for juice production, and so special attention in selection is given to juice quality, notably flavour, colour and ascorbic acid content. Juice colour is due to the presence of several anthocyanins, mainly cyanidin and delphinidin 3-glucosides and 3-rutinosides. The juice processors' preference is for highly coloured juice, rich in ascorbic acid and with a characteristic blackcurrant flavour. Of the new cultivars, Ben Lomond is satisfactory for these qualities, especially for its colour, which is remarkably stable during storage. Selections and parents are routinely assessed for their juice qualities and, at SCRI, we have selected hybrids that have outstanding juice qualities and are capable of producing progenies with substantial improvements.

The superiority of Ben Lomond's juice colour is attributed to three factors: a high total concentration of pigments, an unusually high ratio of delphinidin to cyanidin pigments and the acylation of some of these pigments. Acylated pigments have not previously been reported in blackcurrants, and in Ben Lomond they account for some 3% of the total anthocyanins present. The advantage is that they retain their colour over a wide pH range and do not become colourless when the pH rises. Hence there is less need for synthetic colour additives to augment the natural colour in the pH range 2.5–4.0 (Ford, 1985, pers. comm.).

### 7.2.3 Resistance to diseases and pests

American gooseberry mildew (*Sphaerotheca mors-uvae* (Schw.) Berk.) and leaf spot (*Drepanopeziza ribis* (Klebahn) von Höhnel) are the two most important fungal diseases of blackcurrants; sources of resistance are known and routine screening is practised.

Resistance-breaking strains of mildew occur (Trajkovski and Pääsuke, 1876; Keep, 1983), but the diversity in the form and genetic origin of the resistances gives reason to believe that some of them will be durable, particularly the non-sporulating form of resistance derived from the Swedish cultivars Sunderbyn II and Matkakoski, where resistance is probably governed by a single dominant allele or a block of closely linked loci (Trajkovski, 1976, 1982, 1986). Other sources of strong resistance are Öjebyn, which is heterozygous for a dominant resistance gene (Keep, 1977), and *R. dikuscha* where resistance is determined by alleles at two loci with complementary effects (Trajkovski and Pääsuke, 1976; Temmen et al., 1977, 1980). Additional forms of resistance occur in closely related species such as *R. hudsonianum* (Anderson, 1974), *R. petiolare*, which has a single dominant gene for resistance (Melekhina and Eglite, 1978; Sergeeva, 1981), *R. pauciflorum* and *R. americanum* (Smirnov and Shatilova, 1979), and in two non-hardy flowering currants (Calobotrya species) *R. glutinosum*, which carries the major gene *Sph3*, and *R. sanguineum* (Keep, 1973, 1981).

Among numerous hardy sources of resistance to blackcurrant leaf spot, *R. dikuscha* and *R. pauciflorum* have proved the most useful in breeding (Ravkin and Litvinova, 1976; Ogol'tsova and Sedova, 1979). In *R. dikuscha*, resistance is controlled by two complementary genes, *Pr1* and *Pr2* (Anderson, 1972), and in the USSR, Golubka, a *R. dikuscha* hybrid with resistance to reversion, mildew and leaf spot, has given many highly resistant cultivars and hybrids.

In susceptible cultivars, mildew and leafspot are controlled effectively by fungicides, and so the advantage of resistant cultivars is in reduced production costs rather than improved stability of production. However, sprays to control infection of the flowers and immature fruits with *Botrytis cinerea* are still required in most commercial plantations. Stem infections with *B. cinerea* through wounds caused by machine harvesting may also be a source of yield loss.

Resistance to reversion virus and to its mite vector (*Cecidophyopsis ribis* Westw.) has become more important because growers wish to increase the productive life of their plantations, and the number of reverted bushes present is a major factor determining the time when a plantation ceases to give economical returns. Resistance to the virus occurs in *R. dikuscha* and its hybrids, for example, the USSR-bred cultivar Golubka, and in hybrids of *R. nigrum sibiricum* (Anderson, 1973; Keep, 1985b; Knight, 1985, pers. comm.). Segregation in graft-inoculated progenies of Golubka suggests that the cultivar is heterozygous for a dominant gene which confers resistance to reversion disease (Keep, 1985b; Anderson, unpubl.).

Different forms of resistance to the mite vector occur in blackcurrant × *R. grossularia* hybrids, in which resistance to infestation is controlled by gene *Ce* (Knight *et al.*, 1974; Keep *et al.*, 1982), and in *R. ussuriense* and *R. nigrum sibiricum* hybrids, in which resistance to galling is controlled by gene *P* (Anderson, 1971). Genetic studies show that the genes for gall mite resistance (*Ce*), cytoplasmic male sterility (*Rf1*, *Rf2*), season of leafing out (*Lf1*) and mildew resistance (*Sph2*, *Sph3*) are linked in the order *Sph2–Ce–Rf1–Rf2–Lf1–Sph3* (Keep, 1985c). Austin *et al.* (1983) showed that mite resistant genotypes could be identified with a probability of about 70% by metabolic profiling of plant tissues. This method may reduce the prolonged field exposure at present required to identify resistant plants.

Blackcurrant midge (*Dasyneura tetensi* Rübs.) and vine weevil (*Otiorhynchus sulcatus* (F.)) have recently become increasingly widespread on fruiting plantations. No strong resistance to midge is known among European cultivars, and there is no known resistance to the vine weevil. However, effective midge resistance occurs in the Scandinavian cultivars Sunderbyn II, Kangosfors and Hedda, in the USSR cultivars Naryadnaya (K-7636) and Nochka (K-15510) (Agafonova, 1974), and in the blackcurrant species *R. ussuriense*, *R. dikuscha* and their hybrids (Keep, 1985d). A resistance gene conferring midge resistance in *R. dikuscha* and designated *Dt* is probably linked with a gene(s) controlling resistance to American gooseberry mildew. Resistance in Sunderbyn II and, probably, in *R. ussuriense* is oligogenic and dominant (Keep, 1985d).

## 7.3 CONCLUSIONS

Improvements in yield have been achieved in both raspberries and blackcurrants, and in each the highest yields have been obtained from plants of only moderate vegetative vigour and high harvest index. Reduction of vegetative vigour has been achieved in raspberries by plantation management, and in blackcurrants by selection of plants with a compact growth form.

In blackcurrants the most promising improvement is in reliability of cropping, achieved by combining frost and cold tolerance at flowering with late flowering and improved fruit setting. Potential for further yield improvements is possible from cold-hardy compact blackcurrants with a high harvest index grown in high-density plantations, and production costs would be reduced by the resulting ease of management. In raspberries, better marketing opportunities are facilitated by improvements in resistance to fruit moulds and in firmness of fruit, and in extensions of the season of ripening.

For both crops, selection for durable resistances to the major diseases and pests has been successful.

# REFERENCES

Agafonova, Z. Ya. (1974). [Resistance of blackcurrants to pests.] *Trudy Po Prikladnoi Botanike Genetike I Selektsii* **53**, 243–244.
Anderson, M. M. (1971). Resistance to gall mite (*Phytoptus ribis* Nal.) in the Eucoreosma section of *Ribes. Euphytica* **20**, 422–426.
Anderson, M. M. (1972). Resistance to blackcurrant leaf spot (*Pseudopeziza ribis*) in crosses between *Ribes dikuscha* and *R. nigrum. Euphytica* **21**, 510–517.
Anderson, M. M. (1973). Breeding blackcurrants resistant to gall mite (*Cecidophyopsis ribis* West.) and reversion disease. *Jugoslovensko Vocarstvo* **7**, 61–66.
Anderson, M. M. (1974). *Rep. Scott. Hort. Res. Inst. for 1973*, p. 40.
Anderson, M. M. (1975). *Rep. Scott. Hort. Res. Inst. for 1974*, pp. 49–50.
Anthony, V. M., Williamson, B., Jennings, D. L., and Shattock, R. C. (1986). Inheritance of resistance to yellow rust (*Phragmidium rubi-idaei*) in red raspberry. *Ann. appl. Biol.* **109**, 365–374.
Austin, D. K., Hall, K. J., Keep, E., and MacFie, H. J. H. (1983). Metabolic profiling as a potential aid to blackcurrant breeding. *Proc. 10th Int. Congr. Pl. Prot.* **2**, 835.
Barritt, B. H., and Torre, L. C. (1980). Red raspberry breeding in Washington with emphasis on fruit rot resistance. *Acta Hort.* **112**, 25–30.
Barritt, B. H., Torre, L. C., Pepin, H. S., and Daubeny, H. A. (1980). Fruit firmness measurements in red raspberry. *HortScience* **15**, 38–39.
Dale, A. (1977). Yield responses to cane vigour control. *Bull. Scott. Hort. Res. Inst. Assoc.* **13**, 12–18.
Dale, A. (1981). The tolerance of blackcurrant flowers to induced frosts. *Ann. appl. Biol.* **99**, 99–106.
Dale, A. (1982a). *Rep. Scott. Crop Res. Inst. for 1981*, pp. 59–60.
Dale, A. (1982b). *Rep. Scott. Crop Res. Inst. for 1981*, pp. 62–65.
Daubeny, H. A. (1980). Red raspberry cultivar development in British Columbia with special reference to pest response and germplasm exploitation. *Acta Hort.* **112**, 59–66.
Daugaard, H. (1981). [Factors affecting flower and fruit drop in blackcurrant (*Ribes nigrum* L.): a review.] *Statens Planteavlsforsog Kobenhavn Tidsskift for Planteavls Specialserie*, Beretning No. S1559.
Helliar, M. V., and Turner, E. A. (1984). *National Fruit Trials, Brogdale EHS, Annual Review for 1983*, pp. 30–34.
Hiirsalmi, H. (1985). Winter hardiness in small fruit breeding. *Acta Hort.* **168**, 57–62.
Jennings, D. L. (1979). Resistance to *Leptospheria coniothyrium* in the red raspberry and some related species. *Ann. appl. Biol.* **93**, 319–326.
Jennings, D. L. (1980). Recent progress in breeding raspberries and other *Rubus* fruits at the Scottish Horticultural Research Institute. *Acta Hort.* **112**, 109–116.
Jennings, D. L. (1982a). *Rep. Scott. Crop Res. Inst. for 1981*, p. 59.
Jennings, D. L. (1982b). Resistance to *Didymella applanata* in red raspberry and some related species. *Ann. appl. Biol.* **101**, 331–337.
Jennings, D. L. (1983). Inheritance of resistance to *Botrytis cinerea* and *Didymella applanata* in canes of *Rubus idaeus*, and relationships between these resistances. *Euphytica* **32**, 895–901.
Jennings, D. L., and Dale, A. (1982). Variation in the growth habit of red raspberries with particular reference to cane height and node production. *J. hort. Sci.* **57**, 197–204.

Jennings, D. L., and McGregor, G. R. (1987). Resistance to cane spot (*Elsinoë veneta*) in the red raspberry and its relationship to resistance to yellow rust (*Phragmidium rubi-idaei*). *Euphytica* (in press).

Jennings, D. L., and Williamson, B. (1982). Resistance to *Botrytis cinerea* in canes of *Rubus idaeus* and some related species. *Ann. appl. Biol.* **100**, 375–381.

Jones, A. T., Murant, A. F., Jennings, D. L., and Wood, G. A. (1982). Association of raspberry bushy dwarf virus with raspberry yellow disease; reaction of *Rubus* species and cultivars, and the inheritance of resistance. *Ann. appl. Biol.* **100**, 135–147.

Keep, E. (1973). *Ribes sanguineum* and related species as donors in currant and gooseberry breeding. *Jugoslovensko Vocarstvo* **7**, 3–7.

Keep, E. (1977). North European cultivars as donors of resistance to American gooseberry mildew in blackcurrant breeding. *Euphytica* **26**, 817–823.

Keep, E. (1981). *Ribes glutinosum* and *R. sanguineum* as donors of resistance to American gooseberry mildew in blackcurrant breeding. *Euphytica* **30**, 197–202.

Keep, E. (1983). Powdery mildews of temperate fruit crops. *Appl. Genet.: Proc. 15th Int. Congr. Genet.* **4**, 105–118.

Keep, E. (1985a). Resistance to the gall mite and American gooseberry mildew in blackcurrants in relation to season of leafing out. *Euphytica* **34**, 509–519.

Keep, E. (1985b). *Rep. East Malling Res. Stn for 1984*, pp. 167–168.

Keep, E. (1985c). *Rep. East Malling Res. Stn for 1984*, p. 168.

Keep, E. (1985d). The blackcurrant leaf curling midge, *Dasyneura tetensi* Rübs.: its host range, and the inheritance of resistance. *Euphytica* **34**, 801–809.

Keep, E., Knight, V. H., and Parker J. H. (1982). Progress in the integration of characters in gall mite (*Cecidophyopsis ribis*)—resistant blackcurrants. *J. hort. Sci.* **57**, 189–196.

Keep, E., Parker, J. H., and Knight, V. H. (1980). Recent progress in raspberry breeding at East Malling. Acta Hort. **112**, 117–125.

Keep, E., Parker, J. H., and Knight, V. H. (1984). *Rep. East Malling Res. Stn for 1983*, p. 191.

Kichina, V. (1976). Raspberry breeding for mechanical harvesting in Northern Russia. *Acta Hort.* **60**, 89–94.

Knight, R. L., Keep, E., Briggs, J. B., and Parker, J. H. (1974). Transference of resistance to blackcurrant gall mite, *Cecidophyopsis ribis*, from gooseberry to blackcurrant. *Ann. appl. Biol.* **76**, 123–130.

Knight, V. H. (1980). Screening for fruit rot resistance in red raspberries at East Malling. *Acta Hort.* **112**, 127–134.

Knight, V. H. (1986). Recent progress in raspberry breeding at East Malling. *Acta Hort.* **183**, 67–76.

Larsson, G. (1959). Variety trials of blackcurrant in Norrland, 1944–58. *Meddelanden från Alnarps Trädgardars försöksverksamhet*.

Lawrence, F. J. (1980). Breeding primocane fruiting red raspberries at Oregon State University. *Acta Hort.* **112**, 145–149.

Levitt, J. (1978). An overview of freezing injury and survival, and its interrelationships to other stresses. *In* "Plant Cold Hardiness and Freezing Stress" (P. H. Li and A. Sakai, eds), pp. 3–16. Academic Press, New York.

McNicol, R. J., Williamson, B., Jennings, D. L., and Woodford, J. A. T. (1983). Resistance to raspberry cane midge (*Resseliella theobaldi*) and its association with wound periderm in *Rubus crataegifolius* and its red raspberry derivatives. *Ann. appl. Biol.* **103**, 489–495.

Marshall, H. G. (1982). Breeding for tolerance to heat and cold *In* "Breeding

Plants for Less Favourable Environments", (M. N. Christiansen and C. F. Lewis, eds), pp. 47–70. Wiley Interscience, New York.

Melekhina, A. A., and Eglite, M. A. (1978). [*Ribes petiolare* Dougl.—a new source of resistance to *Sphaerotheca mors-uvae* (Schw.) Berk. in breeding blackcurrant.] *Pl. Breed. Abstr.* **49**, 1252.

Ogol'tsova, T. P., and Sedova, Z. A. (1979). [Study of Far Eastern forms of blackcurrant as initial material for breeding in the central zone of the RSFSR.] *Pl. Breed. Abstr.* **50**, 8784.

Oydvin, J. (1974). [Mildew resistance of 17 cultivars and 4 progenies of blackcurrant.] *Forskning og Forsk i Landbruket* **25**, 239–256.

Pepin, H. S., and MacPherson, E. A. (1980). Some possible factors affecting fruit rot resistance in red raspberry. *Acta Hort.* **112**, 205–207.

Quamme, H. A. (1978). Breeding and selecting temperate fruit crops for cold hardiness. In "Plant Cold Hardiness and Freezing Stress" (P. H. Li and A. Sakai, eds), pp. 313–332. Academic Press, New York.

Ravkin, A. S., and Litvinova, V. M. (1976). [Manifestation of resistance to leaf spot and rust in the hybrid progeny of blackcurrant.] *Pl. Breed. Abstr.* **46**, 11467.

Sergeeva, K. D. (1981). [Blackcurrant breeding at a new stage.] *Sadovodstvo* **1**, 41–42.

Smirnov, A. G., and Shatilova, S. D. (1979). [Resistance of blackcurrant to fungal diseases.] *Pl. Breed. Abstr.* **52**, 3336.

Stushnoff, C. (1972). Breeding and selection methods for cold hardiness in deciduous fruit crops. *HortScience* **71**, 10–13.

Stushnoff, C. (1974). Cold hardiness of woody plants. *Proc. Int. Hort. Congr., Warsaw* **XIX**(II), 13–24.

Temmen, K. H., Gruppe, W., and Schlösser, E. (1977). Basis for resistance of blackcurrant (*Ribes nigrum* L.) against powdery mildew (*Sphaerotheca mors-uvae*). *Phytopath. Z.* **88**, 184–187.

Temmen, K. H., Gruppe, W., and Schlösser, E. (1980). Investigations on the resistance of plants to powdery mildew: III. Basis for the horizontal resistance of *Ribes* cv. to *Sphaerotheca mors-uvae*. *Z. PflKrankh. PflSchutz.* **87**, 129–136.

Trajkovski, V. (1976). Resistance to *Sphaerotheca mors-uvae* (Schw.) Berk. in *Ribes nigrum* L. 6. The mechanism of resistance of *Ribes nigrum* to *Sphaerotheca mors-uvae* (Schw.) Berk. *Swed. J. Agric. Res.* **6**, 215–223.

Trajkovski, V. (1982). *Rep. 1980–81, Division of Fruit Breeding, Balsgard, Alnarp, Sweden*, pp. 29–37.

Trajkovksi, V. (1986). *Rep. 1984–85, Division of Fruit Breeding, Balsgard, Alnarp, Sweden*, pp. 117–120.

Trajkovski, V., and Pääsuke, R. (1976). Resistance to *Sphaerotheca mors-uvae* (Schw.) Berk. in *Ribes nigrum* L. 5. Studies on breeding blackcurrants for resistance to *Sphaerotheca mors-uvae* (Schw.) Berk. *Swed. J. Agric. Res.* **6**, 201–214.

Weiser, C. J., Quamme, H. A., Proebsting, E. L., Burke, M. J., and Yelensky, G. (1979). Plant freezing injury and resistance. In "Modification of the Aerial Environment of Plants" (B. J. Barfield and J. F. Gerber, eds), pp. 55–84. American Society of Agricultural Engineering, St Joseph, Miss., USA.

Williamson, B., and Jennings, D. L. (1986). Common resistance in red raspberry to *Botrytis cinerea* and *Didymella applanata*, two pathogens occupying the same ecological niche. *Ann. appl. Biol.* **109**, 581–593.

Williamson, B., and McNicol, R. J. (1986). Pathways of infection of flowers and fruits of red raspberry by *Botrytis cinerea*. *Acta Hort.* **183**, 137–141.

# 8 Strawberry Breeding in the United Kingdom

D. W. SIMPSON and M. G. BEECH

*Institute of Horticultural Research, East Malling, Kent*

Over the past decade the value of the strawberry crop in the UK has risen relative to other fruit crops to a point where it now rivals that of dessert apples (Fig. 8.1). Despite this, however, the crop has received relatively little attention from research workers over the years and the industry still relies very heavily on one cultivar, Cambridge Favourite (Fig. 8.2), which has now been grown on a wide scale for 30 years. Until recently there were three breeding programmes in the UK although all were operating on rather a small scale. The Scottish Crops Research Institute (SCRI) programme at Auchincruive had been in existence since 1930 and had concentrated on breeding for resistance to red core disease. The John Innes Institute (JII) programme at Norwich was also long established and had concentrated on breeding dual-purpose cultivars, suitable for dessert and processing use. The Institute of Horticultural Research (IHR) programme began in 1965 at Long Ashton Research Station (LARS) near Bristol and had initially been primarily concerned with producing new cultivars for early season production. It is now the only remaining breeding programme in the UK and is currently based at East Malling in Kent.

The reason why these breeding programmes have not provided the industry with a succession of new, constantly improving cultivars, in the way that has occurred in the USA and elsewhere, is hard to establish. Cambridge Favourite is certainly very good and there is a certain reluctance among growers, particularly in eastern England, to change to a different one. It has the advantage of being dual-purpose, which means the growers have the choice of sending the fruit to market when the price is right, or to the processing factories. Nevertheless, this is not the only reason, and many growers for dessert and pick-your-own markets are continually trying new cultivars, mainly from Europe. Gorella, from Holland, is

IMPROVING VEGETATIVELY PROPAGATED CROPS
ISBN 0-12-041410-4

*Copyright © 1987 Academic Press Limited*
*All rights of reproduction in any form reserved*

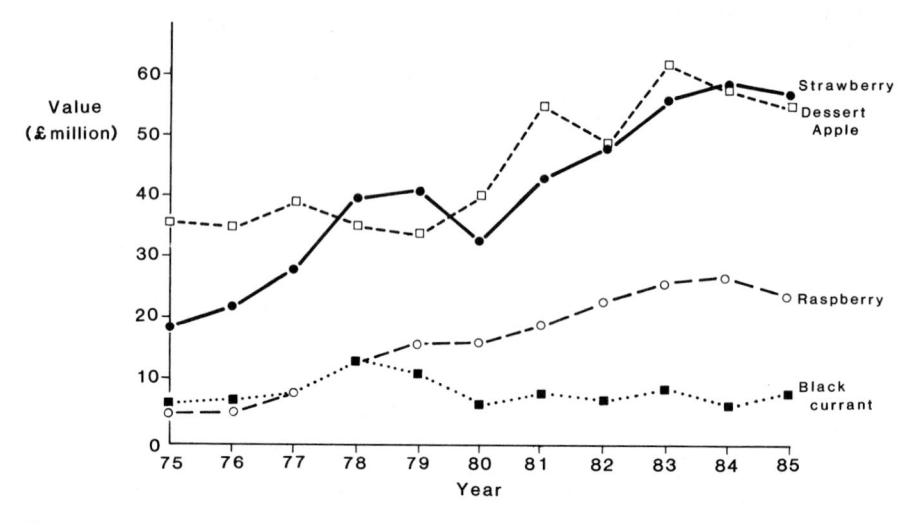

**Figure 8.1.** Annual value of the UK strawberry crop relative to that of dessert apple, raspberry and blackcurrant during 10 years up to 1985.

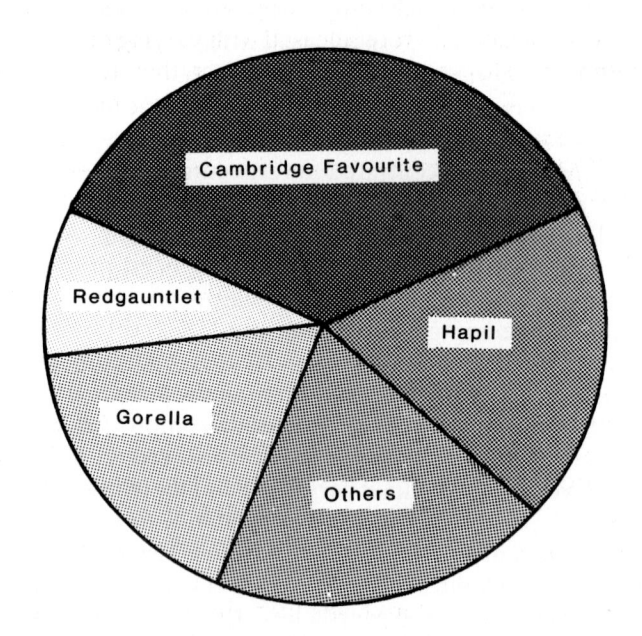

**Figure 8.2.** Relative land area in the UK occupied by the main strawberry cultivars in 1985.

established as a high-quality, second-early while the Belgian cvs Domanil and Hapil are popular because of their large fruit size. More recently, the Dutch cvs Bogota and Elsanta have also become popular.

So why have the British breeding programmes failed to provide the new cultivars which the growers will accept? The answer to this question is crucial to the future of British strawberry breeding and to the British strawberry industry, but it is not a straightforward one. A problem with breeding any crop is that the breeder, when deciding on his approach to the problem, has to predict what the industry will require in several years' time when the cultivar will become available. With strawberries it takes 10–12 years to produce a new cultivar, which is a much shorter time than with most top fruit species, but the strawberry industry has changed considerably over the last 10 years, with the increasing popularity of pick-your-own and a decline in the traditional processing industries of jamming and canning. Changes of this type are difficult to predict and this can often lead to cultivars being released which are not suitable for the current market requirements. This, in turn, can lead to the growers losing confidence in the breeding programmes, resulting in less co-operation, and having a damaging effect on the industry in the long term. Since its commencement, the SCRI breeding programme had concentrated primarily on producing cultivars with resistance to red core disease (*Phytophthora fragariae*); several cultivars were released with varying degrees of resistance to this pathogen. However, despite the fact that red core is virtually endemic in the UK these cultivars have not become established. The reason is that they have been found wanting in other respects, and while the disease can be adequately controlled by chemical and cultural means growers will continue to grow susceptible cultivars which have the right characteristics in terms of yield and fruit quality. The cv. Hapil, for example, is very susceptible to *Verticillium* wilt, red spider mite and some races of red core and yet it is being grown on an increasing scale because of its large berry size which leads to reduced picking costs. Similarly, Elsanta is very susceptible to *Verticillium* wilt and red core but is nevertheless popular with growers because the fruit quality is outstanding.

In contrast, the JII programme had concentrated mainly on increasing yields with the primary objective of producing a dual-purpose cultivar to replace Cambridge Favourite. This goal has proved a very difficult one to achieve and while the cultivars released showed improvements in yield, the fruit quality, particularly with regard to firmness, colour and shape, was found to be unacceptable.

The IHR programme started comparatively recently but has already been subject to marked changes in emphasis. Initially, the main objective was to produce an early fruiting cultivar suitable for growing under polyethylene tunnels. Now, due to increasing pressures from cheap imported fruit in the early part of the season, the pendulum has swung comletely so that one of our principal aims is to breed cultivars which produce large volumes of fruit in August.

The lesson to be learned from the recent history of the strawberry industry is that any breeding programme must remain flexible in its approach. Priorities have changed and will continue to change, although the present demand for high yields of large, firm, high-quality berries seems likely to remain of great importance for some time to come. Large fruit is perhaps the most important single character at the moment because of its effect in reducing picking costs, but it would be folly to concentrate too much on this one character, as a large increase in pick-your-own or move towards mechanical harvesting could change the situation quite dramatically.

As the strawberry is an octoploid species, the inheritance of all the important characters is inevitably complex. The system used for breeding new strawberry cultivars is fundamentally one of recurrent mass selection, with the parents from each year's crossing programme being chosen from the advanced selections derived from earlier crosses. In addition, new cultivars from other programmes may be included if they have shown promise in the trial plots. Families normally comprise 150–200 seedlings, but the most interesting crosses are often repeated with a family size of 1000–4000. This method results in a steady improvement in the gene pool while allowing new germplasm to be added (from other programmes), thus preventing a narrowing of the genetic base.

A good deal of additive variation is available for exploitation by the breeder, but specific combining ability is also important for many of the characters of interest, particularly yield, fruit size and firmness of the berries (Aalders and Craig 1968, Hansche *et al.* 1968 and Spangelo *et al.* 1971a,b). This is exploited by the procedure of the large scale repeat crosses. At IHR the normal practice is to grow 13 000 seedlings each year of which approximately 70% will be small scale crosses.

The seedlings are raised in a glasshouse and planted in the field in mid-July so that they can become established and be assessed as mature plants in the following year. No reliable estimate of yield can be made from single plants at this stage so selection is on the basis of fruit quality characters. The two most important are fruit size and firmness and in most families the majority of seedlings can be rejected because they are not up to standard for one or both of these characters. Other important characters are flavour, flesh and skin colour, shape of the berries, ease of calyx removal, plant habit and runner production. The number of plants selected in any one year will vary according to the families but normally the selection rate is approximately 1%. Selected seedlings are multiplied vegetatively for a 10-plant yield trial where they are compared with the standard cultivars. Here they are again assessed for the fruit quality characters but in addition all the fruit is picked and graded so that the yield of marketable fruit can be compared with the standard cultivars. Any selections which do well in this trial are then multiplied for planting in larger scale replicated trials at Brogdale Experimental Horticulture Station (EHS), Kent. Following these trials the most promising selections are propagated for planting in trials at six EHS's in different parts of the UK and with selected growers.

This method of breeding new cultivars is similar to that employed in many other breeding programmes around the world and gives a good chance of success but there are still many points at which good seedlings could be wrongly rejected. At the seedling population stage the very large numbers of plants involved means that the amount of attention which can be given to any individual is very small. The fruiting season for strawberries is short and an individual plant can only be accurately assessed over a period of 4 or 5 days, sometimes less. This means, in practical terms, that each plant will only be seen at a stage when it can be judged on two occasions, since it is normally only possible to assess the complete seedling population twice a week. Inevitably, many errors will be made but improvements in the system can only be made by either increasing the number of people assessing the plants or decreasing the number of seedlings. Strawberries are very heterozygous and it is necessary to work with large populations to increase the chance of any plants with the right genetic combinations being present. It would thus be counter-productive to work with smaller families, even if it did allow for each individual plant to be more accurately assessed.

The ideal situation would be to reduce the number of plants at the seed-tray stage by some sort of preliminary screen, thereby reducing the number of plants which would ultimately be planted in the field. The most obvious solution would be to screen for disease resistance at this stage, but it has been shown that the relationship between seedling resistance and mature plant resistance to *Phytophthora fragariae*, the most important fungal root disease in the UK, is not a straightforward one. There are many different races of this pathogen and, although sources of major gene resistance to each race probably exist, it is unlikely that these could ever be satisfactorily combined in a single genotype. A more realistic aim is to produce a cultivar which has a tolerance to the disease under field conditions, as have Cambridge Favourite and Saladin. Unfortunately, however, it appears that resistance of this type can only be reliably identified by a visual assessment of mature plants growing in infected soil

(Gooding, 1972). A similar situation exists with *Verticillium* wilt, and the screening of small seedlings was found to be unsatisfactory with this disease also (Wilson, pers. comm.)

Apart from disease resistance all the other important characters in strawberries can only be assessed on fruiting plants and, thus, cannot be screened directly at the small-seedling stage. However, if a strong correlation could be found between an easily measured juvenile character and an important character in mature plants then this would allow a preliminary screening to be done. Recent work at Long Ashton by Guttridge *et al.* (1983) has been investigating this possibility and early results have been encouraging. A significant correlation exists between the vigour of the seedling, as measured by the size of the second emerging leaf, and total yield of the mature plants. This correlation is stronger with selfed progenies than with crosses, as would be expected, and hence may be more useful in the development of parental lines than in crosses aimed at producing new cultivars. Even when a correlation is small, however, it can be valuable, provided that the juvenile character can be measured sufficiently accurately. To ensure this it is important to have synchronous germination of the seeds and a completely uniform environment for the young seedlings, to make certain that, as far as possible, all the variation that is observed is genetic and not environmental. Also, since between 30 and 50 crosses are usually investigated each year, it is important that the correlation should be fairly general if it is to be used effectively, and not fluctuate too greatly from one family to another. The work is currently at too early a stage to know if this will be the case.

A further method of reducing the size of the seedling population would be to reduce the amount of variation present in a family, thus eliminating the need to grow very large progenies from the most promising crosses. This can be done by increasing the homozygosity of the parents, which has the additional benefit of also allowing particular characters to be concentrated in certain parental lines. This approach was first used at Long Ashton several years ago after straightforward crosses between cultivars had failed to make any significant progress in advancing the fruiting season of strawberries in this country. As a new approach, four early fruiting cultivars were crossed in all combinations and the earliest fruiting seedlings with acceptable fruit quality were selected (Fig. 8.3). These plants were then selfed and the earliest fruiting ones selected again from the progenies. The next step was to self these selections again and to cross with one another those which had a different pedigree. This crossing restored the heterozygosity and resulted in a population of 10 000 seedlings which were assessed in 1981. With few exceptions these families were very early fruiting and showed reduced variation for other important

|              | C. Vigour | Glasa | Gorella |
|---|---|---|---|
| Glasa        | ★         |       |         |
| Gorella      | ★         | ★     |         |
| Pantagruella | ★         | ★     | ★       |

↓

900 seedlings assessed in glasshouse

↓

41 earliest fruiting seedlings selected and selfed

↓

820 seedlings assessed in glasshouse

↓

32 earliest fruiting seedlings selected

↓

unrelated selections crossed in all combinations

↓

10 000 seedlings comprising 48 families assessed in field

↓

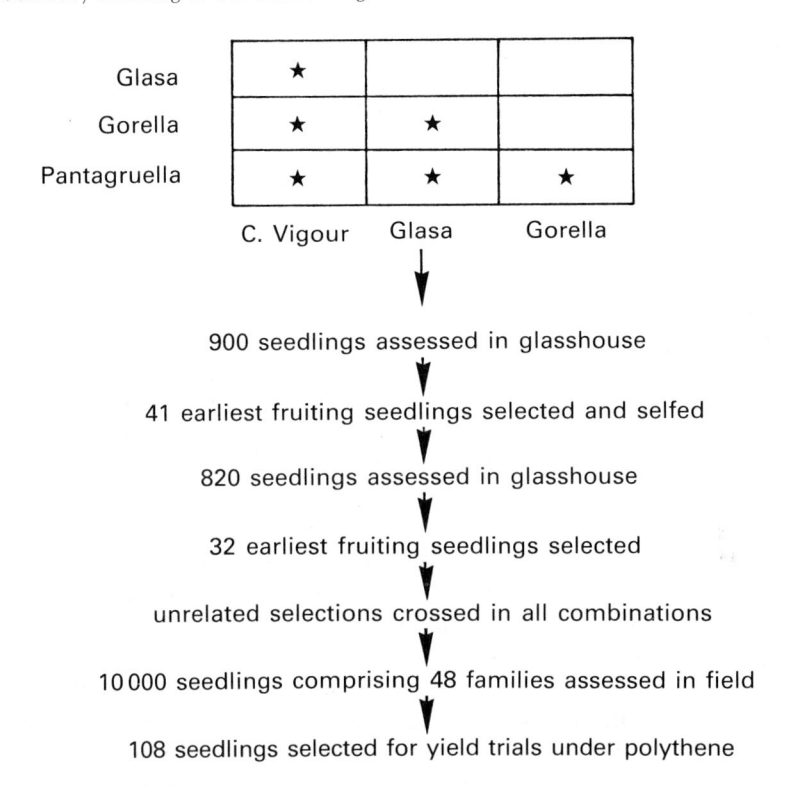

108 seedlings selected for yield trials under polythene

**Figure 8.3.** A recurrent selection programme for earlier fruiting strawberries.

characters. From this population 110 selections were made, the majority of which had a 50% harvest date between 10 and 14 days ahead of Cambridge Favourite. These selections were trialled in 1983, 1984 and 1985 and two have now been identified which may be released as new cultivars if they continue to perform well in large-scale trials in different areas of the country. Both are very early and well adapted to cropping under protection. They produce high-quality fruit suitable for the dessert market.

This technique has been reasonably successful in achieving the objective of advancing the season, but unfortunately many of the families were found lacking in certain fruit quality characteristics. Softness was the main problem in this respect, but in addition some families had very small fruit and others white flesh. Thus, while there is no doubt that limited inbreeding can be a useful tool for producing breeding lines which show the desired expression for a particular character, care must be taken not to fix other undesirable characters in a breeding line.

An interesting observation from this work was the great uniformity and vigour of the families produced when these first generation inbreds were crossed, suggesting that it may be possible to make great improvements by further increasing the homozygosity of the parental lines. Strawberries show extreme inbreeding depression which suggests there are many deleterious genes present in all the cultivars currently being grown. If, by successive generations of inbreeding, the proportion of deleterious genes could be drastically reduced, then there should be the potential for a big improvement on the current cultivars in terms of yield. Furthermore, if sufficiently homozygous lines could be produced to allow the production of $F_1$ hybrids, then there is no reason why strawberries should not be commercially grown from seed. This is not a new idea (Aalders and Craig, 1964, 1968) but it is one which deserves more serious consideration as there would be considerable advantages in this country. First, the problem with viruses would be virtually eliminated and with it the costly and time-consuming procedure of producing virus indexed mother-plants and then certified runners in isolation beds. Secondly, it would allow growers to plant at any time of the year according to their needs. Currently, most strawberries are planted in October or November because this is when runners are plentiful and cheap. This means that plants are in the ground a full 20 months before a worthwhile crop is taken from them, which is obviously an uneconomical use of land. Seedlings could easily be raised in unheated polyethylene tunnels to be ready for planting in July. They would then give a heavy crop the following June. It is already possible to achieve this by the use of cold-stored runners but these have to be hand de-blossomed in the autumn and are generally not as satisfactory as fresh runners. Even more economical use of land could be made where the requirement is for day-neutral cultivars; in this case, seedlings raised under heated glass could be planted in early May and be fruiting by August of the same year. Again, this is possible by using cold-stored runners but cool spring weather often retards the early growth, resulting in small plants and low yields in the first year. If $F_1$ hybrid seedlings were used, then the extra vigour and more rapid growth of the plants would go a long way towards eliminating this problem.

   There would thus be several advantages in converting strawberries from a vegetatively propagated to a seed-propagated crop, but the problem lies in the production of homozygous parental lines. Previous work with repeated selfing has shown that often, by the third generation, the vigour of the plants is reduced to such an extent that it is difficult to keep them alive (Hulewicz and Hortynski, 1979). Therefore, it will be necessary to work with large populations in the early generations of this scheme in order to ensure that acceptable vigour and fertility can be maintained.

Severe selection at the seed-tray stage should enable the number of plants for the field to be drastically reduced, thereby making large families a practical proposition. It is also likely that considerable advantage would be gained from some controlled crossing after the first generation of selfing, to "reshuffle" the gene pool.

Production of specialised parental lines by this method is a long-term process and it is likely to take 10 years to produce a line with the required level of homozygosity. An attractive alternative to this method, which potentially could be much quicker, would be to employ induced haplogenesis followed by chromosome doubling to produce completely homozygous lines. The production of polyhaploid strawberries has already been achieved by the method of intergeneric crossing (Barrientos and Bringhurst, 1973; Hughes and Janick, 1974) using *Potentilla anserina* ($2n = 28$) and *Potentilla fruticosa* ($2n = 14$) as pollen parents and *Fragaria ananassa* ($2n = 56$) as the seed parent. The plants produced were weak, however, and the flowers both pollen and pistil sterile. The authors did not indicate whether a doubling of the chromosome complement by colchicine treatment would restore fertility.

Research at the IHR involving crosses between *Fragaria ananassa* and several species of *Potentilla* and *Geum* has resulted in over 300 hybrids being produced but, so far, no haploid plants. This method does not, therefore, seem a suitable one for the routine production of large numbers of haploid plants.

Two alternative methods are available, and the one which is probably most likely to succeed is the technique of another culture. Niemirowicz-Szczytt *et al*. (1983) produced plants with ploidy of $2x$, $4x$ and $6x$ by this method, but the origin of the plants is not certain and the fact that they were all fully fertile suggests that they may not have derived from microspores. Haploid or polyhaploid plants would be expected to be sterile or, at best, show a very low level of fertility. Investigations at East Malling have so far resulted only in callus, and subsequently some plants, derived from the filament but none from the microspores.

The third method which we are currently investigating at East Malling is the use of ionising radiation. Pandey and Phung (1982) demonstrated that the application of pollen, which has been subjected to high levels of ionising radiation, to the flowers of four different *Nicotiana* species resulted in progeny containing parthenogenetic haploid and diploid maternal individuals. This approach has been applied with strawberries at East Malling and a large number of plants have been produced but their chromosome complement has yet to be determined.

In the long term, the production of $F_1$ hybrid strawberry plants is a very attractive proposition which could radically change the way strawberries are grown in this country. However, in the more immediate future conventional breeding techniques will provide new cultivars to meet the needs of the industry and will extend the season of production to allow new markets to be developed.

## REFERENCES

Aalders, L. E., and Craig, D. L. (1964). The inbred line $F_1$ hybrid approach to strawberry breeding, as practised at Kentville. *Can. J. Genet. Cytol.* **6**, 237.

Aalders, L. E., and Craig, D. L. (1968). General and specific combining ability in seven inbred strawberry lines. *Can. J. Genet. Cytol.* **10**, 1–6.

Barrientos, F., and Bringhurst, R. S. (1973). A haploid of an octoploid strawberry cultivar. *HortScience* **8**, 44.

Gooding, H. J. (1972). Studies on field resistance of strawberry varieties to *Phytophthora fragariae*. *Euphytica* **21**, 63–70.

Guttridge, C. G., Anderson, H. M., and Woodley, S. (1983). Seedling selection in strawberry. *Rep. Long Ashton Res. Stn for 1981*, pp. 19–20.

Hansche, P. E., Bringhurst, R. S., and Voth, V. (1968). Estimates of genetic and environmental parameters in the strawberry. *Proc. Am. Soc. hort. Sci.* **92**, 338–345.

Hughes, H. G., and Janick, J. (1974). Production of tetrahaploids in the cultivated strawberry. *HortScience* **9**, 442–444.

Hulewicz, T., and Hortynski, J. A. (1979). Effect of inbreeding and its use in strawberry breeding. *Genet. Polon.* **20**, 541–546.

Niemirowicz-Szczytt, K., Zakrzewska, Z., Malepszy, S., and Kubicki, B. (1983). Characters of plants obtained from Fragaria × ananassa in anther culture. *Acta Hort.* **131**, 231–237.

Pandey, K. K., and Phung, M. (1982). 'Hertwig effect' in plants: induced parthenogenesis through the use of irradiated pollen. *Theoret. appl. Genet.* **62**, 295–300.

Spangelo, L. P. S., Hsu, C. S., and Fejer, S. O. (1971a). Combining ability analysis in the cultivated strawberry. *Can. J. Pl. Sci.* **51**, 377–383.

Spangelo, L. P. S., Hsu, C. S., Fejer, S. O., Bedard, P. R., and Rouselle, G. L. (1971b). Heritability and genetic variance components for 20 fruit and plant characters in the cultivated strawberry. *Can. J. Genet. Cytol.* **13**, 443–456.

# 9 Breeding for Improved Ornamental Plants

F. A. LANGTON

*Institute of Horticultural Research, Littlehampton, Sussex*

On a worldwide basis, three cut-flower species are pre-eminent: the carnation, the chrysanthemum and the rose. Table 9.1 gives the 1985 values of these and of the next seven most important cut flowers in The Netherlands, Europe's leading glasshouse crops producer. All are vegetatively propagated, although gerbera and freesia are also grown from seed. Other important species include gladiolus and alstroemeria. The chrysanthemum is by far the most important cut flower grown under protection in the UK but narcissus ranks first when the values of outdoor-grown flowers and dry bulbs are taken into account.

The rankings of the top ten pot plant species in The Netherlands in 1985 are given in Table 9.2. All are vegetatively propagated, although cyclamen is mainly grown from seed. Some of those listed are actually groupings of related species, exemplified by ficus, represented by *F. pumila*, *F. benjamina*, *F. elastica* and several others of lesser importance. Many other species are important as pot plants and those listed in Table 9.2 accounted for only 30% of total turnover. There are great differences between countries in the popularity of a given species as is illustrated by the pot chrysanthemum which dominates in the UK, but which fails to reach the top ten in The Netherlands.

Characters of ornamental plants which can be objectively assessed, e.g. those affecting crop productivity, transportation, vase life, etc., are of unquestioned importance and of real concern to the breeder, but aesthetic appeal is at least equally important. This can only be judged subjectively and sets the breeding of ornamentals apart from that of most other crops. The volatility of the ornamentals market is such that one new introduction can completely transform the commercial potential of an ornamental species—as happened, for example, with the introduction of Rieger's Schwabenland elatior begonia (*B.* × *hiemalis*) in Germany in 1955 (see

**Table 9.1.** The top ten cut flowers sold at the Dutch auctions in 1985 (data from Flower Council of Holland)

|  | Value (million Dutch guilders) |
|---|---|
| 1. Rose | 499 |
| 2. Chrysanthemum | 422 |
| 3. Carnation | 250 |
| 4. Tulip | 181 |
| 5. Freesia | 149 |
| 6. Gerbera | 143 |
| 7. Lily | 132 |
| 8. Cymbidium | 94 |
| 9. Gypsophila | 82 |
| 10. Iris | 50 |

**Table 9.2.** The top ten pot plants sold at the Dutch auctions in 1985 (data from Flower Council of Holland)

|  | Value (million Dutch guilders) |
|---|---|
| 1. Ficus | 39 |
| 2. Dracaena | 31 |
| 3. Begonia | 31 |
| 4. Saintpaulia | 26 |
| 5. Yucca | 24 |
| 6. Azalea | 23 |
| 7. Poinsettia | 22 |
| 8. Kalanchoë | 21 |
| 9. Dieffenbachia | 20 |
| 10. Cyclamen | 19 |

Doorenbos, 1973). On the other hand, however well conceived a new introduction, the capricious nature of consumer acceptance can totally thwart the breeder's efforts. This appears to have been the case with the naturally dwarf, free-branching Wye pot chrysanthemums bred in the UK by Allan Jackson in the late 1960s and described by Lovelidge (1969).

Novelty value can itself be sufficient to warrant the marketing of a new ornamental cultivar, and for this reason the release of genetic variation can be an end in itself as opposed to a means to an end as in most other types of crop. The use of procedures to generate variants is thus of especial importance to the breeder of ornamental crops and these will be considered before turning to more specific objectives.

## 9.1 THE GENERATION OF GENETIC VARIANTS

### 9.1.1 Interspecific hybridisation

As in many other crops, the genetic base in ornamentals can be very narrow. For example, 120 or so species of *Rosa* are known (Rehder, 1940), but only seven or eight have featured in the ancestry of modern rose cultivars (Wylie, 1954). The contributions of several of these wild species are still clearly apparent: the recurrent flowering habit which is so important in modern crop roses is due to a recessive single-gene character (Semeniuk, 1971; De Vries and Dubois, 1978) introduced *via* a small number of derivatives of *R. chinensis* and *R. gigantea* brought from China at the end of the eighteenth century. The yellow flower colour in modern roses can be traced back *via* a single cultivar, Soleil d'Or, to *R. foetida* Persian Yellow used by the French breeder, Pernet-Ducher in 1898. There is clearly scope for broadening the genetic base by interspecific hybridisation, both to introduce novel morphological and physiological characters and to improve specific traits of horticultural performance.

Some species which have been important in the past are being re-examined by modern rose breeders. As an example, De Vries and Dubois (1978) have used two *R. foetida* accessions (one of which was Persian Yellow) to re-introduce yellow flower colour together with, presumably, other previously unsampled *R. foetida* genes. Species not formerly of great importance are also being surveyed and *R. rugosa*, for example, may prove to be a useful source of black spot resistance.

Black spot caused by *Diplocarpon rosea* is a most serious disease of roses, particularly of outdoor-grown plants, leading to premature defoliation and a reduction in both flower number and quality. Resistance in *R. rugosa* was demonstrated by Palmer *et al.* (1966) and preliminary results using this species and a black spot resistant accession of *R. multiflora* have been reported (Semeniuk and Arisumi, 1968; Svejda, 1975). An obstacle to rapid implementation, however, is that the black spot resistant species are diploid whereas modern commercial roses are tetraploid. Doubling the chromosome number of diploid roses, although difficult, has been achieved (Semeniuk and Arisumi, 1968) and three tetraploid black spot resistant breeding lines, Spotless Gold, Spotless Yellow and Spotless Pink were released in the late 1970s (Semeniuk, 1979). Two other resistant tetraploid lines originating *via* unreduced gametes have also been described (Svejda and Bolton, 1980). Whether resistance will prove to be long lasting is uncertain, however, since a new race of *Diplocarpon rosae* has been reported on a diploid *R. rugosa* hybrid, Martin Frobisher, following 8 years of cultivation without infection (Bolton and Svejda, 1979). More recently, an accession of *R. rugosa*, Rugosa Ottawa, has been described which has resistance not only to black spot but also to mildew caused by *Sphaerotheca pannosa*, the two-spotted spider mite (*Tetranychus urticae*) and the strawberry aphid (*Chaetosiphon fragaefolii*) (Svejda, 1984).

There must be few commercially important ornamental species which have not been improved by deliberate interspecific hybridisation at one time or another. Indeed, many have a very recent hybrid origin. Amongst these is the alstroemeria where the importance of interspecific hybridisation is reflected in a rapidly increasing crop acreage. Modern cultivars are vigorous, easily propagated vegetatively, have large, showy flowers in an attractive range of colours and have good lasting qualities. Of particular importance, however, the alstroemeria is an "energy-saving" glasshouse crop which is grown cool and gives successive flushes of flowers over a cropping period of several years.

The cultivar, Orchid, which is only now becoming eclipsed, originated as an interspecific hybrid in the 1940s (Robinson, 1963). However, the main impetus to alstroemeria growing was probably the release of the Parigo hybrids in the early 1960s. Three species were involved but only one, *A. aurantiaca*, has been identified (Goemans, 1962). Wilkins *et al.* (1980) speculated that two other species which feature in the parentage of modern cultivars are *A. pelegrina* and *A. violacea*; *A. ligtu* may also be involved (Broertjes and Verboom, 1974). Whatever their origin, there is no doubt that today's cultivars are a great improvement for commercial use on anything known prior to the 1940s.

### 9.1.2 Mutation breeding

Mutation breeding has probably had its greatest impact in the vegetatively propagated ornamentals. This is largely due to the value of novelty *per se* in these species.

Spontaneous mutations releasing hitherto unknown genetic variations are rather uncommon but an example of great horticultural importance led to the production of the pigment, pelargonidin, in the rose, unknown prior to about 1930 and responsible for the brilliant orange hues that are familiar today (Wylie, 1955). The finding of a double-flowered mutant of *Gypsophila paniculata* in the mid-1930s by Alex Cumming of the Bristol Nursery in Bristol, Connecticut (Wilfret *et al.*, 1983) is another example of a spontaneous mutation of considerable horticultural importance. The mutant was named Bristol Fairy and is still the most widely grown gypsophila in commerce today.

Most mutations are less dramatic, and generally their horticultural importance for ornamentals is in giving small, but commercially significant, improvements in growth and flowering characteristics or in extending the range of flower colour and form in already useful cultivars, to give "families" of sports which share the productivity and quality characteristics of their progenitors.

To exploit the occurrence of spontaneous mutations in pot plant species, a clonal selection programme was initiated in Denmark in 1979 based on plant stocks collected on growers' holdings. This has identified superior clones in such species as *Hedera helix* (Bech, 1983) and *Dieffenbachia maculata* (Bech et al., 1985). However, it cannot be certain in these cases that the selected clones were improvements on the originals. It may have been that the grower stocks from which they were selected were genetically heterogeneous, having deteriorated as a consequence of deleterious mutations. More certain examples of beneficial mutations come from programmes using X-ray or γ-irradiation procedures.

A notable example of a "family" of sports is the "Sim" group of carnations. The progenitor, William Sim, was bred in 1938 and rapidly eclipsed other cultivars then grown because of its high productivity and good flower quality. Spontaneous mutants soon extended the colour range and this further stimulated commercial demand. By 1953, as many as 200 sports (mainly periclinal chimeras) of William Sim had been named (Holley and Baker, 1963) and, even today, the "Sim" group dominates standard carnation production.

All but two of the commercially grown carnation sports appear to have been of spontaneous origin (Broertjes and Van Harten, 1978). Prior to the mid-1960s most chrysanthemum sports were also of natural occurrence. However, since then the systematic induction of mutants by irradiation has become increasingly important in chrysanthemum breeding. The highly heterozygous and hexaploid nature of the chrysanthemum make it suitable for mutation breeding, and few European breeders will now release a new cultivar without first carrying out several cycles of irradiation in an attempt both to improve the original selection and to produce a range of sports. Currently, about half the chrysanthemum applications received for Plant Variety Rights in Europe are for sports of existing cultivars (A.J. George, pers. comm.).

From a pink-flowered chrysanthemum all other colours can be derived, usually in a stepwise progression (Jank, 1957). Some pink cultivars produce colour mutants less readily than others, presumably because of the multiple possession of flower-colour genes of dominant effect. The rapidity with which commercially useful chrysanthemum sports can be accumulated from a pink parent has been documented for the "Horim" family (Broertjes et al., 1980), and Fig. 9.1 traces the origin of the early sports of the now widely grown cultivar, Snapper. It took only 5 years to develop the family

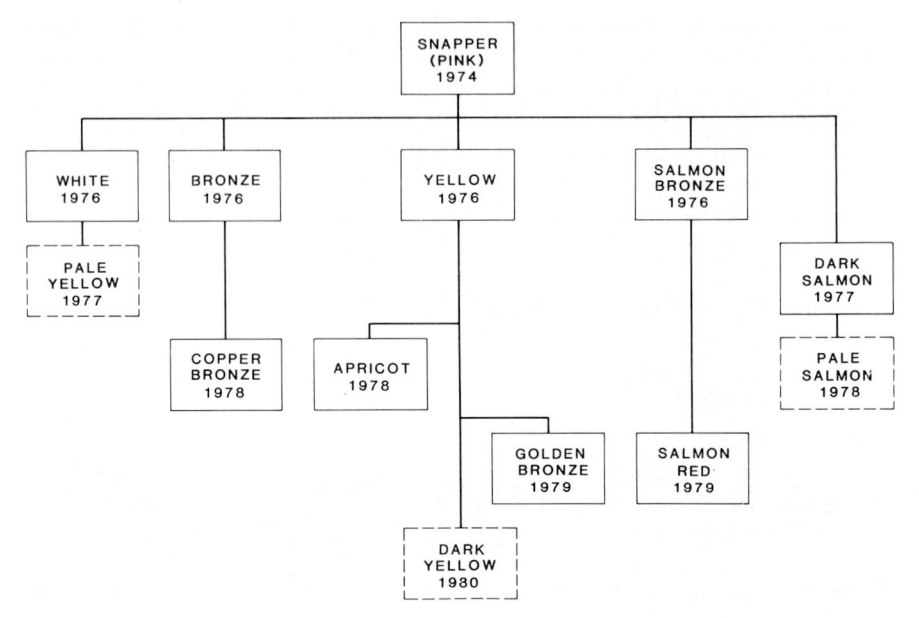

Figure 9.1. Origins of the "Snapper" family of chrysanthemums.

shown and, by 1985, two other sports had received Plant Variety Rights (Dark Pink Snapper and Snappertoo) and applications had been received for 12 others (A. J. George, pers. comm.). The range of flower colours generated from Snapper is shown in an article by Langton (1986). However, far more extensive changes than just flower colour have been induced and Fig. 9.2 shows, for example, lines differing in flower number and peduncle length and, hence, in "spray" form.

Mutations affecting physiological responses are also sought by chrysanthemum breeders. This is a realistic proposition as shown for the Snapper chrysanthemums in the number of leaves that they produce in continuously maintained long days, an important determinant of a disorder, "premature budding" (Cockshull, 1975a). Langton *et al.* (1982) showed that while Snapper averaged only 31 leaves, one of its sports, Golden Bronze Snapper, produced 38 leaves. This sport remained vegetative for 7 days longer than Snapper and ought to be appreciably less prone to premature budding. A further example of variation in a physiological response is shown in the recent induction of low-temperature tolerance in chrysanthemums (Broertjes *et al.*, 1983; Preil *et al.*, 1983; Broertjes and Lock, 1985).

Since most chrysanthemum sports which arise using current methods are periclinal chimeras (Broertjes and Van Harten, 1978), there can be

**Figure 9.2.** Snapper chrysanthemum sports, showing variation in "spray" form (Courtesy of Perifleur Ltd).

problems in using these mutants as breeding lines. Langton (1980) showed, for example, that many of the yellow "sports" grown commercially are changed only in the epidermal layer (LI) and are identical to their white progenitors in breeding behaviour. This was subsequently confirmed by Shibata and Kawata (1986). Methods are therefore required to produce "solid" non-chimeral forms. The growing of plantlets from irradiated and cultured pedicels (strictly peduncles) will achieve this aim (Roest and Bokelmann, 1975; Broertjes *et al.*, 1976), and the use of irradiated or EMS-treated callus or cell-suspension cultures (Jung-Heiliger and Horn, 1980; Preil *et al.*, 1983) could become important.

The successful application of mutation breeding to chrysanthemum improvement is being mirrored in an increasing number of ornamental species, including alstroemeria (Broertjes and Verboom, 1974) and elatior begonia (Doorenbos and Karper, 1975). A novel development which may add to this impact is the recognition of somaclonal variation (Larkin and Scowcroft, 1981) whereby plant cell culture itself generates genetic variants. From callus cultures of the scented geranium, *Pelargonium graveolens*, for example, Skirvin and Janick (1976a) derived numerous plantlets which differed from the original cultivar in leaf and flower morphology, pigmentation, pubescence, essential oil constituents and plant stature. In some cases the production of variants was clearly related to chimerism in the tissues from which the callus had been derived; in other cases the tissue-culture procedure itself appeared to have induced variants. One of these, a clone having the general leaf morphology characteristic of the internal tissues of the chimeral cultivar, Rober's Lemon Rose, but with double the chromosome number, was subsequently released as Velvet Rose (Skirvin and Janick, 1976b).

When Roest and Bokelmann's (1975) pedicel culture technique is used without irradiation, a range of distinct and stable genetic variants can be produced (pers. observation; B. Machin, pers. comm.). It is possible that the method merely "captures" mutant genotypes present but not expressed in the explant tissues, but the extensive range of variants which have been found from a single cultivar could, perhaps, indicate somaclonal variation. A "dwarf" variant of Yellow Snowdon obtained by B. Machin via pedicel culture is shown in Fig. 9.3.

### 9.1.3 Induced autopolyploidy

Induced polyploidy has an obvious value in relation to interspecific hybridisation (see earlier) but, since induced autopolyploids are frequently

**Figure 9.3.** Chrysanthemum stock plants of Snowdon and the dwarf mutant, YS 52 generated by pedicel culture (courtesy of Dr B. J. Machin).

conspicuously different from their progenitors in having, for instance, larger flowers, e.g. snapdragon, *Antirrhinum majus* (Straub, 1940), it is not surprising that breeders of ornamentals turned to colchicine as a rapid means of improving flower quality and of producing novelties. To date, however, this approach has met with only limited commercial success.

In the case of the carnation, autotetraploids are generally found to have larger flowers and thicker stems, but these commercial advantages are outweighed by generally shorter stems and fewer flowers (Stewart, 1951; Howard, 1971; Sparnaaij, 1979). Bearing in mind that colchicine-doubling gives genotypes equivalent to about two generations of inbreeding in tetraploids (Mendoza and Haynes, 1973), further crossing is required to restore heterozygosity before the true value of tetraploidy becomes evident. This has been done with the carnation and promising results have been reported (Holley and Baker, 1963; Howard, 1971). Nevertheless, the carnations currently grown remain diploid.

Several ornamental species are grown both as diploids and as tetraploids. An example is the geranium. In a survey of vegetatively propagated cultivars then available in West Germany, Badr and Horn (1971) found that ten were diploid while 67 others were tetraploid. A similar survey carried out a little earlier in the USA by Knicely and Walker (1966) showed that of 49 cultivars, 27 were diploid and 22 tetraploid. It appears likely that the first tetraploid geraniums appeared around 1880 (Badr and Horn, 1971) and that, initially at least, these were unknowingly selected by breeders for their superior horticultural characteristics.

There has been a similar occurrence in freesia breeding. Goemans (1980) reported that the first tetraploid freesia was introduced by van Tubergen in 1911 but that its tetraploid status was not realised at the time. Sparnaaij *et al.* (1968) surveyed 39 vegetatively propagated cultivars then grown in The Netherlands and found nine to be diploid, one triploid, 17 tetraploid and 12 aneuploid ($4x - 2$, $4x - 1$ and $4x + 1$). As with geranium, tetraploids were rapidly supplanting diploids in commercial horticulture.

On the basis of commercial impact, tetraploidy appears to confer advantages over diploidy in both the geranium and the freesia. Tetraploids have come to the fore in these species and it may be that the optimum ploidy for commercial growing will have already been reached in other species with a long breeding history. Induced autopolyploidy is most likely to have commercial impact in ornamentals as yet little grown and which have not been subjected to long and intensive hybridisation and selection. A case in point is *Archimenes* where successful autotetraploid pot plant cultivars have been obtained from adventitious shoots growing from colchicine-treated, detached leaves (Broertjes, 1972).

## 9.2 BREEDING OBJECTIVES

While breeding objectives are usually dictated by the commercial needs peculiar to the species under consideration, for example, dwarf habit in normally tall pot species (geranium) or bract retention in species which normally drop their flowers (*Bougainvillea*), it is possible to discern common trends spanning the ornamentals as a whole or at least significant groups of ornamentals.

### 9.2.1 Seed propagation

Vegetative propagation gives genetic homogeneity but has the disadvantages of virus build-up in the stocks and high cost of planting material, particularly if certified virus-free. In contrast, seed tends to be virus free, is relatively cheap, but frequently gives genetically variable crops. For several species, therefore, breeders have attempted to improve the uniformity of seed "strains" to give a commercially realistic alternative to vegetative propagation.

Several important ornamental species are already grown commercially from both vegetatively derived planting material and from seed. The freesia, for example, is grown mainly from heat-treated corms in The Netherlands but frequently from seed in the Channel Islands (Goemans, 1980). Due to the freesia's almost complete self-sterility, seed is genetically heterogeneous but crop variability is much reduced when tetraploid seed is used because of the stabilising effect of tetrasomic inheritance (Sparnaaij, 1979). There can be problems of seed production at the tetraploid level, however, and attempts have been made to produce improved lines in this respect (Sparnaaij *et al.*, 1968). More recently self-compatibility has been incorporated into tetraploid lines and $S_1$ seed has been trialled. This is relatively uniform, still vigorous and can be produced using bees rather than by expensive hand cross-pollination (Sparnaaij, 1979).

The gerbera is also propagated vegetatively (40% or more plants are produced by micropropagation) and from seed. However, vegetative multiplication is not without its problems and inflates the cost of planting material; this makes seed propagation particularly attractive. The problem in producing true-breeding lines for commercial growing or for $F_1$ hybrid production lies, in this case, in severe inbreeding depression (Jordan and Reimann-Philipp, 1979). However, this might be circumvented by the *in vitro* production of haploids as suggested by Preil *et al.* (1977) and Meynet and Sibi (1984).

In the geranium, $F_1$ hybrid seed-raised cultivars have been bred to compete successfully with vegetatively propagated cultivars. The modern era of seed geraniums began with the inbred Nittany Lion Red, released in 1963 by the Pennsylvania Agricultural Experiment Station (Craig, 1971). This was rapidly followed by the release of $F_1$ hybrids by commercial seed firms in both the USA and The Netherlands. A problem with the early releases was the extremely long growth period from sowing to flowering, which was largely overcome with the release of the early-flowering Sprinter in 1972 (Horn, 1974). "Early-flowering" is a simply inherited dominant character (Hanniford and Craig, 1982) and is, presumably, present in many of the more than 100 seed cultivars which are currently available to growers.

The trend to seed propagation is likely to continue as evidenced by the recent release of the first commercial seed-raised ivy-leaf geranium (*Pelargonium peltatum*), Summer Showers, by the Pan-American Plant Company. However, the breeding problems can be formidable. Many ornamentals are high polyploids, have genetically determined self-incompatibility systems and show severe inbreeding depression; the chrysanthemum, for example, has all of these problems. In other species the problems are less great and it may be that the kalanchoe will be the next vegetatively propagated ornamental to receive competition from $F_1$ hybrid seed-raised cultivars (Royle, 1982). If this proves to be the case, propagation methods will have gone full cycle from the early days of seed-raised *K. blosfeldiana*.

### 9.2.2 Year-round production

Glasshouse flower crops broadly divide into those which remain in more-or-less continuous production for one to several years after planting and those which are harvested over one to several weeks and are then cleared. In the former category are the rose, carnation, gerbera and alstroemeria; these typically show seasonal fluctuations in flower yield and cultivars are sought which show good productivity under adverse conditions. These would either replace or complement existing cultivars and so stabilise year-round production. In the second category are species such as the chrysanthemum and freesia (and pot plant species) where year-round production is achieved by successional planting, usually combined with environmental manipulation or pre-treatment of the planting material to encourage flowering outside of the "natural" season. Cultivars better adapted to flower at "inclement" times are sought to improve year-round production.

#### 9.2.2.1 Long-duration crops—the rose

Seasonal variation in rose flower production is closely linked with the light integral (Post and Howland, 1946) and winter production is depressed by reduced growth and increased flower abortion (blindness). Supplementary lighting can almost double winter bloom production (Cockshull, 1975b), but is usually judged to be commercially uneconomic. Cultivars with improved winter performance are sought, and Hand and Cockshull (1975) pointed out that the introduction of Sonia in the early 1970s had been of great benefit to growers since it at least doubled the winter bloom production of the long-established Pink Sensation which it superseded.

De Vries (1977) demonstrated the independence of total shoot production and blind shoot production and developed a screening procedure using low-intensity lighting in growth rooms to facilitate selection for high rates of total shoot production combined with low levels of blindness (De Vries and Smeets, 1978). The efficiency of this selection method was subsequently demonstrated by growing selected genotypes under commercial glasshouse conditions (De Vries et al., 1978). In addition, genetic variation for flowering ability under low-irradiance conditions was analysed and found to be mainly additive, and genotypes with high general combining ability were identified for further breeding (De Vries et al., 1980).

#### 9.2.2.2 Short-duration crops—the chrysanthemum

In Europe and the USA, year-round cropping of the chrysanthemum is achieved by growing naturally autumn-flowering types and manipulating daylength to control flowering date. This is effective but there is a need for cultivars better adapted for winter growing when low light levels in Northern Europe delay flowering and reduce flower quality (Cockshull and Hughes, 1971).

De Jong (1986) demonstrated marked differences in flowering behaviour when cultivars were grown along a light gradient that fell from 35 to 3 W m$^{-2}$ maintained for 9 h per day (PAR integrals of 1.13–0.10 MJ m$^{-2}$ d$^{-1}$). The more sensitive cultivars showed first a reduction in the number of flowers per stem produced, then delayed flowering and, finally, no flowering at all. One commercially grown cultivar, Pink Boston, showed little delay in flowering over the whole range of irradiances sampled but, nevertheless, produced progressively fewer flowers. Genetic variation does exist for the capacity to produce a reasonable number of flowers at the lowest irradiances and one of De Jong's selections, IVT 80089-4, produced 14 flowers per stem at 15 W m$^{-2}$, while Accent, one of the better winter cultivars currently grown, produced only four.

A quite different approach to year-round growing is seen in Japan where photoperiod manipulation is rarely used and a succession of cultivars (in different locations) is grown, each suited to flower naturally at a different time of year (Kawata, 1969). With a view to integrating the contrasting approaches of East and West, Langton and Cockshull (1976) suggested that naturally summer-flowering cultivars might be introduced into year-round schedules in the UK. Flower buds would be initiated by short-day treatment but, since these cultivars differ from those currently grown in being quantitative rather than qualitative short-day plants for flower development (Langton, 1977) and are able to produce open flowers rapidly in long days, blackout treatment in summer would not, thereafter, be needed. This is highly desirable to reduce high-temperature flowering delays and to lower the incidence of disease due to high humidities. Blackouts also lower the total amount of light reaching the crop and reduce potential quality. A difficulty still to be overcome if this approach is to be commercially realistic is that breeding lines with the long-day flower development character also initiate buds rapidly in long days (Langton, 1981). Consequently, it is difficult to maintain stock plants in the vegetative condition. Nevertheless, the concept of broadening the range of chrysanthemum response types to complement environmental control methods remains an attractive proposition.

### 9.2.3 Low-temperature tolerance

As fuel costs have soared there has been an increased effort to develop cultivars which can be successfully grown at temperatures lower than those currently used. This applies equally to pot plant species such as the African violet (*Saintpaulia ionantha*) (Englert, 1980) and cut-flower species such as the chrysanthemum. In the former case, a range of cool-tolerant cultivars, the Endurance series, has recently been bred by interspecific hybridisation between *S. ionantha* Georgia and *S. shumensis* with further back-crossing to commercial cultivars (Anon, 1983). In the case of chrysanthemums, there is already considerable variation between existing cultivars in response to temperature.

Cathey (1954, 1955) classified chrysanthemums into three response types: thermopositive, in which flowering is inhibited more by low temperatures than by high; thermonegative, in which high temperatures are more inhibitory than low temperatures; and thermozero, in which effects of high and low temperatures are similar. De Jong (1978a) has shown that cultivars which are currently grown in year-round programmes are mainly of the thermozero type. He further showed that cultivars differ in their optimum temperature to achieve rapid flowering and that, even within the thermozero types, some such as Early Yellow are relatively temperature insensitive while others such as Star Stream are very sensitive to temperature. There appeared to be no relationship between the time to flowering at the optimum temperature and the delay caused by low-temperature growing. De Jong concluded, therefore, that initial selection should be practised at low temperatures. Those cultivars retained would be expected to flower rapidly at higher temperatures around the optimum and, in most cases, due to thermozero response, would not be unduly delayed by high summer temperatures. This was later confirmed (De Jong, 1984) when parents and progeny that flowered quickly at one temperature were generally fast-flowering at all temperatures.

An alternative strategy dictated by the high costs of winter trialling in glasshouses is to select against slow-flowering types and those with other obvious commercial defects in an initial summer trial with rapid throughput and then to trial the survivors at low temperatures in the following winter (Poisson, 1978). It remains to be seen how successful these strategies can be in meeting this most important challenge and, indeed, that of the mutation breeding approach referred to earlier.

### 9.2.4 Pest and disease resistance

Pest damage or merely the obvious presence of pests on ornamentals greatly reduces their commercial value and can render crops unsaleable. The problem of pest control is exacerbated by instances of pesticide resistance, and breeders increasingly are turning their attention to the possibilities of utilising genetic resistance to pests in their breeding programmes. The possible value of Rugosa Ottawa as a source of resistance in rose breeding to the two-spotted spider mite and the strawberry aphid (Svejda, 1984) has already been referred to. For chrysanthemum breeding, *C. pacificum* may be a useful source of resistance to the leafminer fly, *Liryomyza trifolii*, and genetic variation affecting the development of spider mites has been identified (Anon, 1985).

Ornamental species are in no way exceptional amongst vegetatively propagated plants in being subject to virus "degeneration". For example, at least 15 viruses are recorded as infecting the chrysanthemum (Hollings, 1968). Breeding has so far played little part in combating this problem. Instead, meristem-tip culture methods combined with indexing procedures have proved successful, if expensive, and the major diseases of economic importance are now fungal and bacterial.

Black spot of roses has already been mentioned. Chrysanthemum white rust caused by *Puccinia horiana* can prove devastating and is a notifiable disease in the UK where the compulsory destruction of affected plants is rigorously enforced. Resistance has been identified in commercial cultivars and breeding is under way to transfer this into further types (van Veen, 1978; Yamaguchi, 1981; Anon., 1985). Recent work has shown that resistance to white rust can be of three types (Rademacher and De Jong, 1987). The first, complete resistance, occurs in the cvs Guilderland, Renine, Rewilo and Lameet, and gives protection with no macroscopic symptoms of fungal attack. Its mechanism of action appears to be similar to "early hypersensitivity" in cereals and, although it is controlled by only a single dominant gene (Renine is duplex for the gene), no instances have been reported of new races of the fungus arising which can overcome the resistance. The second, incomplete resistance, is possessed by the cvs Refour and Ready. Under heavy infection pressure, sporulation is delayed but not prevented. Nevertheless, it is judged that this form of resistance offers sufficient protection under commercial conditions. It appears to be based on the presence of several genes and is expected to be durable. The third type, present in Redemine, is associated with the development of visible necrotic spots on the leaves at points of infection. The mechanism of action is referred to as "late hypersensitivity" and is proving difficult to transfer to progeny.

The disease which has probably claimed as much attention as any from plant breeders in recent years is carnation wilt caused by *Fusarium oxysporum* f. sp. *dianthi* and *F. oxysporum* var. *redolens*. These cause crop losses as high as 50% (Matthews and Arthur, 1979) and many growers have had to abandon the traditional two-year cropping cycle and grow shorter duration crops instead. While disease-free stock plants and hygiene measures such as soil sterilisation are important for control, resistant cultivars with good growth and flowering characters are urgently needed.

Resistance (or tolerance) to carnation wilt has been identified in some wild species (Matthews and Arthur, 1979), in some older cultivars such as Orchid Beauty, and in breeding lines resistant to *Phialophora* wilt (Sparnaaij and Demmink, 1977). A common defence mechanism appears to operate (Anon., 1985), which is determined by an additive genetic system with many genes involved (Arthur, 1984). Resistance is not absolute but is at a level sufficient to give satisfactory control in commercial trials (Sparnaaij and Demmink, 1977). A complication to breeding, however, is that two, and possible three, distinct races of *F. oxysporum* f. sp. *dianthi* appear to exist in Europe (Garibaldi, 1977; Matthews, 1979) and at least three in the USA (Hood and Stewart, 1957). In the UK, race 2, as defined by Garibaldi (1977), predominates and most of the commonly grown cultivars including the "Sims" (see earlier) are extremely susceptible to it (Matthews, 1979). The emphasis in the UK, therefore, has been to develop "Sim" types resistant to race 2 (and preferably with resistance to race 1 also). Promising selections are currently being trialled (Arthur, 1984). Two resistant cultivars have been released in the USA (Carrier, 1977).

*Fusarium oxysporum* can also cause great losses in narcissus and limits the export trade of dry bulbs from the UK. In this case the causal organism is f. sp. *narcissi* and the disease is bulb "basal rot". Resistance, presumably polygenic in inheritance, is present in cultivars such as St. Keverne and has been transferred to a range of newly released cultivars better suited to commercial needs (Tompsett, 1984). Attempts to induce resistance by X-ray irradiation in otherwise good cultivars of narcissus have recently been initiated (Squires *et al.*, 1986). Other important diseases of bulbous ornamentals which are the subjects of active breeding programmes include bulb rot disease of tulip caused by *F. oxysporum* f. sp. *tulipae* (Van Eijk and Eikelboom, 1983), bulb rot of nerine caused by *F. sacchari* var. *elongatum* (Van Tuyl and Kwakkenbos, 1986), and "yellow" disease of hyacinth caused by the bacterium, *Xanthomonas campestris* pv. *hyacinthi* (Van Tuyl, 1984).

### 9.2.5 Longevity

The potential longevity of cut flowers (vase life) and flowering pot plants after they have left the grower's holding is a character of concern to the ornamentals breeder since it is a major factor determining consumer satisfaction and potential market penetration. This is a particularly difficult character to assess because it is markedly affected by conditions before harvest, during distribution and after sale.

The gerbera is an example of a species in which cultivars differ in their potential vase life (Serini and De Leo, 1979; De Jong, 1978b) and also in their proneness to irreversible stem bending ("folding") which reduces longevity (De Jong, 1978b). The ideal cultivar has high structural stem strength and rapidly regains turgor when placed in water after a period of dry storage (De Jong, 1978b) although all genotypes will regain turgor if the bases of stems are cut after storage (De Jong and Garretsen, 1985). Harding *et al.* (1981) calculated broad- and narrow-sense heritabilities for longevity in the gerbera. Their estimates were low but it was concluded that a slow response to selection can be expected. De Jong and Garretsen (1985) have also carried out a genetic analysis of longevity and have put forward a more optimistic view, stating that a 14-day vase life is a realistic breeding goal. Selection on the basis of stem curvature after dry storage may give cultivars less prone to "folding".

The choice of ornamental subjects for improvement by breeding, especially where there is state involvement, is largely dictated by commercial value. On the other hand, there can be no doubt that the relative importance of a species can be strongly influenced by the breeding inputs to which it is subjected. The tenfold increase in the area of gerbera grown in The Netherlands between 1970 and 1980 must, in part, be attributable to improved cultivars, as must the recent rise in the importance of "spray" carnations to eclipse the traditional "standard" type. The imagination of the ornamentals breeder is a key element in determining his commercial success. In the final analysis, however, the ornamentals breeder has the same basic objectives as other breeders—to improve yield and quality while, at the same time, bearing in mind the need to reduce inputs of labour, fuel and growing materials.

**REFERENCES**

Anon. (1983). Cooler saintpaulias. *Grower* **99** (13), 29.
Anon. (1985). Breeding research ornamental crops. *Jaarverslag, Instituut voor de Veredeling van Tuinbouwgewassen*, pp. 71–95.
Arthur, A. E. (1984). Carnation breeding: Scope for the future. *Sci. Hort.* **35**, 78–83.
Badr, M., and Horn, W. (1971). Cytologische Untersuchungen bei *Pelargonium zonale*—Hybriden. *Z. PflZücht.* **66**, 158–174.
Bech, A.-M. (1983). Homogeneity and heterogeneity in vegetatively propagated pot plants. *Acta Hort.* **147**, 135–142.
Bech, A.-M., Christensen, O.V., and Ottosen, C.-O. (1985). Udvaelgelse af kloner af *Dieffenbachia maculata* (Lodd.) G. Don. *Tidsskr. Planteavl.* **89**, 185–189.
Bolton, A. T., and Svejda, F. J. (1979). A new race of *Diplocarpon rosae* capable of causing severe black spot on *Rosa rugosa* hybrids. *Can. Pl. Dis. Surv.* **59**, 38–41.
Broertjes, C. (1972). Mutation breeding of *Achimenes*. *Euphytica* **21**, 48–62.
Broertjes, C., and Lock, C. A. M. (1985). Radiation-induced low-temperature tolerant solid mutants of *Chrysanthemum morifolium* Ram. *Euphytica* **34**, 97–103.
Broertjes, C., and Van Harten, A. M. (1978). "Application of Mutation Breeding Methods in the Improvement of Vegetatively Propagated Crops". Elsevier, Amsterdam, 316pp.
Broertjes, C. and Verboom, H. (1974). Mutation breeding of *Alstroemeria*. *Euphytica* **23**, 39–44.
Broertjes, C., Roest, S., and Bokelmann, G. S. (1976). Mutation breeding of *Chrysanthemum morifolium* Ram. using *in vivo* and *in vitro* adventitious bud techniques. *Euphytica* **25**, 11–19.
Broertjes, C., Koene, P., and Van Veen, J. W. H. (1980). A mutant of a mutant of a mutant of a . . .: Irradiation of progressive radiation-induced mutants in a

mutation-breeding programme with *Chrysanthemum morifolium* Ram. *Euphytica* **29**, 525–530.

Broertjes, C., Koene, P., and Pronk, Th. (1983). Radiation-induced low-temperature tolerant cultivars of *Chrysanthemum morifolium* Ram. *Euphytica* **32**, 97–101.

Carrier, L. E. (1977). Breeding carnations for disease resistance in Southern California. *Acta Hort.* **71**, 165–168.

Cathey, H. M. (1954). Chrysanthemum temperature study. B. Thermal modifications of photoperiods previous to and after flower bud initiation. *Proc. Am. Soc. hort. Sci.* **64**, 492–498.

Cathey, H. M. (1955). Temperature guide to chrysanthemum varieties. *N.Y. St. Flow. Grow. Bull.* **119**, 1–4.

Cockshull, K. E. (1975a). Premature budding in year-round chrysanthemums. *Rep. Glasshouse Crops Res. Inst. 1974*, pp. 128–136.

Cockshull, K. E. (1975b). Roses II: The effects of supplementary light on winter bloom production. *J. hort. Sci.* **50**, 193–206.

Cockshull, K. E., and Hughes, A. P. (1971). The effects of light intensity at different stages in flower initiation and development of *Chrysanthemum morifolium*. *Ann. Bot.* **35**, 915–926.

Craig, R. (1971). Cytology, genetics and breeding. *In* "Geraniums. A Penn State Manual" (J. W. Mastalerz, ed.), pp. 315–346. Pennsylvania Flower Growers.

De Jong, J. (1978a). Selection for wide temperature adaptation in *Chrysanthemum morifolium* (Ramat.) Hemsl. *Neth. J. agric. Sci.* **26**, 110–118.

De Jong, J. (1978b). Dry storage and subsequent recovery of cut gerbera as an aid in selection for longevity. *Scientia Hort.* **9**, 389–397.

De Jong, J. (1984). Genetic analysis in *Chrysanthemum morifolium*. I. Flowering time and flower number at low and optimum temperature. *Euphytica* **33**, 455–463.

De Jong, J. (1986). Adaptation of *Chrysanthemum morifolium* to low light levels. *Scientia Hort.* **28**, 263–270.

De Jong, J., and Garretsen, F. (1985). Genetic analysis of cut flower longevity in gerbera. *Euphytica* **34**, 779–784.

De Vries, D. P. (1977). Shoot production in cut roses with reference to breeding for winter flowering. *Euphytica* **26**, 85–88.

De Vries, D. P., and Dubois, L. A. M. (1978). On the transmission of the yellow flower colour from *Rosa foetida* to recurrent flowering hybrid tea-roses. *Euphytica* **27**, 205–210.

De Vries, D. P., and Smeets, L. (1978). Hybrid tea-roses under controlled light conditions: 2. Flowering of seedlings as dependent on the level of irradiance. *Neth. J. agric. Sci.* **26**, 128–132.

De Vries, D. P., Dubois, L. A. M., and Smeets, L. (1978). Hybrid tea-roses under controlled light conditions: 3. Flower and blind shoot production in the glasshouse of seedlings selected for flowering or flower bud abortion at low irradiances in a growth room. *Neth. J. agric. Sci.* **26**, 399–404.

De Vries, D. P., Smeets, L., and Dubois, L. A. M. (1980). Hybrid tea-roses under controlled light conditions. 4. Combining ability analysis of variance for percentage of flowering in $F_1$ populations. *Neth. J. agric. Sci.* **28**, 36–39.

Doorenbos, J. (1973). Breeding 'elatior'-begonias (*B.* × *hiemalis* Fotsch). *Acta Hort.* **31**, 127–131.

Doorenbos, J., and Karper, J. J. (1975). X-ray induced mutations in *Begonia* × *hiemalis*. *Euphytica* **24**, 13–19.

Englert, U. (1980). *Saintpaulia*. Neuheiten aus Vaihingen. *Gb und Gw* **80**, 131–132.
Garibaldi, A. (1977). Race differentiation in *Fusarium oxysporum* f. sp. *dianthi* and varietal susceptibility. *Acta Hort.* **71**, 97–101.
Goemans, J. A. M. (1962). Breeding of alstroemerias. *Jl R. hort. Soc.* **87**, 282–284.
Goemans, R. A. (1980). The history of the modern *Freesia*. In "Petaloid Monocotyledons: Horticultural and Botanical Research" (C. D. Brickell, D. F. Cutler and M. Gregory, eds), pp. 161–170. Academic Press for Linnean Society of London.
Hand, D. W., and Cockshull, K. E. (1975). Roses I: the effects of $CO_2$ enrichment on winter bloom production. *J. hort. Sci.* **50**, 183–192.
Hanniford, G. G., and Craig, R. (1982). Genetics of flowering—geraniums for the future. *Sci. Agric.* **29**, 9.
Harding, J., Byrne, T., and Nelson, R. L. (1981). Heritability of cut-flower vase longevity in gerbera. *Euphytica* **30**, 653–657.
Holley, W. D., and Baker, R. (1963). "Carnation Production". Brown, Dubuque, Iowa, 142pp.
Hollings, M. (1968). "Virus Diseases of Chrysanthemums". Advisory Leaflet 555. Ministry of Agriculture, Fisheries and Food, Her Majesty's Stationery Office, Edinburgh, 6pp.
Hood, J. R., and Stewart, R. N. (1957). Factors affecting symptom expression in *Fusarium* wilt of *Dianthus*. *Phytopathology* **47**, 173–178.
Horn, W. (1974). The glasshouse environment and selection for continuously varying characters. In "Eucarpia Meeting on Ornamentals", John Innes Institute, Norwich, pp. 17–25.
Howard, G. S. (1971). Polyploid carnations: five quantitative character estimates for nine hybrid populations. *J. Am. Soc. hort. Sci.* **96**, 673–674.
Jank, H. (1957). Experimentelle Mutationsauslösung durch Röntgenstrahlen bei *Chrysanthemum indicum*. *Züchter* **27**, 223–231.
Jordan, C., and Reimann-Philipp, R. (1979). Breeding of *Gerbera jamesonii* at the Federal Research Center for Horticultural Plant Breeding. In "Eucarpia Meeting on Genetics and Breeding of Carnation and Gerbera, Alassio, 1978" (L. Quagliotti and A. Baldi, eds), pp. 223–225. Institute of Plant Breeding and Seed Production, Turin.
Jung-Heiliger, H., and Horn, W. (1980) Variation nach mutagener Behandlung von Stecklingen und in vitro—Kulturen bei *Chrysanthemum*. *Z. PflZücht.* **85**, 185–199.
Kawata, J. (1969). Year-round production of chrysanthemums in Japan. *Jap. Agric. Res. Quart.* **4**, 23–27.
Knicely, W. W., and Walker, D. E. (1966). Chromosome numbers and crossability studies in the genus *Pelargonium*. *Proc. XVII Int. hort. Congr.* **1**, 209.
Langton, F. A. (1977). The responses of early-flowering chrysanthemums to daylength. *Scientia Hort.* **7**, 277–289.
Langton, F. A. (1980). Chimerical structure and carotenoid inheritance in *Chrysanthemum morifolium* (Ramat.). *Euphytica* **29**, 807–812.
Langton, F. A. (1981). Breeding summer-flowering chrysanthemums: The relationship between flower bud development and vegetative duration in long days. *Z. PflZücht.* **86**, 254–262.
Langton, F. A. (1986). Mutation breeding and its role in the improvement and commercialization of vegetatively propagated crops. In "Infraspecific Classification of Wild and Cultivated Plants" (B. T. Styles, ed.), pp. 263–276.

Oxford University Press, Oxford.

Langton, F. A., and Cockshull, K. E. (1976). An ideotype of chrysanthemum (*C. morifolium* Ramat.). *Acta Hort.* **63**, 165–175.

Langton, F. A., Royce, H. M., and Cockshull, K. E. (1982). Chrysanthemums: a check-list of long-day leaf numbers. "*Tests of Agrochemicals and Cultivars*" (*Ann. appl. Biol.* **100** *Suppl.*) **3**, 110–111.

Larkin, P. J., and Scowcroft, W. R. (1981). Somaclonal variation—a novel source of variability from cell cultures for plant improvement. *Theoret. appl. Genet*, **60**, 197–214.

Lovelidge, B. (1969). On the threshold of a new era in pot plants. *Grower* **172**, 1023–1026.

Matthews, P. (1979). Variation in English isolates of *Fusarium oxysporum* f. sp. *dianthi*. *In* "Eucarpia Meeting on Genetics and Breeding of Carnation and Gerbera, Alassio, 1978" (L. Quagliotti and A. Baldi, eds), pp. 115–126. Institute of Plant Breeding and Seed Production, Turin.

Matthews, P., and Arthur, A. E. (1979). Resistance and selecting for resistance to *Fusarium* wilt in the carnation. In "Eucarpia Meeting on Genetics and Breeding of Carnation and Gerbera, Alassio, 1978" (L. Quagliotti and A. Baldi, eds), pp. 127–139. Institute of Plant Breeding and Seed Production, Turin.

Mendoza, H. A., and Haynes, F. L. (1973). Some aspects of breeding and inbreeding in potatoes. *Am. Potato J.* **50**, 216–222.

Meynet, J., and Sibi, M. (1984). Haploid plants from *in vitro* culture of unfertilized ovules in *Gerbera jamesonii*. *Z. PflZücht.* **93**, 78–85.

Palmer, J. G., Semeniuk, P., and Stewart, R. N. (1966). Roses and blackspot: 1. Pathogenicity to excised leaflets of *Diplocarpon rosae* from seven geographic locations. *Phytopathology* **56**, 1277–1282.

Poisson, C. (1978). Selection of chrysanthemums for growth and flowering at low temperatures. *In* "Eucarpia Meeting on Chrysanthemums, Littlehampton", (F. A. Langton, ed.), pp. 23–31.

Post, K., and Howland, J. E. (1946). The influence of nitrate level and light intensity on the growth and production of greenhouse roses. *Proc. Am. Soc. hort. Sci.* **47**, 446–450.

Preil, W., Engelhardt, M., and Walther, F. (1983). Breeding of low temperature tolerant poinsettia (*Euphorbia pulcherrima*) and chrysanthemum by means of mutation induction in *in vitro* culture. *Acta Hort.* **131**, 345–351.

Preil, W., Huhnke, W., Engelhardt, M., and Hoffmann, M. (1977). Haploids in *Gerbera jamesonii* from *in vitro* cultured capitulum explants. *Z. PflZücht.* **79**, 167–171.

Rademaker, W., and De Jong, J. (1987). Types of resistance to *Puccinia horiana* in chrysanthemum. *Acta Hort.* (in press).

Rehder, A. (1940). "Manual of Cultivated Trees and Shrubs Hardy in North America," 2nd edn. MacMillan, New York, 930pp.

Robinson, G. W. (1963). Alstroemeria. *Jl R. hort. Soc.* **88**, 490–494.

Roest, S., and Bokelmann, G. S. (1975). Vegetative propagation of *Chrysanthemum morifolium* Ram. *in vitro*. *Scientia Hort.* **3**, 317–330.

Royle, D. (1982). Floranova breed fuel economy into pot plants. *Grower* **98**, (10), 21–25.

Semeniuk, P. (1971). Inheritance of recurrent blooming in *Rosa wichuraiana*. *J. Hered.* **62**, 203–204.

Semeniuk, P. (1979). Spotless Gold, Spotless Yellow, and Spotless Pink rose: Blackspot resistant breeding lines. *HortScience* **14**, 764–765.
Semeniuk, P., and Arisumi, T. (1968). Colchicine-induced tetraploid and cytochimeral roses. *Bot. Gaz.* **129**, 190–193.
Serini, G., and De Leo, V. (1979). Phenotypic characters and preservability of gerbera flowers. *In* "Eucarpia Meeting on Genetics and breeding of Carnation and Gerbera, Alassio, 1978" (L. Quagliotti and A. Baldi, eds), pp. 269–277. Institute of Plant Breeding and Seed Production, Turin.
Shibata, M., and Kawata, J. (1986). Chimerical structure of the Marble sports series in chrysanthemums. In "Development of New Technology for Identification and Classification of Tree Crops and Ornamentals" (K. Kitaura *et al.*, eds), pp. 47–52. Ministry of Agriculture, Forestry and Fisheries, Japan.
Skirvin, R. M., and Janick, J. (1976a). Tissue culture-induced variation in scented *Pelargonium* spp. *J. Am. Soc. hort. Sci.* **101**, 281–290.
Skirvin, R. M., and Janick, J. (1976b). 'Velvet Rose' *Pelargonium*, a scented geranium. *HortScience* **11**, 61–62.
Sparnaaij, L. D. (1979). Polyploidy in flower breeding. *HortScience* **14**, 496–499.
Sparnaaij, L. D., and Demmink, J. F. (1977). Progress towards *Fusarium* resistance in carnations. *Acta Hort*, **71**, 107–113.
Sparnaaij, L. D., Kho, Y. O., and Baer, J. (1968). Investigations on seed production in tetraploid freesias. *Euphytica* **17**, 289–297.
Squires, W., Bowes, S., and Langton, A. (1986). Work to keep the bulb industry ahead. *Grower* (Supp.) **105**, (3), 10–13.
Stewart, R. N. (1951). Colchicine-induced tetraploids in carnations and poinsettias. *Proc. Am. Soc. hort. Sci.* **57**, 408–410.
Straub, J. (1940). Quantitative and qualitative Verschiedenheiten innerhalb von polyploiden Pflanzenreihen. *Biol. Zbl.* **60**, 659–669.
Svejda, F. (1975). New approaches in rose breeding. *HortScience* **10**, 564–567.
Svejda, F. (1984). 'Rugosa Ottawa', a source for insect and disease resistance in roses. *HortScience* **19**, 896–897.
Svejda, F. J., and Bolton, A. T. (1980). Resistance of rose hybrids to three races of *Diplocarpon rosae*. *Can. J. Pl. Path.* **2**, 23–25.
Tompsett, A. A. (1984). Narcissus varieties for the future. *Sci. Hort.* **35**, 84–87.
Van Eijk, J. P., and Eikelboom, W. (1983). Breeding for resistance to *Fusarium oxysporum* f. sp. *tulipae* in tulip (*Tulipa* L.): 3. Genotypic evaluation of cultivars and effectiveness of pre-selection. *Euphytica* **32**, 505–510.
Van Tuyl, J. M. (1984). Veredeling op resistentie tegen geelziek bij hyacinth. *Bloembollencultuur* **94**, 818–819.
Van Tuyl, J. M., and Kwakkenbos, A. A. M. (1986). Veredelingsonderzoek *Nerine bowdenii*. *Fusarium* veroorzaakt grote verliezen. *Bloembollencultuur* **97**, 18.
Van Veen, J. W. H. (1978). Japanese chrysanthemums. *In* "Eucarpia Meeting on Chrysanthemums, Littlehampton" (F. A. Langton, ed.), pp. 49–59.
Wilfret, G. J., Weiler, T. C., Harbaugh, B. K., and Hammer, P. A. (1983). Floriana Mist and Floriana Cascade. New Bristol Fairy type *Gypsophila* for cut flowers. Circular S-299, Florida Agricultural Experiment Stations. University of Gainesville, Florida.
Wilkins, H. F., Healy, W. E., and Gilbertson-Ferris, T. L. (1980). Comparing and contrasting the control of flowering in *Alstroemeria* 'Regina', *Freesia* ×

*hybrida* and *Lilium longiflorum*. *In* "Petaloid Monocotyledons: Horticultural and Botanical Research" (C. D. Brickell, D. F. Cutler and M. Gregory, eds), pp. 51–63. Academic press for the Linnean Society of London.

Wylie, A. P. (1954). The history of garden roses, Part I. *Jl R. hort. Soc.* **79**, 555–571.

Wylie, A. P. (1955). The history of garden roses, Part II. *Jl R. hort. Soc.* **80**, 8–24.

Yamaguchi, T. (1981). Chrysanthemum breeding for resistance to white rust. *Jap. J. Breed.* **31**, 121–132.

# 10 Selecting and Breeding for Better Potato Cultivars

G. R. MACKAY

*Scottish Crop Research Institute (Pentlandfield), Roslin*

In common with other scientific disciplines, there is a considerable and increasing quantity of publications on the subject of the breeding of potatoes and associated research (Ross, 1986). A review of this literature might lead one to conclude that there have been substantial advances in the state of this particular art, since the turn of the century. Since the ultimate objective of breeding programmes is the production of new and improved cultivars, one might then assess the progress of modern breeders by their cultivars and the state of their art by comparing the methods employed to produce them with those of their predecessors.

In the UK, breeders may take credit for having finally ousted the cv. Majestic, introduced in 1911 (Salaman, 1926), which dominated the British maincrop ware acreage until the 1960s. However, the ware potato acreage continues to be dominated by rather few varieties introduced more than a decade ago and all of which have substantial weaknesses, which could be improved upon by breeding. Désirée, introduced into the UK in 1962, is a high-yielding cultivar with excellent table qualities. It is, however, very susceptible to common scab, moderately susceptible to late blight and gangrene, susceptible to potato cyst nematode (PCN) and has only a moderate spectrum of resistance to the common viruses (Anon., 1986a); under some conditions it can also exhibit secondary growth, gemmation, which can result in unattractive tuber shapes. Maris Piper (1963), currently the most widely grown cultivar in the UK, is also high yielding, of good quality and is resistant to pathotypes $Ro_1$ and $Ro_4$ of the golden cyst nematode, *Globodera rostochiensis*, but it is also very susceptible to common scab and, while field-immune to potato virus X (PVX), is susceptible to virus Y (PVY) and leafroll virus (PLRV) (Anon., 1986a). Pentland Crown (1958), now declining in popularity, is an extremely high-yielding cultivar combining high levels of resistance to PVY and PLRV

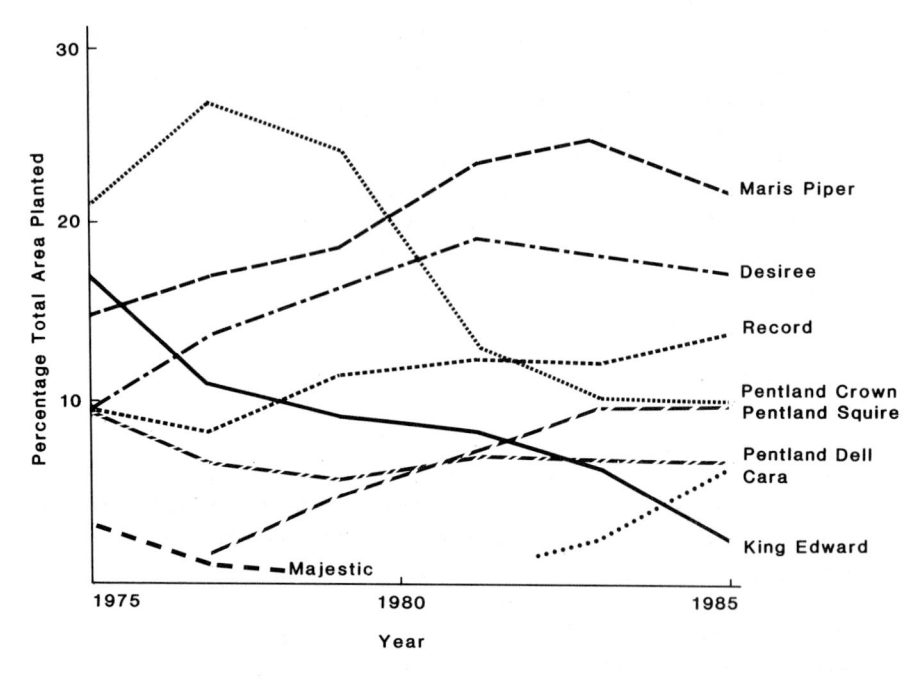

**Figure 10.1.** Trends in the areas planted to maincrop potato cultivars in the UK (1975–1985).

with resistance to common scab, but it is susceptible to PCN and gangrene (Anon., 1986a). Despite its obvious merits as a grower's cultivar it lacks the quality that an increasingly discerning public demands. King Edward, raised by a Northumberland gardener and first marketed in 1902 (Salaman, 1926), still continues to provide a bench mark for quality in the eyes of the British consumer and makes a significant contribution despite its susceptibility to wart, late blight, skin spot and the common viruses (Anon., 1986a) (Fig. 10.1). The needs of the British processing industry, which utilises approximately 25% of the crop, continue to be met by two rather old cultivars, Record (1944) for crisping and Pentland Dell (1960) for French-fry production. The situation in North America and Europe is broadly similar; old cultivars such as Russet Burbank (pre-1890) and Bintje (1910), respectively, continue to predominate (Beukema and van der Zaag, 1979).

An examination of the methods employed by breeders of commercial cultivars also suggests that though breeders now understand a great deal more about the genetics and cytogenetics, physiology and pathology of the potato, conventional potato breeding has changed little in essence, save in scale and technology since the turn of the century. Howard *et al.* (1970, 1978) describe the principal basis of breeding a potato cultivar as crossing a cultivar with high yield and poor quality or low disease resistance with another of low yield and high quality or disease resistance in order to select a new cultivar which has a yield as high or higher than the former, but with the quality or disease resistance of the latter. Phenotypic recurrent selection as implied here remains the principal basis of most cultivar breeding programmes.

## 10.1 HISTORICAL PERSPECTIVE AND CYTOGENETIC BACKGROUND

The modern cultivated potato, *Solanum tuberosum* ssp. *tuberosum* is a tetraploid, $2n = 4x = 48$, probably derived from introductions of *Solanum tuberosum* ssp. *andigena* into Europe at the time of and since the Spanish conquest of South America. Its precise evolutionary history is a matter of conjecture, but it is likely that the potato was brought into cultivation in South America during the period 2000–5000 BC, by human selection for less-bitter-tasting clones from the wild. Several wild species are believed to have contributed to the tetraploid form and diploid, triploid and pentaploid cultivated clones are also grown in the centre of origin. The diploid species of the genus are obligate outbreeders with a gametophytic self-incompatibility system which becomes inoperative at higher levels of ploidy. Cultivated tetraploids, if pollen fertile, are self-fertile but will exhibit the usual symptoms of inbreeding depression on selfing and the potato is regarded and treated as an outbreeder. Consequently, potato cultivars and breeders' clones are usually highly heterozygous.

The original introductions into Europe were not adapted to the long days of the northern temperate zone growing season but, within 200 years, the potato had become an important staple food crop. This will have been achieved by a combination of conscious human phenotypic selection aided by natural selection for clones capable of initiating tubers and producing above-average yields under the long-day conditions of the European growing season (Dodds, 1965; Howard, 1970; Simmonds, 1974).

The great proliferation of cultivars during the nineteenth century seems largely to have been due to the observation that all cultivars "degenerated" after several years of clonal reproduction in agriculture. This "degeneration" was believed by many to be an intrinsic feature of clonal reproduction, and passage through a sexual phase was necessary to re-invigorate stocks (Robb, 1921). Many cultivars were simply selections from amongst seedlings obtained from self-set, naturally pollinated berries (Salaman, 1926). There are, however, early reports of controlled hybridisation and the use of wild species is reported as long ago as 1850 (Simmonds, 1974). Many early breeders did not disclose the pedigrees of their cultivars (Salaman, 1926) or may have invented them in order to mislead their contemporaries (Glendinning, pers. comm.). However, there is little doubt that the principal basis of breeding was then, as now, phenotypic selection.

## 10.2 GENERAL CONSIDERATIONS

Many authorities are of the opinion that European cultivars and their derivatives are founded on a narrow genetic base. Much effort has been devoted to expanding this in order to exploit sources of resistance to diseases and pests not available within the *tuberosum* group and also, hopefully, to benefit from an anticipated heterotic effect, with an increase in yield (Simmonds, 1969). There are, however, many constraints placed upon the breeder of new cultivars which must be at least of the same quality, yield and all-round disease resistance, or better, than existing cultivars. Table 10.1 lists some of the characteristics which the breeder has to consider in making his selections for European agriculture.

While it is unreasonable to expect a single clone to exhibit all these attributes to a high degree, a high expression of resistance to one disease will not compensate for extreme susceptibility to another. Also, whereas a high-yielding clone of acceptable quality but an indifferent disease-resistance spectrum may become a successful cultivar, no amount of disease resistance will compensate for a serious agronomic defect. Consequently, conventional varietal breeding schemes tend to be rather conservative. Breeders are reluctant to move outside the *tuberosum* group unless they are obliged to do so because a specific character, usually resistance to a particular disease, is not available within it. Nevertheless, despite this innate conservatism, in a recent survey Ochoa and Schmiediche (1983) reported that of 595 cultivated European cultivars, more than 200 had at least one wild species in their pedigrees.

The selective introgression of a number of simply inherited characteristics from wild species and primitive cultivars has been achieved very successfully. The $H_1$ gene from the *andigena* group which provides resistance to pathotypes $Ro_1$ and $Ro_4$ of *Globodera rostochiensis*, discovered by Ellenby (1952), is now available in numerous cultivars (Anon., 1986a). Many new cultivars are field-immune to PVX by virtue of genes also introgressed from *andigena* and *Solanum acaule*. Comprehensive resistance to all

*Selecting and Breeding Potato Cultivars*

strains of PVY, from *Solanum stoloniferum* and other species, is becoming available in cultivars and breeders' advanced selections (Cockerham, 1970; Davidson, 1980).

More complex polygenically controlled characters, such as resistance to potato leafroll virus (PLRV) or horizontal resistance to late blight, *Phytophthora infestans*, have proved more difficult to introgress from wild species or primitive cultivars. Nevertheless, where the objective is well defined and fairly straightforward to screen for, this too is being achieved. More comprehensive forms of resistance to the white cyst nematode, *G. pallida*, from *andigena*, and to both *G. pallida* and *G. rostochiensis*, from *Solanum vernei*, have been incorporated into breeders' clones and are

**Table 10.1.** A sample of some of the characteristics which a potato breeder has to consider in selecting new cultivars.

| | |
|---|---|
| Yield | Tuber number, tuber size, bulking rate, drought resistance, storageability (marketable vs. total yield) |
| Conformity | Tuber shape, regularity and uniformity |
| Absence of growth defects | Gemmation, hollow heart, growth cracks |
| Quality | Table and processing: Enzymic browning, after-cooking blackening, sloughing, texture, dry-matter content, sugars (crisp colour), storage characteristics (dormancy) |
| Resistance to mechanical damage | External: shatter cracks, scuffing<br>Internal: bruising (black spot) |
| Eye appeal | Consumer preferences: skin colour, flesh colour |
| Miscellaneous disorders | Internal rust spot, wind damage, sensitivity to herbicides |
| Disease and pest resistances | Late blight—tuber and foliage<br>Common viruses, PVX, PVY, PLRV<br>Cyst nematodes, *rostochiensis* and *pallida*<br>Common scab<br>Gangrene<br>Wart<br>Skinspot<br>Powdery scab<br>Spraing (tobacco rattle virus)<br>Soft rot<br>Dry rot |

becoming available in new cultivars (Dunnett, 1960; Howard *et al.*, 1970). This progress has been greatly aided by the development of screening techniques capable of testing large numbers of clones rapidly and repeatably, so enabling the breeder to exert positive selection pressure at an early stage in his programme and to more accurately estimate the breeding value of parental clones (Phillips and Dale, 1982).

Given such suitable screening techniques, there is no doubt that conventional breeding can, and will, continue to make progress.

## 10.3 A CONVENTIONAL BREEDING PROGRAMME

Most conventional potato breeding programmes follow a broadly similar pattern and, whereas their emphases may differ and the scale of their operations will vary, that of the Scottish Crop Research Institute (SCRI) a few years ago afforded a fairly typical example. Table 10.2 provides an

**Table 10.2.** Outline of SCRI Potato Variety Breeding Scheme, 1982.

| Year | Number of clones | |
|---|---|---|
| 1 | 115 000 | Glasshouse seedlings: PVX, PVY and PCN screen on selected progenies |
| 2 | 40 000 | Singles (1 site) |
| 3 | 4 000 | Three-plant plots (1 site): PCN tests on selected progenies |
| 4 | 1 000 | Two sites (seed and ware): yield, trials, quality and routine disease tests, storageability, agronomic assessment etc. |
| 5 | 500 | |
| 6 | 200 | |
| 7 | 60 | Regional trials, 6–7 sites (UK), 2–3 sites (overseas): further disease tests, samples to collaborators for independent assessment, production of Approved Stocks |
| 8 | 10 | |
| 9 | 5 | |
| 10 | 5 | National List Trials (2 years, 1981–1982); maintenance of Approved Stock (SCRI), VTSC initiated (DAFS)[a] |
| 11 | | |
| 12 | Named cultivars | Commercialisation (NSDO)[a] |

[a] VTSC, Virus Tested Stem Cuttings (nuclear stock); DAFS, Department of Agriculture and Fisheries for Scotland; NSDO, National Seed Development Organisation.

outline of the SCRI varietal breeding programme as of 1982. Historically, Pentlandfield has concentrated its breeding policy on three important areas: breeding for resistance to late blight; breeding for resistance to the common viruses; breeding for resistance to PCN (Holden, 1977; Mackay, 1982). In common with other such programmes, parental clones which express high degrees of resistance to one or other or several of these major diseases were hybridised with other clones or cultivars whose characters complement them, or they were intercrossed in order to produce segregants with a higher expression of resistance. Each year, approximately 100 000 seedlings were grown in glasshouses to produce tubers for field evaluation. A single tuber was taken from each selected seedling and planted in progeny blocks at a high grade seed site. These "singles" were dug by hand and selected in the field, the selected clones were then planted back the next year as unreplicated three-plant plots which were also selected in the field at harvest. From this stage on, until submission to National List Trials of perhaps one or two potential cultivars, decreasing numbers of selected clones were grown in increasingly sophisticated trials at various sites within the UK and overseas. They were also subjected to a battery of tests for their cooking and processing characteristics, and their resistance to diseases and pests.

Table 10.3 illustrates the routine disease and pest screening procedure at SCRI. In the course of passage through such a system more than 60 variates are recorded or scored and used as criteria in deciding whether to continue assessment or discard a clone. In common with other similar programmes, more than 90% of the variation in the original seedling population may be eliminated by the third clonal year. In The Netherlands, where the structure of the plant breeding industry is substantially different from that of the UK and production of finished cultivars is entirely in the hands of the private sector, about 180 potato breeders produce approximately 1 000 000 seedlings each year. Approximately one in 200 000 of these may become a cultivar, following a sequence of selection not dissimilar to that described at SCRI. On average, less than 2% of these seedlings survive selection beyond the third clonal year stage and more than 30% are discarded at the glasshouse seedling stage (K. Louwes, pers. comm., 1986). Until recently, with a few exceptions, the principal information gained on a progeny of a particular cross during these stages was the percentage survival rate.

Recent developments in the use of electronic data capture systems and the computerisation of data storage, retrieval and analysis now provide the means to greatly improve the quantity and quality of data in these early stages. A data base management system (DBMS) developed at the SCRI (Brown, 1984) permits the recording, collation and analysis of data

**Table 10.3.** Approximate numbers of SCRI Commercial Breeding Department clones routinely screened annually.

| Disease tests Total: 115 000 | Year: 1 | 2 | 3 | 4 | 5 | 6 | 7 | 8–12 incl. 20–40 | Yearly totals 165 000 |
|---|---|---|---|---|---|---|---|---|---|
| PVX and/or PVY | 15 000[a] | — | 150[b] | 1000 | 500 | 200 | 60[c] | 30[c] | 15 300 |
| PCN (Ro and Pa) | 200[d] | — | 300 | 300 | — | — | — | 30 | 600 |
| Foliage blight | 200[d] | — | — | — | 500 | 200 | 60 | 30 | 600 |
| Tuber blight | — | — | — | — | 500 | 200 | 60 | 30 | 600 |
| Gangrene | — | — | — | — | 500 | 200 | 60 | 30 | 600 |
| Common scab | — | — | — | — | 500 | 200 | 60 | 30 | 600 |
| Skin spot | — | — | — | — | — | 200 | 60 | 30 | 300 |
| Wart | — | — | — | — | 300 | 200 | 60 | 30 | 590 |
| PLRV | — | — | — | 200 | 200[e] | — | — | — | 200 |
| Soft rot | — | — | — | — | — | — | — | 30 | 30 |
| TRV | — | — | — | — | — | — | 60[f] | 30[g] | 180 |
| PMTV | — | — | — | — | — | — | — | 30[g] | 30 |

[a] Glasshouse seedling screen by spray and hand inoculation with PVX and PVY.
[b] X and/or Y by sap inoculation.
[c] X, Y, A, B, C and Vn by sap inoculation and/or graft.
[d] Progeny tests of seedlings; up to 200 progenies per annum to date.
[e] Two-year field trial, includes all advanced clones reaching 6th year of selection.
[f] Glasshouse pot trials—under development.
[g] Field trials.

on many more clones and entire progenies than hitherto, including those discarded in the early stages of a programme.

Until recently, there was very little published information on the efficacy of selection at these early stages of a potato breeding programme, and such as there was supported the view that it is an area requiring a great deal more study (Howard, 1963; Anderson and Howard, 1981). Opinions differ as to whether or not selection at the glasshouse seedling stage is relevant at all in a clonally reproduced crop, but under glasshouse conditions a proportion of seedlings fail to tuberise at all and are thus self-eliminating, others produce such small, poorly shaped tubers that experience dictates little purpose would be served in planting them for field evaluation.

The proceedings of a symposium on early generation selection methods sponsored by the Potato Association of America (PAA) in 1984 clearly highlighted weaknesses in this current methodology of potato breeding (Martin, 1984; Plaisted *et al.*, 1984; Sanford *et al.*, 1984; Swiezynski, 1984; Tai and Young, 1984). At the SCRI, all clonal selection for agronomic traits has now ceased at the glasshouse seedling stage, as recent research has clearly confirmed the inefficiency of attempting clonal selection at these early stages (Mackay *et al.*, 1984; Brown *et al.*, 1987b).

However, evaluation of clones even in these early stages and selection on the basis of progeny means is proving potentially very worthwhile. The ranking of progeny means is remarkably consistent over a wide diversity of environments including the glasshouse and field (Brown *et al.*, 1987a). Identification of superior progenies, with a greater likelihood of their containing superior clones than contemporaries is feasible even on the basis of assessment of tubers produced in pots from true seed. Comparisons between progeny means and variances of breeders' preference scores based on visual appraisal offer the possibility of efficient cross prediction (Caligari and Brown, 1986). Similar developments for other traits, such as resistance to late blight (Caligari *et al.*, 1984) are possible. Progeny tests, for a variety of disease and pest resistances and quality characteristics are being developed, and incorporated into the routine selection procedures at the SCRI; thus providing the means to increase selection pressure for desirable traits and/or to estimate the breeding value of parents, so increasing the emphasis on genotypic versus phenotypic selection. Results of some of this work have been published but much is ongoing and, clearly, further research is needed, as evidenced by the series of papers introduced by Martin (1984).

## 10.4 NON-CONVENTIONAL APPROACHES

The view that the genetic base of the *tuberosum* breeding-pool is rather narrow has led to a number of breeding programmes designed to broaden it. The *andigena* germplasm of South American cultivated tetraploids represents a very diverse range of potentially useful variation, and has already been the source of such useful characteristics as resistance to PCN and PVX immunity. It is the first logical choice of the breeder when he is obliged to move outside the *tuberosum* group (Howard, 1970). However, *andigena* cultivars are adapted to the short-day conditions of Central and South America. When used to introgress specific characters, it is necessary to backcross for several generations to *tuberosum* in order to regain the maturity, yield, quality and disease resistance characteristics demanded of a modern cultivar. By subjecting a population of *andigena* to cyclical recurrent selection under long-day conditions it has proved possible to emulate in a comparatively short time the process that is presumed to have occurred since their original introduction in the sixteenth century (Simmonds, 1969; Glendinning, 1979). These neo-*tuberosum* populations now contain clones which are capable of producing yields comparable to *tuberosum* cultivars in experimental trials in the UK (Glendinning, pers. comm.). A recent evaluation of the SCRI neo-*tuberosum* population has identified within it some potentially useful levels of resistance to several major diseases (Glendinning, pers. comm.). Several neo-*tuberosum* × *tuberosum* hybrids, the results of speculative crosses made in the early 1970s, reached an advanced stage of selection in the commercial breeding programme at SCRI, and one has been National Listed as the named cultivar Shelagh (Anon., 1986b, 1987). The neo-*tuberosum* experiment has therefore reached a point where neo-*tuberosum* clones can begin to feature in the breeding of commercial cultivars.

An alternative approach to the neo-*tuberosum* experiment has been a similar attempt to produce adapted populations based on the cultivated diploid species, *Solanum phureja* and *S. stenotomum*. While this is not as advanced as the former, progress has been made and promising results from experimental trials are now forthcoming (Carroll, 1982). Ochoa and Schmiediche (1983) have also reported that recurrent selection for tuber yield and conformity in a population based on *S. sparsifolium*, *S. phureja* and *S. chacoense* had elicited a marked response in yield and tuber conformity. Such populations may also provide the commercial breeder with a novel source of breeding material, but as yet no cultivars have emerged and this promising work must still be considered as in the development phase (Peloquin, 1983).

Dihaploids induced by interspecific pollination of *tuberosum* tetraploids have long been considered a means of circumventing the complexities of dealing with a tetraploid (Chase, 1963). Such techniques allied to novel methods of *in vitro* culture of microspores (pollen grains) and protoplast fusion offer exciting prospects for the production of unique genotypes for breeding and/or research purposes (Wenzel *et al.*, 1982). The exploitation of somaclonal variation is currently being investigated as a technique capable of "engineering" improved variants from existing successful cultivars, without the problems associated with the massive release of unwanted hidden variation when the normal sexual reproductive process is employed.

The use of irradiated pollen also offers the possibility of transferring specific genes or limited genetic information from the paternal genome in intra- and inter-specific crosses (Jinks *et al.*, 1981). While such a technique may be more immediately attractive in an inbreeding crop such as barley or wheat, in avoiding the time-consuming process of a conventional backcrossing programme, the use of irradiated pollen could offer the potato breeder a more rapid means of transferring specific characteristics from wild species and primitive cultivars into an adapted *tuberosum* background, than hitherto. The mechanism whereby matromorphic seedlings arise following pollination with irradiated pollen is not yet known, but clones from seedlings produced by pollination of selected cultivars, with irradiated pollen of others, have been produced and may provide this information (Caligari, Perryman and Swan, pers. comm.).

All these techniques and others offer the breeder the means of inducing or increasing the range of variation upon which he can practise his art. However, it is essential that the creation of such variation is paralleled by the development of techniques to permit the positive identification of desirable segregants amongst heterogeneous populations. The probability of breeding a perfect potato is very low (Simmonds, 1969); it will be impossible if the breeder cannot recognise it when he has it.

The development of screening tests capable of testing representative progeny samples of many combinations of parents is therefore necessary in order to identify superior genotypes. Otherwise, breeding will continue to rely heavily on phenotypic selection and it will be difficult to supersede old-established cultivars which are now protected from "degeneration" by heat treatment and meristem culture, and seed certification schemes (Mackay and Wastie, 1981).

## 10.5 FUTURE STRATEGIES

Research into breeding has led to a greater understanding of the underlying cytogenetics, physiology and pathology and is contributing indirectly to the production of new cultivars, but has had little direct impact on the actual methods of breeding cultivars. Less conventional approaches have still to demonstrate that they will be more successful to this end than conventional breeding techniques.

The potato breeder is fortunate that a number of economically important diseases and pests can be combated by simply inherited major genes, such as $H_1$ and others conferring PVX immunity and PVY comprehensive resistance (Cockerham, 1970) as well as the $H_2$ gene ex *S. multidissectum* conferring resistance to at least one pathotype of *G. pallida* (Dunnett, 1961). However, most cultivars and breeders' clones possessing these characters are genetically simplex. Consequently, when they are hybridised with a susceptible clone there is only a 50% chance that any individual clone of the ensuing progeny will inherit that character. In a traditional scheme where more than 90% of the variation can be eliminated in the first two or three cycles of selection, there is a high probability that such useful genes as $H_1$ can be lost. The cv. Drayton from the Plant Breeding Institute, Cambridge, is a hybrid between Maris Piper and Red King Edward, but it has failed to inherit the $H_1$ gene. Baillie, a second early from SCRI, is also a Maris Piper seedling which has not inherited the $H_1$ gene (Anon., 1986a). Screening techniques capable of positively identifying desirable phenotypes at the seedling stage are, in this respect, of great value. In a programme at SCRI designed specifically to breed for resistance to the common viruses, some 15 000 seedlings are screened each year for their reaction to PVX and/or PVY. Only resistant clones are retained for further selection. It would not be possible with current resources to screen all seedlings raised each year for PVX and PVY resistance, and since PVX and PVY resistance are not prerequisites for a successful cultivar it would be difficult to justify such a course of action. However, by including fully susceptible clones or cultivars as "testers" in the resistance breeding crossing programmes and by monitoring the segregation ratios of the ensuing progenies, a number of clones duplex for their PVY resistance loci and others for PVX and $H_1$ have now been identified (Anon., 1981). Using such clones as parents, the frequency of resistant segregants amongst the resultant progenies is increased substantially. By carefully planned crossing schedules, the intercrossing, and/or where possible selfing, of these duplex clones has also permitted the isolation of putative triplex or quadruplex segregants in a comparatively short time (Anon., 1986c). Such clones, when used as parents, would guarantee that all their progenies inherit the desired characteristic, the need for screening would be eliminated and the breeder and his pathologist colleagues could then concentrate their combined efforts on other more complex problems.

Crossing programmes designed to produce parental material such as is described need not compromise a varietal breeding programme, and what is true of major genes is also true of polygenes.

Breeders well recognise the ability of particular clones to "nick" and produce higher frequencies of superior segregants in their progenies than would be predicted by reference to their phenotypes. In order to capitalise on this phenomenon the breeder needs progeny tests; data on representative samples of progenies at early stages of breeding programmes are essential. Genotypic selection must eventually supersede phenotypic selection and future potato breeding strategies should be devoted to this end by whatever means the actual variation is produced.

## 10.6 TRUE POTATO SEED

At a conference held at Lima, Peru (1982) a great deal of discussion revolved around the potential of true botanic potato seed (TPS) as an alternative to seed tubers. Dr Li Jing Hua reported that true seed crops of potatoes have been grown in China for 15 years in order to produce healthy seed tubers for ware production, where clonal reproduction of healthy stocks was climatologically, technologically or economically not possible (Li, 1983). In his opinion the growing of ware potatoes from directly sown true seed was unlikely to succeed, but there was a substantial body of opinion that supported the possibility of growing ware crops from transplants of TPS seedlings, particularly in the Third World. Research in Japan has demonstrated that induced apomixis is experimentally possible (Iwanaga, 1983) and might be a means of overcoming the lack of uniformity amongst true seed cultivars from heterozygous parents. Investigations into this area are rapidly expanding, with potentially profound implications for the potato crop (Anon., 1983).

Acknowledgements

I wish to thank Miss Elizabeth Elliott and Mrs Audrey Sinclair for typing my manuscript, my colleague Dr Peter Caligari for his useful comments on my draft manuscript, and my other colleagues who have permitted me to quote from their, as yet, unpublished work.

## REFERENCES

Anon. (1981). *Rep. Scott. Pl. Breed. Stat. for 1980–81*.
Anon. (1983). Bibliography on true potato seed, potato flowering, pollen and pollination. Bibliog. no. 17, CIP, Lima, Peru, 17 pp.
Anon. (1986a). "Classified List of Potato Varieties, England and Wales, 1987" NIAB.
Anon. (1986b). *Plant Varieties and Seeds Gazette* No. 260, September, 11 pp.

Anon. (1986c). *Rep. Scott. Crop Res. Inst. for 1985.*
Anon. (1987). *Rep. Scott. Crop Res. Inst. for 1986.*
Anderson, J. A. D., and Howard, H. W. (1981). Effectiveness of selection in the early stages of potato breeding programmes. *Potato Res.* **24**, 289–299.
Beukema, H. P., and van der Zaag, D. E. (1979). "Potato Improvement: Some Factors and Facts." I.A.C., Wageningen, The Netherlands.
Brown, J. (1984). A new data base computer package for plant breeders. *Euphytica* **33**, 935–942.
Brown, J., Caligari, P. D. S., and Mackay, G. R. (1987a). The repeatability of progeny means in the early generations of a potato breeding programme. *Ann. App. Biol.* **110** 365–370.
Brown, J., Caligari, P. D. S., Mackay, G. R., and Swan, G. E. L. (1987b). The efficiency of seedling selection by visual preference in a potato breeding programme. *Ann. appl. Biol.* **110**, 357–363.
Caligari, P. D. S., and Brown, J. (1986). The use of univariate cross prediction methods in the breeding of a clonally reproduced crop (*Solanum tuberosum*). *Heredity* **57**, 395–401.
Caligari, P. D. S., Mackay, G. R., Stewart, H. E., and Wastie, R. L. (1984). A seedling progeny test for resistance to potato foliage blight (*Phytophthora infestans* (Mont.) de Bary). *Potato Res.* **27**, 43–50.
Carroll, C. P. (1982). A mass-selection method for the acclimatization and improvement of edible diploid potatoes in the United Kingdom. *J. agric. Sci., Camb.* **99**, 631–640.
Chase, S.E. (1963). Analytical breeding of *Solanum tuberosum*. *Can. J. Genet. Cytol.* **5**, 359–363.
Cockerham, G. (1970). Genetic studies on resistance to Potato Viruses X and Y. *Heredity* **25**, 309–348.
Davidson, T. M. W. (1980). Breeding resistance to virus diseases of the potato (*Solanum tuberosum* L.) at the Scottish Plant Breeding Station. *Rep. Scott. Pl. Breed. Sta. for 1979–80*, pp. 100–108.
Dodds, K. S. (1965). The history and relationships of cultivated potatoes. In "Crop Plant Evolution" (Sir J. Hutchinson, ed.) pp. 119–141. Cambridge University Press.
Dunnett, J. M. (1960). The role of *Solanum vernei* Bitt. et Wittm. in breeding for resistance to potato root eelworm (*Heterodera rostochiensis* Woll.). *Rep. Scott. Pl. Breed. Sta. for 1960*, pp. 39–44.
Dunnett, J. M. (1961). Inheritance of resistance to potato root eelworm in a breeding line stemming from *Solanum multidissectum* Hawkes. *Rep. Scott. Pl. Breed. Sta. for 1961*, pp. 39–46.
Ellenby, C. (1952). Resistance to the potato root eelworm. *Nature* **170**, 1016.
Glendinning, D. R. (1979). Enriching the potato gene-pool using primitive cultivars. In "Proceedings of Eucarpia Conference, Wild Species and Primitive Forms Section, Wageningen, 1978", pp. 27–33. Pudoc, Wageningen.
Holden, J. H. W. (1977). Potato breeding at Pentlandfield. *Rep. Scott. Pl. Breed. Sta. for 1976–77*, pp. 66–97.
Howard, H. W. (1963). Some potato breeding problems. *Rep. Pl. Breed. Inst. Camb. for 1961–62*, pp. 5–21.
Howard, H. W. (1970). "Genetics of the Potato". Logos Press, London.
Howard, H. W., Cole, C. S., and Fuller, J. M. (1970). Further sources of resistance to *Heterodera rostochiensis* Woll. in the Andigena potato. *Euphytica* **19**, 210–219.

Howard, H. W., Cole, C. S., Fuller, J. M., Jellis, G. J., and Thomson, A. J. (1978). Potato breeding problems with special reference to selecting progeny of the cross Pentland Crown × Maris Piper. *Rep. Pl. Breed. Inst. Camb. for 1977*, pp. 22–50.
Jinks, J. L., Caligari, P. D. S., and Ingram, N. R. (1981). Gene transfer in *Nicotiana rustica* using irradiated pollen. *Nature* **291**, 586.
Li Jing Hua (1983). Prospects for the use of true seed to grow potato. *In* "Proceedings of Conference: Research for the Potato of the Year 2000", pp. 17–18. CIP, Lima, Peru.
Iwanaga, M. (1983). Chemical induction of aposporous apomictic seed production. *In* "Proceedings of Conference; Research for the Potato of the year 2000", pp. 104–105. CIP, Lima, Peru.
Mackay, G. R. (1982). Breeding for resistance to pests and diseases. *In* "Producing Quality Seed Potatoes in Scotland, Bulletin No. 1", *Proc. Scott. Soc. Crop Res.*, pp. 27–36.
Mackay, G. R., and Wastie, R. L. (1981). Problems and prospects in progeny testing for disease and pest resistance in a commercial potato breeding programme. *In* "Abstracts of Conference Papers, 8th Triennial Conference of EAPR 1981", pp. 30–31. EAPR, Wageningen.
Mackay, G. R., Caligari, P. D. S., Brown, J., Dale, M. F. B., Torrance, C. J. W., Swan, G. E. L., and Spence, J. (1984). Potato Breeding Department. *Rep. Scott. Crop Res. Inst. for 1983*, pp. 62–69.
Martin, M. W. (1984). Early generation selection methods for resistance and horticultural factors. *Am. Pot. J.* **61**, 383.
Ochoa, C., and Schmiediche, P. (1983). The systematic exploitation and utilisation of potato germ plasm. *In* "Proceedings of Conference, Research for the Potato in the Year 2000", pp. 142–144. CIP, Lima, Peru.
Peloquin, S. J. (1983). New approaches to breeding for the potato for the year 2000. *In* "Research for the Potato in the Year 2000", pp. 82–84. CIP, Lima, Peru.
Phillips, M. S., and Dale, M. F. B. (1982). Assessing potato seedling progenies for resistance to the white potato cyst nematode. *J. agric. Sci., Camb.* **99**, 67–70.
Plaisted, R. L., Thurston, H. D., Brodie, B. B., and Hoopes, R. W. (1984). Selecting for resistance to diseases in early generations. *Am. Pot. J.* **61**, 395–403.
Robb, W. (1921). Breeding selection and development work in Britain. *In* "Report of the International Potato Conference, London 1921" (W. R. Dykes, ed.), pp. 27–34. Royal Horticultural Society, London.
Ross, H. (1986). Potato breeding—problems and perspectives. *In* "Advances in Plant Breeding" (W. Horn and G. Röbbelen, eds), Suppl. 13 to *J. Pl. Breed.*, Verlag Paul Pavey, Berlin.
Salaman, R. N. (1926). "Potato Varieties". Cambridge University Press.
Sanford, L. L., Ladd, T. L., Sinden, S. L., and Cantelo, W. W. (1984). Early generation selection of insect resistance in potato. *Am. Pot. J.* **61**, 405–418.
Simmonds, N. W. (1969). Prospects of potato improvement. *Rep. Scott. Pl. Breed. Sta. for 1968–69*, pp. 18–38.
Simmonds, N. W. (1974). Potatoes, *Solanum tuberosum* (*Solanaceae*). *In* "Evolution of Crop Plants" (N. W. Simmonds ed.), pp. 279–283. Longman, London.
Swiezynski, K. M. (1984). Early generation selection methods used in Polish potato breeding. *Am. Pot. J.* **61**, 385–394.

Tai, G. C. C., and Young, D. A. (1984). Early generation selection for important agronomic characteristics in a potato breeding population. *Am. Pot. J.* **61**, 419–434.

Wenzel, G., Bapart, V. A., and Uhrig, H. (1982). New strategies to tackle breeding problems on potato. *In* "Plant Cell Culture in Crop Improvement" (K. L. Giles and S. K. Sen, eds), pp. 337–349. Plenum Press, New York.

# Part III Improving Forest Trees

# 11 Genetic Resources and Variation in Forest Trees

W. J. LIBBY

*Department of Forestry, University of California, Berkeley 94720, USA*

During the past few decades, several major approaches to the genetic manipulation of our forest gene-pools have been developed. At one end of the scale, there is an attempt to maintain the *status quo*, either through certain kinds of natural regeneration or by artificial regeneration from seeds of trees selected in areas similar to the plantation site.

As an interim measure, above-average stands have been selected for management as seed-production areas after some roguing of undesirable trees. The most common strategy has been to concentrate highly selected genotypes or families into seed orchards for open cross-pollination and economic seed production. A few programmes have used interracial or even interspecies hybrids, as $F_1$s or as the start of land-race formation.

At the other end of this range of approaches is the clonal option, possible for a few favourable species. As new techniques become available, more programmes are beginning to explore the use of clonal material. However, the testing and widespread use of new or exotic genetic variation is conservative (Libby, 1986), and properly so. Our present knowledge is very limited in comparison with the much more considerable information available on, for example, crop plants.

Consider the total amount of effort and knowledge that has gone into the planting and growing of potatoes, compared with the amount that has been applied to the planting and growing of redwood. Now consider a similar comparison of the poetry written about potatoes and redwood. The forest-tree breeder envies the amount of knowledge available to the crop breeder, and ignores the public attention directed to the aesthetic qualities of our forests at some peril. These are two of several differences between forest trees and most other crops. They often affect overall breeding strategy (Roulund, 1981), and they may also be interesting in their own right.

IMPROVING VEGETATIVELY PROPAGATED CROPS  
ISBN 0–12–041410–4

*Copyright © 1987 Academic Press Limited*  
*All rights of reproduction in any form reserved*

Among economic plants, forest trees were very late in being domesticated. Even where civilisation has existed for millennia, except for the selections of a few interesting cultivars such as narrow-crowned poplar or cypress, the domestication of forest trees has tended to be passive. The gene-pools of our forests have responded to repeated entry by humans, and the response has often been dysgenic. The parents of the next generation were too often the trees least useful to humans, which, being left by the woodcutters and loggers, were available to seed nearby logged sites. As the forestry enterprise shifted from a hunting-and-gathering mode, which depended on natural regeneration, to its early neolithic phase, many plantations were made using seeds gathered from overly-sexy, easy-to-reach trees. Alternatively, the seeds were collected in other populations, which were frequently not as well adapted to the plantation site as seeds from the populations they replaced. In both cases, dysgenic elements were thus increased during this early period of domestication.

As we have come to appreciate these problems, we have devised at least five strategies to stop the dysgenic trend. We have begun also to manipulate our forest gene-pools so that our forests can be managed more efficiently to produce products in greater quantity and/or of greater value. However, before this could happen, it was necessary for our decision makers to change their perception of the forest as a non-renewable resource to be mined, to that of a renewable resource to be managed.

## 11.1 ALTERNATIVE STRATEGIES FOR FOREST TREES

### 11.1.1 The *status quo*

The forest is large and complex; forest geneticists cannot begin to serve all areas at once. Until we can get to them, many productive forests or their species components are being placed under management prescriptions designed merely to halt the dysgenic elements of human activities. Furthermore, due to poor soils or climates, the potential productivity of many other forests is too low to ever expect their economic management as a renewable resource. For these forests, the *status quo* level of management will be permanent. The *status quo* strategy may be applied to plantation forestry, or to natural regeneration systems. Management systems employing small clearings, with natural seeding from adjacent edges of standing forest, probably do it best. For plantations, attention is given to the source of the seed.

Seeds are collected near the plantation site or from trees at several more distant sites that are similar to the plantation site. The seed-donor trees are widely spaced to avoid founder effects or later inbreeding.

## 11.1.2 Seed production areas

An intermediate strategy has been to select natural stands or plantations that appear to be of above-average genetic quality, and that are on sites that physically allow their management for seed production. These, and their nearby surrounding forest, usually have their genetic quality further upgraded by removing trees with undesirable qualities. (There are a few indications that, if the natural stand has family structure, such roguing may increase the degree of relatedness of the remaining trees and thus increase inbreeding among their offspring. This is not a problem in plantations used as seed-production areas.) Fertilising and other techniques then increase the quantity and frequency of seed production above that of the rest of the forest, to make seed harvest more efficient and more reliable. Good-quality seed of known origin is thus better assured.

## 11.1.3 Hybrids

The reputation of hybrid corn stimulated some programmes to produce interracial and even interspecific hybrids of forest trees. This has led to a good understanding of species relationships, but little of practical value has come of this approach. Expensive control-pollination is necessary, and seed yields of the interspecific crosses are usually only a small fraction of those obtained with similar effort within species. In a few cases, usually where one or both parents are not native to the area in which the hybrid is to be used, hybrids are outperforming both parent populations. In these cases, it is likely that breeding through $F_2$ or backcross generations will develop new land races for use in such areas.

## 11.1.4 Classical tree improvement: seed orchards

From the 1930s onwards, programmes in Scandinavia (closely followed by programmes in Australia, New Zealand and the southeastern United States) pioneered a more aggressive approach to tree improvement (see Anon., 1957–1986), soon copied and put into practice by most of the world's forest enterprises.

First, breeding areas, or client areas, were delineated. Each was supposed to include physiographically similar sites. It was calculated, or hoped or assumed, that trees that had evolved on these sites would be adapted to survive and grow well on all client sites included in a designated breeding or client area. A few provenance (common-garden) tests were available for a few species. These indicated that substantial genetic differences exist among populations of the same tree species, and that these genetic differences were in large part associated with gradients or differences of elevation, latitude, soil, rainfall, and other elements of the forest environment. Furthermore, variations in pattern were found among different species. A crucial question was, how large should a given breeding or client area be? Biological evidence argued for small areas. But seed orchards proved to be expensive and subject to economies of scale. Thus, management considerations argued for large seed orchards with large client areas. Guesses and compromises were made.

The second step was to select trees from within the breeding area. Selection was frequently intense. Using a phenotypic index, selection ranged from the best tree in about 2 ha to the best of about one million trees observed. Usually, only one tree per neighbourhood was selected (a decision that has generally proven wise and useful, as it virtually eliminated inbred offspring from plantations of seed-orchard origin). The selected genotypes were then concentrated and multiplied in seed orchards. The common technique was grafting, but rooted cuttings were used when graft incompatibility was a problem or in a few cases where rooting of mature cuttings was easier than grafting. These mature propagules generally retained enough sexual activity to permit adequate breeding at the seed-orchard site soon after its establishment.

Such breeding was for progeny testing and to establish second-generation families for further breeding (note that these are two different purposes and require different mating designs). Less commonly, the genes of the select trees were concentrated and multiplied in seed orchards by using their offspring. Open-pollinated seeds were generally quickly obtainable and cheap, but their use diluted the effect of selection. Control-pollinated seeds, using select trees as males, are more expensive and result in some delay. Although not as sexually precocious as the grafted or rooted mature parent clones, these juvenile seedling families frequently produce more abundant seed than their mature cloned parents. Commercial seed is generally obtained by unrestricted open-pollination of the orchard trees.

As we gained experience with seed orchards, some problems became apparent. Severe graft incompatibility is common, and in some species occurs after 10 or more years. Pollen and seed production is not uniform among select clones, and there is some evidence that the best wood-growing families are under-represented. Seed production is sensitive to soil and climate, and many of our seed orchards are incorrectly sited. Pollen contamination from adjacent seed orchards, or from the surrounding forest, is a significant problem. Seed-destroying pests build up to damaging population levels. Cone collection and other time-dependent seed-orchard activities often conflict with other management demands on available personnel. The client areas are too large, and different families within the seed orchard are adapted to different sub-areas of the client area.

Finally, the amount of within-population genetic variability, including a substantial non-additive component, is proving to be large, and seed orchards capture only a fraction of it.

### 11.1.5 The clonal option

This strategy solves all of the problems noted with the other four strategies above, and it provides substantial additional advantages not available with the other four.

For taxa in which cloning was easily done by rooting cuttings, clonal forestry has been practiced for decades and even centuries (Roulund, 1981). These taxa include cottonwoods (*Populus deltoides*, *P. nigra*, and their hybrid), willows (most species of *Salix*) and tsugi (*Cryptomeria japonica*). We can profit from the accumulated experience of clonal forestry with such species, particularly by recognising some of the early mistakes. One of the most serious errors appears to be the use (and certification for use) of too few clones within each region.

Since the 1960s, two events have occurred to make, in my opinion, the clonal option the strategy of choice for most forest species of major importance as a renewable resource. Managers have begun to recognise the shortcomings of the other strategies and the advantages of the clonal option. Equally important, the effect of maturation state had been realised and solutions to this biological problem are becoming available (see Chapters 12 and 14, this volume).

## 11.2  SOME PARTICULAR CHARACTERISTICS OF FOREST TREES AND TREE BREEDING

### 11.2.1 Test extent and duration

Forest trees grown as a renewable resource must survive with acceptable levels of damage and grow well in a varied, uncertain and minimally managed environment. They must survive and grow in this manner for two or more years as short-rotation biomass plantations, for several decades to over a century if grown for wood and fibre, and even for millennia if used in park and amenity plantings. Foresters and tree breeders thus, and properly so, tend to be conservative. As a general principle, the more alike newly selected families or clones are to previously tested trees, or to the native populations they are replacing, the more quickly they can be used for large-scale reafforestation. Conversely, if the new trees are radically different, they should be tested in many conditions and for many decades before their widespread use is appropriate.

### 11.2.2 Levels of genetic variability

Forest trees are long-lived, possibly because they are highly heterozygous. Alternatively perhaps, their long-lived sessile lifestyle requires them to be highly heterozygous, and their high fecundity and wind-pollinated outcrossing mating systems allow them to maintain a great deal of genetic variability within populations. Whichever is the case, with a few interesting exceptions, forest trees seem to be among the most genetically variable of all organisms thus far studied (Hamrick *et al.*, 1979).

The patterns of genetic architecture in forest trees are also interesting. For example, among North American pines, *Pinus resinosa* and *P. torreyana* have little genetic variability. *Pinus radiata* has much variability between distant, disjunct, similarly sited populations, but little between stands on different soils, aspects, etc. within populations. *Pinus taeda* has much variation between local stands, but populations hundreds of kilometres apart are similar enough to be in the same breeding areas. Southern populations of *P. muricata* are highly variable, while northern populations are remarkably uniform. *Pinus monticola* has important regional variation, little genetic variation between populations in a region or between stands in a population, but much genetic variation within stands. *Pinus lambertiana* and *P. ponderosa* appear to have substantial genetic variability at all levels. With few exceptions, most of these species manage to maintain a great deal of genetic variation within stands.

To complicate things, when more than one kind of information is available, we seem to get rather different patterns for the same species. Morphological characteristics tend to load more heavily at the regional and populations-within-region levels, and less at the stands and within-stands levels. Components of growth, survival, and susceptibility to various physical and biotic events are important in all four categories. Isozymes tend to concentrate more than 90% of their variability in the within-stands component. One might speculate that when we learn to study their allelic distributions, regulatory elements will have more geographical or ecological differentiation than do the structural genes being sampled in the isozyme analyses.

The available observations on growth, survival and susceptibility make it clear that most of our forest tree species maintain a great deal of meaningful genetic variation within local populations. Furthermore, it is common to find higher levels of isozyme heterozygosity among mature trees than among their embryonic offspring, or levels higher than those predicted by Hardy–Weinberg algebra using observed allele frequencies.

One interpretation is that the high fecundity of forest trees makes it possible for them to do a great deal of experimentation through genetic recombination. Their long generation times, and their changing and uncertain environments, make such costly experimentation profitable.

If this is true, one expects many failures among the experiments. With natural regeneration, where (typically) thousands of seedlings per hectare are reduced to a few hundred harvestable trees during stand development, the failed experiments have little impact. But given plantation forestry, many genetic recombinants that would (and perhaps should) die in nature are pampered through the nursery and then planted in a cleared area of a few square metres, and are expected to occupy it.

A contrast between extensive forest management and intensive agriculture is that only a small fraction of the seeded or planted forest trees are expected to be harvested, while most planted crop plants are expected to contribute to the harvest. As forest management intensifies, the fraction of planted trees expected to be harvested will probably increase.

Clonal tests are very good for separating the successful experiments in a family of forest trees from the flawed ones.

### 11.2.3 Clonal-option *versus* seed-orchard selection strategies

If sufficient numbers of initial selections have been made, the genetic constitution of a seed orchard can be modified by roguing out parents whose progeny are performing poorly. This is typically a low-intensity selection with reasonably high accuracy. Thus, the typical seed orchard starts with a relatively small number (less than 500) of phenotypically selected parents and is then reduced to a smaller number of the best-performing parents. They are typically heterozygous and their offspring are thus highly variable, with no two alike.

The clonal option usually starts with highly promising families, frequently including sib-tested offspring from a seed orchard or a genetic experiment (Skrøppa, 1981). Selection among the members of these families is not a one-time event, but is incremental (Roulund, 1981). Small tests are established (typically 2–4 ramets per site, on about three sites) using 10 000–100 000 candidate clones. Evaluations are made at 3–5-year intervals. At each evaluation the poorer clones in the test are disqualified and no longer contribute to production plantations. It is anticipated that 10–20% of the clones in use will be disqualified at each evaluation round. After several such rounds, a new set of tests is established on additional sites, using only the better-performing clones.

Note that, in this scenario, the average performance of the unselected set of clones is about equal to the average performance of seedlings from a seed orchard. If the candidate clones are from pedigreed control-pollinated families, their average performance will probably be better because selfs and other inbreds will not be included, and the parents committing relatively less photosynthate to fruit and pollen will not be at a disadvantage. At each round of selection, the average clonal performance is further improved.

When the seed orchard is rogued, entire families are removed. By contrast, as candidate clones are evaluated, the poorer members of even the best families will be disqualified, while it is likely that the best members of many of the poorer families will continue to be used. Furthermore, all surviving clones will remain in the tests and foundation orchard, so that a clone disqualified on the basis of its early performance may be reinstated if its later performance warrants it.

In summary, under a clonal strategy, it will take many evaluation rounds and many years before the number of clones in use is reduced to fewer than 1000, and by that time considerable site-specific information will be available for each of these clones. This strategy is much more consistent with forestry's conservative philosophy than the early and intense selection typical of seed-orchard strategy.

### 11.2.4 Developmental genetic variation

Although many of the advantages of clonal forestry were recognised between the 1930s and 1960s, early attempts to exploit it were generally disappointing or failed entirely. The importance of the maturation state was not then understood, and most cloning attempts were made with propagules from mature trees, whose economic characters could be evaluated. Rooting of mature cuttings is generally difficult. Those mature cuttings that did root generally produced trees with characters very different from, and mostly less desirable than, those of the cutting-donor (Roulund, 1981).

Techniques of hedging, serial propagation, tissue culture and micropropagation are beginning to allow us to arrest and even manipulate the maturation state (Zimmerman, 1976; Franclet, 1979). Thus, we can hold a clone in, or return a clone to, its juvenile state. Cuttings can then be effectively rooted and their development will parallel that of the ortet and of tested juvenile ramets.

Furthermore, it is becoming possible to use a maturation state other than that of a seedling for plantation propagules. For example, when juvenile *P. radiata* is hedged at a height of about 1 m, maturation appears to be arrested at a late-juvenile or early-adolescent state. Compared with fully juvenile ramets of the same clones, ramets from 1 m hedges root more slowly, have better form, smaller and fewer branches and are more resistant to western gall rust. Where western gall rust is a problem, the option of using these late-juvenile cuttings is available.

### 11.2.5 Two theoretical clonal strategies

There are two strategies, largely theoretical, that are attractive for clonal forestry.

One strategy is based on the reasonable hypothesis that neighbouring trees making complementary demands on their common environment will be more productive than those neighbours that make competing demands. The available literature (Harper, 1977; Adams, 1980) indicates that clonal or varietal mixtures often do better than the average of the component genotypes grown in pure stands, and that mixtures sometimes do better than the best of them. It will take large and expensive experiments to find complementary sets of clones that are consistently more productive than either mixtures of random sets or monoclonal plantings of the best of them. Once these complementary sets have been found, planting sequences can be prescribed. It is likely that these will lead to mixtures of relatively few clones (seven being a likely number if hexagonal spacing is used).

The second strategy suggests that cross-adaptation by short-generation insects and diseases may be more difficult in a mixture of relatively few (7–30) unrelated clones than in a continuum of seedling genotypes originating in a seed orchard (Libby, 1982). If this is true, a clonal forest employing this strategy would be safer than a forest of seedling origin.

Both of these strategies require substantial knowledge of the clones employed. The first requires detailed knowledge of specific clonal complementarities. The second requires only pedigree information. But since this strategy requires few clones in order to work, the conservative selection strategy appropriate to forestry requires a level of confidence in the performance of the selected clones attained only by long and extensive testing. Thus, neither is likely to find widespread use early in clonal forestry programmes.

### 11.2.6 The unit of management

Perhaps the most compelling advantage of the clonal option is that the individual clone is the unit of management.

If a mistake was made in designating the seed orchard's breeding or client area, such that some families in the seed orchard do well on one set of client areas but poorly on another, while other families do well on the second set but not on the first, the seed-orchard manager has some hard decisions to make. If seeds from the whole orchard are used for both sets of client environments, only a fraction of them will be adapted to each area; many will be poorly adapted. If the first group of parents is removed from the seed orchard, the second set of client areas is then well served, but the first set no longer has a source of improved seed.

In the above situation, a clonal-propagation manager has no serious difficulty. Clones can be assigned to sites on which they do well and withheld from sites on which they do poorly. Differences in adaptation even among clones of the same full-sib family can be usefully exploited. This advantage is particularly important for species with a fine-grained genetic architecture, i.e. species sensitive to small changes in elevation, aspect, soils, etc. (e.g. see Campbell and Sorensen, 1978).

### 11.2.7 Genetic conservation

Any domestication programme that modifies or replaces substantial areas of the native population should have a genetic conservation component. This is both ethical and practical. Various techniques for conserving alleles, co-adapted gene complexes, genotypes, species genetic architecture, and ecosystems are under active discussion and study (Anon., 1982). All but the ecosystem level would be effectively served by clonal samples of the original population, and of various land races and cultivars derived from it, maintained and periodically regenerated in clonal archives.

Because of the low intensity of management, and because many of our forests are still in a close cause-and-effect evolutionary relationship to their environment, genetic conservation efforts are particularly appropriate and important as we accelerate our early attempts to consciously domesticate our important forest species.

Acknowledgements

Many of the concepts and examples in this chapter are drawn from conversations and correspondence with colleagues and have not formally appeared in the literature. These are acknowledged with thanks.

**REFERENCES**

The references cited do not cover fully the pertinent literature. They are given as reasonable entry points to that larger literature.

Adams, W. T. (1980). Intergenotypic competition in forest trees. *Proc. 6th N. Am. For. Biol. Workshop*, pp. 1–14. University of Alberta, Canada.

Anon. (1957–1986). "Annual Reports". North Carolina State Industry Cooperative Forest Tree Improvement Program. School of Forest Resources, N.C. State University, Raleigh, USA.

Anon. (1982). "Douglas Fir Genetic Resources: an Assessment and Plan for California", National Council on Gene Resources, California, 275 pp.

Campbell, R. R., and Sorensen, F. C. (1978). Effect of test environment on expression of clines and on delineation of seed zones in Douglas fir. *Theoret. appl. Genet.* **51**, 233–246.

Franclet, A. (1979). Rajeunissement des arbres adultes en vue de leur propagation végétative. *In* "Micropropagation D'Arbres Forestiers", pp. 3–18. AFOCEL Etudes et Recherches, No. 12.

Hamrick, J.L., Linhart, Y. B., and Mitton, J. B. (1979). Relationships between life history characteristics and electrophoretically detectable genetic variation in plants. *Ann. Rev. Écol. Syst.* **10**, 173–200.

Harper, J. L. (1977). "Population Biology of Plants", Academic Press, New York, 892 pp.

Libby, W. J. (1982). What is a safe number of clones per plantation? *In* "Resistance to Disease Pests in Forest Trees" (H. M. Heybroek, B. R. Stephan and K. von Weissenberg, eds), Topic 7, pp. 342–360. PUDOC, Wageningen.

Libby, W. J. (1986). Testing and deployment of genetically engineered trees. *In* "Cell and Tissue Culture in Forestry", (J. M. Bonga and D. J. Durzan, eds), 2nd edn, vol. 1, pp. 167–197. Martinus Nijhoff/Dr W. Junk, Dordrecht.

Roulund, H. (1981). Problems of clonal forestry in spruce and their influences on breeding strategy. *For. Abstr.* **42**, 457–471.

Skrøppa, T. (1981). Some results from a 20-year-old cutting experiment with Norway spruce. *In* "Symposium on Clonal Forestry", pp. 105–115. Department of Forest Genetics, Uppsala, Sweden.

Zimmerman, R. H., ed. (1976). "Symposium on Juvenility in Woody Perennials". *Acta Hort.* **56**, 317 pp.

# 12 Vegetative Propagation for Improved Tropical Forest Trees

R. D. BARNES and J. BURLEY

*Oxford Forestry Institute, University of Oxford, Oxford*

Since 1950, tree breeding has passed through various phases of development. First, tree breeders progressively accepted and developed genetic theory while the availability of computers and the widespread teaching of statistics led to the use of more complex environmental and mating designs which, in turn, yielded increasingly precise information on genetic variances and covariances. Secondly, the objectives of breeding changed from simple production traits, such as stem volume, to multiple characteristics of vigour, form, crown and stem quality, wood and chemical properties, and pest resistance, particularly as more rapid and precise analytical methods became available. Thirdly, the location of breeding extended from the temperate regions to the tropics.

While these phases continue, a fourth major innovation is required, namely the initiation of breeding of multipurpose species, particularly for use in the rural development of tropical countries with large human and cattle populations, little forest land and many difficult sites. A recent study (Wood *et al.*, 1982) indicated that there are some two billion hectares of degraded tropical lands, much of which could be afforested (Burley, 1980).

Tropical forestry typically consists of monocultural plantations managed for sawtimber, veneer logs and industrial wood. However, trees for rural development may be required for the production of poles, fuelwood, charcoal, fruit, fodder and exudates, for site amelioration through nitrogen fixation, soil protection and reclamation and for special management characteristics such as agrisilvopastural mixtures, coppicing ability and seed production.

It is almost axiomatic that successful plantation trees in the tropics are colonising species that are gregarious and light-demanding in their natural

IMPROVING VEGETATIVELY PROPAGATED CROPS
ISBN 0–12–041410–4

*Copyright © 1987 Academic Press Limited*
*All rights of reproduction in any form reserved*

state in which they reproduce precociously and profusely by seed and not by vegetative means. Until recently, virtually none had been propagated vegetatively for plantations, even on an experimental scale. From the mid 1950s there has been a need to clone selected trees for preservation and seed production in breeding programmes. Grafting, budding, layering and cutting methods have now been successfully developed for these purposes and their potential for producing plantation propagules have become increasingly attractive for transferring genetic gain from the breeding to the commercial population, for exploiting heterosis in hybrid crosses and for exploiting promising species which have been of restricted use because of seed problems. With the advent of tissue culture, the prospects for the general use of vegetative propagation in the genetic improvement of a wide spectrum of species is becoming a reality and it is timely to consider the values and hazards of using it in domestication, in breeding strategy and in the actual production of tropical forest trees (Burdon, 1982; Libby and Rauter, 1984).

## 12.1 VEGETATIVE PROPAGATION TECHNIQUES

Grafting, using various techniques such as tip cleft, side veneer and budding, was the first method used to propagate tropical forest trees vegetatively. Its purpose has been to preserve selected phenotypes and to multiply them to create clonal orchards to produce seed for progeny tests and commercial use. Incompatibility between a scion and rootstock has been widespread among clones in some species and provided the first incentive for tree breeders to try to produce ramets on their own root systems.

Initially, there was virtually no success with cuttings because of the age of the ortets but various layering techniques were developed with which to induce physiologically old material to root (e.g. Barnes, 1974). Once the importance of juvenility had been realised, rejuvenation techniques (Franclet, 1979) and propagation from young trees increased the success rate from cuttings. At this stage the potential of vegetative propagation became evident and a progression from stem and branch to leaf cuttings occurred in attempts to increase the amount of propagable material from a single ortet or clone. For most tree species, however, propagation by cuttings still requires inherently expensive management techniques to maintain juvenility and to produce useful numbers of propagules. Now considerable resources are being invested in the development of organ (Fossard *et al.*, 1974), tissue (Bonga and Durzan, 1982; John, 1983; Dodds, 1983b; Bajaj, 1985; Hakman and von Arnold, 1985; Naguani and

Bonga, 1985; Hasnain and Cheliak, 1986; Jelaska and Bornman, 1986), cell and protoplast (Dodds, 1983a) culture for the commercial production of trees (Hasnain *et al.*, 1986; Mott and Amerson, 1982). (Two Working Parties of the International Union of Forestry Research Organizations have been formed recently to cover these activities.) Vegetative reproduction by apomixis has also been suggested for cloning some species (Ashton, 1981), and haploid culture is already developed for poplars (Lu *et al.*, 1981).

Such techniques are particularly attractive where a species is planted as an exotic in a large number of countries and could therefore benefit from centrally co-ordinated conservation and breeding. In these cases, a specialised institution could conduct the propagation and disseminate the select material for commercial bulking in the planting site; this is the system employed for oil palm (Corley *et al.*, 1979).

Other uses for tissue culture techniques are likely to emerge (Bonga, 1977) and some may have application for tropical species. These include the following.

(a) *Production of disease-free clones.* This is important for virus and mycoplasma diseases such as sandal spike (Lakshmi Sita, 1985).
(b) *Early selection.* This is only possible if juvenile–mature correlations can be evaluated. At present this seems to be likely only for chemical properties (including media selection by clonal genotypes) and screening for disease (e.g. Diner and Mott, 1982) since the field correlations between seedlings and mature plants are generally poor, and predictions of growth rate from photosynthetic or respiration rates are imprecise (e.g. Ledig, 1974). Nevertheless, cultures could be usable for early screening among very large numbers or for obtaining some idea of genotype–environment interactions and clonal stability.
(c) *Production of mutants.* Mutants induced by radiation or chemical mutagenesis of cultures can be screened immediately and *in situ* (Bottino, 1975). However, there is so much variation in most wild tree species that induced variation is unnecessary.
(d) *Somatic hybridisation.* This is being developed with crop plants but has not been attempted with forest trees.
(e) *Genetic engineering.* This has some long-term potential for tropical trees. The most commonly quoted *desideratum* is the introduction of nitrogen fixation into non-leguminous trees.

In all these applications, eventually, whole propagules would be produced for field planting. A sixth potential application is the use of the culture itself as a biomass energy or chemical source. The former is unlikely to replace normal commercial plantations, simply because of the volumes and costs involved. However, the latter is becoming increasingly attractive as the costs of developing new chemical products rise and as more naturally occurring drugs and pharmaceuticals are discovered.

## 12.2  CYCLOPHYSIS, TOPOPHYSIS AND PERIPHYSIS

The increasing interest in vegetative mass production has led to a greater concern about the differences between vegetative- and seed-derived propagules. Ramets tend to retain the adult characteristics of the ortet and this can enhance, but more often limits, their commercial use. *Cyclophysis* is the process of maturation of the apical meristem; *topophysis* is the phenomenon in which the effects due to position in the ortet are passed on to the ramet; *periphysis* concerns the transmission to the ramet of effects which are due to the environment of the ortet. Cyclophysis is generally irreversible whereas topophysis and periphysis can be reversible (Olesen, 1978).

The most troublesome effect of cyclophysis is the decrease in rooting ability of cuttings that takes place with maturation. The process of maturation is most advanced in the apical meristem of the terminal leader. However, even in a large tree, if shoots can be induced from a young meristem towards the base of the tree, they will be juvenile in nature. This has been the basis of success in rooting *Eucalyptus* species (Chaperon and Quillet, 1977). There have been claims of rejuvenation being brought about by repeated vegetative propagation of the same material, such as *Eucalyptus* and *Cupressus* species (Franclet, 1979).

The most common problem associated with topophysis in forest trees is plagiotropism in propagules taken from the lower branches of certain species both softwood, e.g. *Araucaria cunninghamii* Ait. (Nikles, 1973b), and hardwood, e.g. (potentially) the *Dipterocarpaceae* (Ng, 1981). Beneficial topophytic effects are found in stem form and branching in *Pinus radiata* D. Don. (Thulin and Faulds, 1968) and differences in flower productivity in various parts of the crown of the ortet might be used to maximise productivity, as in pine seed orchards (Barnes and Mullin, 1974). The chemical composition of the wax on the surface of the needles of *P. radiata* changes with age and, in old trees, the stomata become blocked (Wells and Franich, 1977). This is is now thought to prevent penetration by the needle blight fungus, *Scirrhia pini* Funk and Parker, and accounts for the transition from widespread susceptibility when the tree is young to virtual immunity at about 18 years (Anon., 1977) and for the promising resistance of ramets originating from a mature ortet

(Barnes, 1970). Topophytic effects may be advantageous in propagating trees selected for crown fodder production (Burley, 1980).

## 12.3 GENETIC INFORMATION

Where planting stock can be produced economically by vegetative means, phenotypic selection followed by clonal testing to evaluate the genotypes is an obvious procedure by which to produce superior material for commercial use. Clonal replication can also be used as a selection tool and to elucidate genetic architecture (Libby, 1969), both in species which are propagated vegetatively for commercial use and in those where propagation is by seed.

Vegetative propagules could be used in a breeding programme to assess general combining ability as an alternative to polycross or wind-pollinated tests. The advantages would be the early test establishment and the fewer plants required because of low within-clone variation (Shelbourne, 1969), with greater precision of ranking because of release from the uncontrolled influence of the pollen parents; but this information can be affected by such factors as topophysis and non-additive gene action and unless planting stock is produced directly from the material, commercial application of the results could lead to serious errors in estimating general combining ability (Burdon and Shelbourne, 1974).

The more sophisticated full-sib progeny tests in tropical tree breeding programmes almost invariably have three objectives; the first to estimate genetic variances and covariances, the second to identify the best general and/or specific combiners among the parents of the progenies, and the third to provide material for the next generation of selection. Invariably a very high proportion of the variance occurs between trees within plots Wyk, 1976; Barnes and Schweppenhauser, 1979). If trees could be replicated clonally within plots, not only would it be possible to estimate the genetic variance between trees and the genotype–environment interaction at individual tree level, but it would also make selection within the test much more precise (Libby, 1969; McKeand, 1981). The dangers of topophysis could be overcome by cloning seedlings at a very early age. However, it has been shown (Libby, 1969) that, provided genotype–environment interaction at the individual tree level is not high, clonal schemes are not as efficient as uncloned schemes because the selection differential has to be reduced to maintain tests of comparable size and structure.

Progeny testing is difficult to justify if it is used purely to evaluate the parents (Libby, 1973). If it is to give comprehensive information on genetic architecture, it may be inefficient in providing enough material for a satisfactory selection differential in the next generation of selection and the latter is economically the most valid reason for progeny testing. Even if topophysis can be circumvented, it would seem that there is little to be gained by cloning and, when the increased complications of technique and cost are taken into account, it does not appear to be an avenue worth exploring for increased genetic information.

## 12.4 SEED PRODUCTION

Most tree breeding programmes for forest species in the tropics have followed the traditional pattern of selection of superior phenotypes ("plus" trees) with commercial seed production from progeny-tested clonal orchards. Because the "plus" trees are almost invariably mature, cuttings have been difficult to root and there has been widespread use of grafts. Clonal orchards have had the advantage over seedling orchards of early seed production, a cyclophytic effect. A major problem in clonal and seedling orchards has been the extreme variability between genotypes in seed productivity and this has given cause for concern because sparse flowering is often associated with superior vegetative growth (Libby et al., 1969). In this regard, clonal orchards do hold an advantage over seedling orchards in that cyclophysis, topophysis, rootstock influence and the potential to increase production from one genotype indefinitely, can be used to overcome this unfavourable correlation (Barnes and Mullin, 1974). Rootstock influence on ramet morphology and rootstock resistance to disease are also useful aids in seed production from clonal orchards.

Breeding programmes with most tropical species have indicated a large proportion of additive genetic variance for the traits of interest. A key feature of breeding in such populations is the recurrent and cumulative nature of genetic improvement. Fast progress can be made with tropical tree species because of their rapid growth rate and short breeding cycles (Namkoong et al., 1980) and the quality of improved seed that can be passed on to the grower can be upgraded in frequent steps. Clonal orchards are expensive to establish and maintain, as are progeny tests. If the functions of progeny testing, selection and seed production can be combined in a single seedling planting, this would be very advantageous for tropical trees, especially if the concept of multiple populations, i.e. keeping separate distinct populations within a species (Namkoong et al., 1980), is to be followed. It is possible, therefore, that the use of clonal orchards for commercial seed production will become more limited in the future, although they will always have an essential role to play in exploiting specific combining ability and in providing conveniently located material in breeding orchards to make controlled crosses for the development of the breeding population.

## 12.5 COMMERCIAL VEGETATIVE PROPAGATION

When planting stock can be raised by vegetative means, the best elements in the breeding population can be passed on to the industry as soon as they are produced; the genotype and probably its pedigree are known, the clonal constitution of the forest can be precisely controlled and it is possible to take advantage of topophysis. If a clone in use should prove to be susceptible to diseases or pests or otherwise defective, a change could be made immediately by replacement with another selection from the breeding population.

The main objection to clonal forestry is the increased risk of biotic or physical catastrophe which is associated with the narrowing of the genetic base. However, theoretical arguments have recently been put forward that monoclonal plantations are often the best strategy and that mixtures of modest numbers of clones (7–25) apparently provide a robust and perhaps even an optimum strategy (Libby, 1980). In tropical trees, rapid breeding progress will lead to a frequent change to new clones and this itself will increase genetic diversity.

The advantages of vegetative propagation would be likely to outweigh the disadvantages in most cases. Provided that there is sufficient genetic variation and that stock can be produced economically, there will be an overwhelming case for practising clonal forestry, particularly with tropical species where environmental conditions are naturally conducive to rooting, growth is fast and development of the breeding population is rapid. This may apply even to species which produce abundant, easily handled seed and are susceptible to improvement by recurrent selection for general combining ability.

There are circumstances, common in tropical tree species, where vegetative propagation may be the key factor in optimising production or even in domestication. These include situations where (a) there is a high proportion of non-additive genetic variance, (b) there is useful heterosis, (c) there is a useful topophytic effect, (d) only the male or only the female plant is required in dioecious species, (e) male and female flowering is asynchronous, (f) there is large or recalcitrant seed, (g) seed is sporadically produced, (h) there is an unfavourable correlation between vegetative and reproductive growth, (i) segregation in the commercial population presents a problem, and (j) there is a particularly high value per tree.

Vegetative propagation could carry risks where (a) trees that are valuable individually are planted on long rotations from very few clones, (b) trees are being established in the first step of domestication from the indigenous forest when neighbourhood inbreeding could already be depressing growth, and (c) there is a danger of transmission of microorganisms in the plant tissue, although *in vitro* techniques do exist for producing "clean" propagules in some species (Young et al., 1984). In many woody plants, outstanding clones have been discovered by chance and not as a result of systematic breeding. It could take many generations to achieve through breeding the same level of improvements; however, a systematic breeding programme is always justified for long-term advance and to minimise the risks of catastrophe with a single clone. It is perhaps with this in mind that proposals are currently being made to establish legal instruments in some countries to control numbers of clones to be used in a plantation.

Tropical tree species fall naturally into a number of groups within which they have common potential, requirements and problems. These are the tropical pines, the araucarias, other tropical conifers, the eucalypts, the casuarinas, other tropical hardwoods and multipurpose trees.

Among the tropical softwoods, *Pinus* species have been most highly developed as plantation trees. All species seed freely under the right environmental conditions and there has been little incentive to develop vegetative propagation techniques. There are no species yet for which propagules are produced for large-scale operational use.

Many progeny tests of tropical pines have been established and there are indications of high specific combining ability variance, but surprisingly few of the tests are designed to estimate non-additive effects. Among those that have, e.g. for *P. patula* Schiede and Deppe (Barnes and Schweppenhauser, 1979), most show general combining ability to be much higher than specific combining ability variance. However, in *P. taeda* L., a species planted in the tropics, spiral grain, tracheid length and rust resistance have all been found to be under appreciable non-additive genetic control (Stonecypher et al., 1973) and this is likely to occur in traits of interest in other species. Two-clone orchards would be one way of exploiting this variance but problems of asynchronous flowering (Barnes and Mullin, 1974), the needs for self-sterile clones and for compatibility, all increase the selection requirements and further limit the clones that can be used. Vegetative propagation would undoubtedly prove to be a much more reliable means of realising such gains efficiently and quickly.

As more interspecific hybrids are made among pines, useful heterosis is being discovered. Because of differences in flowering times, low seed viability, and segregation in the $F_2$ generation, seed production is impracticable—as shown by the vast programme of controlled pollination necessary in Korea to produce the *P. rigida* Mill. × *P. taeda* hybrid (Wright, 1976). This hybrid vigour will probably only be available for large-scale planting when clonal forests become a reality. *Pinus elliottii* var. *elliottii* Engelm. × *P. caribaea* Mor. var. *hondurensis* Barr. and Golf. has outstanding performance on ill-drained sites (Nikles, 1973a; Sijde and Slabbert, 1980). *Pinus caribaea* var. *hondurensis* × *P. oocarpa* Schiede is exceptionally wind-firm on sites subject to cyclone damage (Nikles *et al.*, 1981). *Pinus elliottii* var. *elliottii* × *P. taeda* hybrids perform well in environments which are marginal for both species in the tropics (Barnes and Mullin, 1978).

Among the pines, greatest advances in the development of operational vegetative propagation techniques have been made in *Pinus radiata*. In New Zealand and Australia the species is on the verge of small-scale commercial propagation by cuttings produced from specially managed clonal hedges (Thulin and Faulds, 1968; Arnold and Gleed, 1985; Anon, in press). More recently, in New Zealand, it has been demonstrated that complete plantlets can be generated from embryonic tissue using micropropagation techniques, and this method is now being used in a pilot scheme to amplify small amounts of valuable controlled-pollinated seed (Aitken-Christie and Gleed, 1984).

Both genera of the Araucariaceae, *Araucaria* and *Agathis*, contain a number of species which are better adapted for growth in the lowest latitudes and lower altitudes of the tropics than are the tropical pines. They have high-quality timber and could be especially useful in reafforestation after destruction of tropical high forest. Seed problems of scarcity, large size, viability and recalcitrance have been major obstacles. Vegetative propagation is possible on a small scale (Whitmore, 1977; Ng and Sabariah, 1979); development of suitable mass production techniques could lead to widespread use of these species in the tropics.

*Cupressus*, *Cunninghamia* and *Cryptomeria* are three more coniferous genera which contain important plantation trees used in the tropics and subtropics. *Cupressus* species are grafted successfully for clonal seed orchards in breeding programmes. Seeds are small and produced early and profusely and therefore there has been little incentive for the mass development of vegetative techniques. There are, however, very distinctive genotypes in these forest species of *Cupressus*, just as there are in those commonly used horticulturally; if the non-additive variation is large, vegetative propagation will be useful as a means of reproducing the desired genotype. Although *Cunninghamia lanceolata* (Lamb.) Hook. and *Cryptomeria japonica* (L.f.) D. Don. produce abundant quantities of small seed, both are commonly propagated vegetatively for afforestation, the former on an enormous scale for centuries in China (Richardson, 1966), the latter in Japan, with the result that plantations of outstanding uniformity and quality have been created.

Among the tropical hardwoods, *Eucalyptus* species are the most widely planted and are rapidly becoming the most intensively domesticated. Seed production in most eucalypts is early and copious, and most tropical plantations have been established from seed. Grafting was successfully developed in breeding programmes for seed production. However, substantial specific combining ability variance has been found in various species (Wyk, 1976; Chaperon, 1977) and the extraordinary propensity of species in the genus to hybridise has, in some instances, led to serious loss of productivity through segregation in the $F_2$ and subsequent generations, e.g. in Brazil where *Eucalyptus urophylla* S.T. Blake has hybridised with several other *Eucalyptus* species (Campinhos and Ikemori, 1977). Both these factors, and the potentially valuable heterosis that has been displayed, have acted as stimulants to develop techniques for vegetative propagation. Recently, mass cutting techniques with specific hybrids in the Congo, *E. urophylla* × *E. alba* Reinw. ex Bl. and *E. tereticornis* Sm. × *E. saligna* Sm. (Chaperon, 1977; Delwaulle *et al.*, 1980), and Brazil, *E. urophylla* × *Eucalyptus* spp. (Campinhos and Ikemori, 1980) have been perfected to the point where extensive clonal plantations of exceptional quality are now being established, in some cases where performance of the pure species was so poor that afforestation originally could not be contemplated. This approach has great potential for the improvement of hundreds of thousands of hectares of the "Mysore hybrid" eucalypt and *E. tereticornis* × *E. camaldulensis* in India (Jacobs, 1979).

Species of the genus *Casuarina* are becoming increasingly important for protection forestry, tree-growing in arid areas and agroforestry. The casuarinas seed freely but several species are dioecious. In Thailand, an importation of all male seedlings of *Casuarina junghuhniana* Miq. of exceptional form (Pravit, 1981) made it necessary to propagate by vegetative means and a cutting technique with commercial application has been developed. *Casuarina decaisneana* F. Muell. has large woody "cones" which themselves might be used for fuel, in which case vegetative propagation of clones which are prolific in this respect would be necessary.

There is a multitude of tropical forest species with potential for domestication. Those which are already successfully used for the establishment of plantations range from small, short-lived to large, long-lived gregarious pioneer species, generally from the monsoonal forests, and reproducing freely from seed, for example *Leucaena leucocephala* (Lam.)

de Wit., *Albizia lebbek* (L.) Benth., *Tectona grandis* L.f., *Gmelina arborea* Roxb., *Campnospermum brevipetiolata* Volk. and *Terminalia* spp. Because of their copious, easily handled seed, their outcrossing breeding systems, wide adaptability and yet comparative phenotypic uniformity, there has been little need for vegetative propagation. There are, however, particularly in the tropical rainforest, many species which have well-known valuable timbers and apparent potential for plantation or enrichment use but which are difficult to propagate by seed (Sheikh, 1977; Yap, 1981). Conspicuous among these are species of the genus *Dipterocarpus* which provide the majority of the world's tropical hardwood sawlogs. Similar problems have been found with seed of important species like *Triplochiton scleroxylon* K. Schum. in West Africa (Leakey *et al.*, 1982). Development of vegetative propagation techniques could mean a breakthrough in the domestication of species such as these, and there has already been some successful work done (Srivastava and Manggil, 1981; Leakey *et al.*, 1982).

Many species have potential for use as multipurpose trees in tropical environments (e.g. NAS, 1979, 1980; Webb *et al.*, 1980); of these the nitrogen-fixing trees, mainly in the family *Leguminosae*, attract the most interest, particularly the genera *Acacia* and *Prosopis*. A major problem is the existence of many local land races, particularly in those species that produce fodder or edible fruit. With the exception of the self-pollinating *Leucaena leucocephala* (NAS, 1977) few have been seriously bred and none have been vegetatively propagated for either breeding purposes or commercial plantation.

Vegetative propagation could be used for the widespread distribution of individual phenotypic selections from local land races while awaiting the release of seed of proven varieties, e.g. the fast-growing K8 cultivar of *L. leucocephala* developed at the Nitrogen Fixing Tree Association in Hawaii or the backcrossed *L. leucocephala* × *L. pulverulenta* (Schlecht.) Benth. Hook. developed for tolerance of acid soils by the Centro Internacional de Agricultura Tropical in Colombia (Brewbaker and Hu, 1981; Hutton, 1981). Even though land races of many legumes may differ from each other as the result of intense, though possibly unconscious, selection, there remains considerable variation within races of which the specific combining ability component may be captured by vegetative propagation.

Most *Leucaena* species are self-pollinating but many other leguminous trees are not. A further specific use of vegetative propagation for *Leucaena* is the propagation of sterile, interspecific hybrids. Mist spray techniques have yielded 52–100% successful rooting of various one-year-old materials and 0–50% for three-year-old materials (Hu and Liu, 1981).

## 12.6 CONSERVATION AND THE GENETIC BASE

Tree breeders continually face a dilemma; narrowing the genetic base through selection achieves maximum gain in the present generation but intensifies the inherent risks of using monocultures and reduces the variation available for future generations of selection. Multipopulation breeding strategies defer the onset of this problem and the associated inbreeding depression (see Namkoong *et al.*, 1980). Considerable efforts are currently being made with several tropical species to explore and conserve natural variation, e.g. Central American tropical pines and arid-zone species (OFI, Oxford, UK and CAMCORE, North Carolina State University, USA); eucalypts (CSIRO, Canberra, Australia); Asian pines, teak and *Gmelina*, (DTSC, Humlebaek, Denmark); arid-zone species (FAO/IBPGR, Rome, Italy).* Many of these natural populations are threatened with genetic impoverishment and *in situ* ecosystem conservation is not possible politically, administratively or financially. Thus, some form of *ex situ* conservation is desirable (Wood and Burley, 1980).

In formulating genetic conservation policy it is necessary to decide what is to be conserved—genes, gene frequencies or genotypes. For most species it is probable that virtually all genes occur somewhere in most populations (Burley and Namkoong, 1980; Namkoong *et al.*, 1980). Generally, sampling is best distributed over a few seeds per tree from many trees per population for heterozygous outcrossing species, and over few trees from many populations for small inbred populations; in view of the higher cost per propagule compared with seed and because scion materials are difficult to transport internationally, clonal propagation is not feasible for conservation of genes or gene frequencies *ex situ* unless a very sophisticated field-to-laboratory system were developed for tissue culture. For genotype conservation, however, vegetation propagation is ideal.

When genetic conservation is practised through tissue culture (Wilkins and Dodds, 1983) the chances of genotype loss (through culturing accidents) and genetic change (through mutation of active cultures) are expected to be higher than in refrigerated seed. However, even seed of species with long viability has to be grown out eventually (say 10–20 years) while tissue cultures may be preservable in liquid nitrogen for longer periods. For species with low seed-storage ability, e.g. the dipterocarps, clonal conservation has obvious advantages.

---

* OFI, Oxford Forestry Institute: CAMCORE, Central America and Mexico Coniferous Resources Cooperative: CSIRO, Commonwealth Scientific and Industrial Research Organization: DTSC, Danish Tree Seed Centre; FAO/IBPGR, Food and Agriculture Organization/International Board for Plant Genetic Resources.

## 12.7 EXPERIMENTS ON ENVIRONMENTAL EFFECTS

Uncontrolled genetic variation obscures the results from forestry experiments set up to investigate the effects of environment. This variation has led to the necessity for bigger plots and more replications and, therefore, more costly trials which, in turn, have increased the site variation.

New cultural methods from establishment to harvesting are being developed and new genetic strains of trees are being produced. Because of increasing pressures on land for growing food crops, the site types available for forestry in the tropics are constantly changing and there are new needs for intimate integration of trees and agriculture on the same land. A sound understanding of climatic, edaphic and biotic effects on trees and their interactions is, therefore, crucial if productivity is to be improved, or even maintained.

The use of clonal material entirely removes the genetic source of variation, and comparatively small, precise and efficient experiments can be designed to test environmental factors, including the effects of within-stand competition at the individual genotype level, a notorious source of bias in experiments but an important source of variation to use in maximising site productivity (Libby *et al.*, 1969; Burdon and Shelbourne, 1974). Even if vegetative propagation is not practicable on a commercial scale, it may well be worth a considerable investment to produce clonal material for experiments such as these, particularly in tropical forestry where growth rates, strains and cultural techniques are changing so fast and have to interact favourably with each other.

## 12.8 THE FUTURE

Successful techniques of vegetative propagation have been developed for many tropical forest tree species. Until recently, these have been used just as tools in the breeding process to clone selected individuals for preservation and seed production. Now, breakthroughs in technique are being made for a few species where propagules can be vegetatively produced cheaply on an operational scale; the enormous variability in the still almost-wild tree populations, and the even greater variability in their interspecific hybrid swarms, is suddenly being released for immediate commercial exploitation. This is leading to quantum increases in productivity and in the immediate future it will probably be the most significant factor in the improvement of many tropical forest tree species.

However, there is a danger of looking on this as genetic improvement rather than as the development of a technique, and consequently of neglecting the breeding back-up to clonal forestry. It is essential to maintain variability; without it, there will be no potential to improve and respond to changing requirements and there will be no advance. If genetic gain is to be cumulative, broadly based breeding populations must be maintained with flexible strategies which include a conservation element to preserve variability. The new techniques of tissue culture can contribute significantly to this conservation and allow rapid exploitation of the variability produced by recombination in the breeding programmes.

**REFERENCES**

Aitken-Christie, J., and Gleed, J. A. (1984). Uses for micropropagation of juvenile radiata pine in New Zealand. *In* "Proceedings of International Symposium: Recent Advances in Forest Biotechnology", (J. Hanover, D. F. Kamosky and D. Keathley, eds), pp. 47–57. Michigan Biotechnology Institute, East Lansing, USA.

Anon. (1977). The wax coating on radiata pine needles. "What's New in Forest Research." No. 52, 4 pp. Forest Research Institute, Rotorua, New Zealand.

Anon. (in press). Proceedings of Radiata Pine Cuttings Workshop. Rotorua, May 1986. *NZ For. Res. Inst. Bull.*

Arnold, R., and Gleed, J. A. (1985). Raising and managing radiata pine cuttings for production forests. *Austral. For.* **48**, 199–206.

Ashton, P. S. (1981). Future directions in dipterocarp research. *Malay. For.* **44**, 193–196.

Bajaj, Y. P. S. (ed.) (1985). "Biotechnology in Agriculture and Forestry, 1. Trees I." Springer-Verlag, Berlin, 515 pp.

Barnes, R. D. (1970). The prospects for re-establishing *Pinus radiata* as a commercially important species in Rhodesia. *S. Afr. For. J.* **72**, 17–19.

Barnes, R. D. (1974). Air-layering of grafts to overcome incompatibility problems in propagating old pine trees. *N.Z. J. For. Sci.* **4**, 120–126.

Barnes, R. D., and Mullin, L. J. (1974). Flowering phenology and productivity in clonal seed orchards of *Pinus patula, P. elliottii, P. taeda* and *P. kesiya* in Rhodesia. Forestry Research Paper No. 3, 81 pp. Research Division, Rhodesia Forestry Commission.

Barnes, R. D., and Mullin, L. J. (1978). Three-year height performance of *Pinus elliottii* Engelm. var. *elliottii* × *P. taeda* L. hybrid families on three sites in Rhodesia. *Silvae Genet.* **27**, 217–223.

Barnes, R. D., and Schweppenhauser, M. A. (1979). *Pinus patula* Schiede and Deppe progeny tests in Rhodesia: Genetic control of 1.5-year-old traits and a comparison of progeny test methods. *Silvae Genet.* **28**, 156–167.

Bonga, J. M. (1977). Applications of tissue culture in forestry. *In* "Applied and Fundamental Aspects of Plant, Cell, Tissue and Organ Culture" (J. Reinert and Y. P. S. Bajaj, eds), pp. 93–108. Springer-Verlag, New York.

Bonga, J. M., and Durzan, D. J. (1982). "Tissue Culture in Forestry". Martinus Nijhoff/Dr W. Junk, Dordrecht, 420 pp.

Bottino, P. J. (1975). The potential of genetic manipulation in plant cell cultures for plant breeding. *Radiat. Bot.* **15**, 1–16.

Brewbaker, J. L., and Hu, T. W. (1981). Nitrogen fixing trees of importance in the tropics. Summary of Papers of Workshops, Biological Nitrogen Fixation, Colombia

and Taiwan. Mimeo, NFTA, Hawaii, USA, 8 pp.
Burdon, R. D. (1982). The roles and optimal place of vegetative propagation in tree breeding strategies. Symposium IUFRO Subject Group S2.04 (Genetics), Escherode, Germany, September, 17 pp.
Burdon, R. D., and Shelbourne, C. J. A. (1974). The use of vegetative propagules for obtaining genetic information. *N.Z. J. For. Sci.* **4**, 418–425.
Burley, J. (1980). Choice of tree species and possibility of genetic improvement for smallholder and community forests. *Commonw. For. Rev.* **59**, 311–326.
Burley, J., and Namkoong, G. (1980). Conservation of forest genetic resources. 11th Commonwealth Forestry Conference, Trinidad, 25 pp.
Campinhos, E., and Ikemori, Y. K. (1977). Tree improvement program of *Eucalyptus* spp. Preliminary results. *In* "Proceedings of the Third World Consultation on Forest Tree Breeding", pp. 717–738. Canberra, Australia. FAO, Rome.
Campinhos, E., and Ikemori, Y. K. (1980). Mass production of *Eucalyptus* spp. by rooted cuttings. *Silvicultura* **8**, 770–775.
Chaperon, H. (1977). Amélioration génétique des *Eucalyptus* hybrides au Congo–Brazzaville. *In* "Proceedings of the Third World Consultation on Forest Tree Breeding", pp. 1055–1069. Canberra, Australia. FAO, Rome.
Chaperon, H., and Quillet, G. (1977). Résultats des travaux sur le bouturage des *Eucalyptus* au Congo–Brazzaville. *In* "Proceedings of the Third World Consultation on Forest Tree Breeding", pp. 835–855. Canberra, Australia. FAO, Rome.
Corley, R. H. B., Wooi, K. C., and Wong, C. Y. (1979). Progress with vegetative propagation of oil palm. *Planter, Kuala Lumpur* **55**, 337–380.
Delwaulle, J. C., Laplace, Y., and Quillet, G. (1980). Production massive de boutures d'*Eucalyptus* en République Populaire de Congo. *Silvicultura* **8**, 779–781.
Diner, A. M., and Mott, R. L. (1982). A rapid assay for hypersensitive resistance of *Pinus lambertiana* to *Cronartium ribicola*. *Phytopathology* **72**, 864–865.
Dodds, J. H. (1983a). The use of protoplast technology in tissue culture of trees. *In* "Tissue Culture of Trees" (J. H. Dodds, ed.), pp. 103–112. Croom Helm, London.
Dodds, J. H. (1983b). Tissue culture of hardwoods. *In* "Tissue Culture of Trees" (J. H. Dodds, ed.), pp. 22–28. Croom Helm, London.
Fossard, R. A. de, Nitsch, C., Cresswell, R. J., and Lee, E. C. M. (1974). Tissue and organ culture of *Eucalyptus*. *N.Z. J. For. Sci.* **4**, 267–278.
Franclet, A. (1979). Rejeunissement des arbres adultes en vue de leur propagation végétative. *Afocel Etudes et Recherches* **12**, 3–18.
Hakman, I., and Arnold, S. von (1985). Plantlet regeneration through somatic eurkaryogenetics in *Picea agies* (Norway Spruce). *J. Plant Physiol.* **121**, 149–158.
Hasnain, S., and Cheliak, W. (1986). Tissue culture in forestry: Economic and genetic potential. *For. Chronicle* **62**, 219–225.
Hasnain, S., Pigeon, R., and Overend, R. P. (1986). Economic analysis of the use of tissue culture for rapid forest improvement. *For. Chronicle* **62**, 240–245.
Hu, T. W., and Liu, C. C. (1981). Vegetative propagation of *Leucaena* by leafy cuttings under mist spray. *Leucaena Res. Reports* 2, NFTA, Hawaii, USA. 50 pp.
Hutton, E. M. (1981). *Leucaena* and acid soils in tropical America. *ICRAF Newslett.* **6**, 1–2.
Jacobs, M. R. (1979). "Eucalypts for Planting". FAO, Rome, 677 pp.
Jelaska, S., and Bornman, C. H. (1986). Application of cell culture methods in forestry. *In* "Proceedings 18th IUFRO World Congress, Division 2, Vol. II", pp. 554–564.
John, A. (1983). Tissue culture of coniferous trees. *In* "Tissue Culture of Trees" (J. H. Dodds, ed.), pp. 6–21. Croom Helm, London.

Lakshmi Sita, G. (1985). Sandalwood (*Santalum album* L.). *In* "Biotechnology in Agriculture and Forestry, 1. Trees I" (Y. P. S. Bajaj, ed.), pp. 363–374. Springer-Verlag, Berlin.

Leakey, R. R. B., Last, F. T., and Longman, K. A. (1982). Domestication of tropical trees: an approach securing future productivity and diversity in managed ecosystems. *Commonw. For. Rev.* **61**, 33–42.

Ledig, F. T. (1974). Photosynthetic capacity: a criterion for the early selection of rapidly growing trees. *Yale Univ. Sch. For. Bull.* **85**, 19–39.

Libby, W. J. (1969). Some possibilities of the clone in forest genetics research. *In* "Genetics Lectures", vol. 1 (R. Bogart, ed.), pp. 121–136. Oregon State University Press, Oregon.

Libby, W. J. (1973). Domestication strategies for forest trees. *Can. J. For. Res.* **3**, 265–276.

Libby, W. J. (1980). What is a safe number of clones per plantation? Paper of Workshop on the Genetics of Host–Parasite Interactions in Forestry, Wageningen, 14–21 September, pp. 342–361.

Libby, W. J., and Rauter, R. M. (1984). Advantages of clonal forestry. *For. Chronicle* **60**, 145–149.

Libby, W. J., Stettler, R. F., and Seitz, F. W. (1969). Forest genetics and forest tree breeding. *Ann. Rev. Gen.* **3**, 469–494.

Lu, C. H., Chiang, H. F., and Liu, V. H. (1981). Induction and cultivation of pollen plants from poplar pollen. *In* "Plant Tissue Culture: Proceedings of the Beijing (Peking) Symposium, 1978". Pitman, London.

McKeand, S. E. (1981). Loblolly pine tissue culture: present and future uses in southern forestry. North Carolina State University School of Forestry Resources, Tech. Rep. No. 64, 50 pp.

Mott, R. L., and Amerson, H. V. (1982). A tissue culture Process for the clonal production of Loblolly pine plantlets. *N. Carolina Agric. Res. Ser. Tech. Bull.* No. 271, 14 pp.

Namkoong, G., Barnes, R. D., and Burley, J. (1980). A philosophy of breeding strategy for tropical forest trees. Tropical Forestry Paper No. 16. Dept. of Forestry, Commonwealth Forestry Institute, Oxford, England, 67 pp.

NAS (National Academy of Sciences) (1977). "*Leucaena*: Promising forage and tree crop for the tropics". National Academy of Sciences, Washington, D.C. 115 pp.

NAS (1979). "Tropical legumes: resources for the future". National Academy of Sciences, Washington, D.C., 331 pp.

NAS (1980). "Firewood crops: shrub and tree species for energy production". National Academy of Sciences, Washington, D.C., 237 pp.

Naguani, R., and Bonga, J. M. (1985). Embryogenesis in subcultured callus of *Larix decidua*. *Can. J. For. Res.* **15**, 1088–1091.

Ng, F. S. P. (1981). Vegetative and reproductive phenology of dipterocarps. *Malay. For.* **44**, 197–221.

Ng, F. S. P., and Sabariah, B. A. (1979). Vegetative propagation of *Araucaria hunsteinii*. *Malay. For.* **42**, 71–72.

Nikles, D. G. (1973a). Progress in breeding *Pinus caribaea* Morelet in Queensland, Australia. *In* "Selection and Breeding to Improve some Tropical Conifers" (J. Burley and D. G. Nikles, eds), vol. II, pp. 245–266. Commonwealth Forestry Institute, Oxford.

Nikles, D. G. (1973b). Biology and genetic improvement of *Araucaria cunninghamii* Ait. in Queensland, Australia. *In* "Selection and Breeding to Improve some

Tropical Conifers" (J. Burley and D. G. Nikles, eds), vol. II, pp. 305–334. Commonwealth Forestry Institute, Oxford.
Nikles, D. G., Boyer, P. C., Matthews-Frederick, D., Spidy, T., and Rider, E. J. (1981). Breeding of Honduras Caribbean pine and its hybrids in Queensland. Seminar Paper, Department of Forestry, Queensland, Australia, 14 pp.
Olesen, P. O. (1978). On cyclophysis and topophysis. *Silvae Genet.* **27**, 173–178.
Pravit, C. (1981). Silviculture of *Casuarina junghuhniana* in Thailand. Paper presented at Casuarina Workshop, CSIRO, Division Forest Research, Canberra, ACT, Australia, 5 pp.
Richardson, S. D. (1966). "Forestry in Communist China". The Johns Hopkins Press, Baltimore, Maryland, 238 pp.
Sheikh, I. (1977). Problems of seed production in moist tropical climates. *In* "Proceedings of the Third World Consultation on Forest Tree Breeding", pp. 807–819. Canberra, Australia. FAO, Rome.
Shelbourne, C. J. A. (1969). Tree Breeding Methods. Technical Paper No. 55, 43 pp. Forest Research Institute, New Zealand Forest Service.
Sijde, H. A. van der, and Slabbert, R. G. (1980). Performance of some pine hybrids in South Africa. *S. Afr. For. J.* **112**, 23–26.
Srivastava, P. B. L., and Manggil, P. (1981). Vegetative propagation of some dipterocarps by cuttings. *Malay. For.* **44**, 301–313.
Stonecypher, R. W., Zobel, B. J., and Blair, R. (1973). Inheritance patterns of Loblolly pines from a non-selected natural population. *N. Carolina Agric. Exp. Sta. Tech. Bull.* No. 220, 60 pp.
Thulin, I. J., and Faulds, T. (1968). The use of cuttings in the breeding and afforestation of *Pinus radiata*. *NZ J. For.* **13**, 66–77.
Webb, D. E., Wood, P. J., and Smith, J. (1980). A guide to species selection for tropical and sub-tropical plantations. Tropical Forestry Paper No. 15, 342 pp. Commonwealth Forestry Institute, Oxford.
Wells, L. G., and Franich, R. A. (1977). Morphology of epicuticular wax on primary needles of *Pinus radiata* seedlings. *N.Z. J. Bot.* **15**, 525–529.
Whitmore, T. C. (1977). A first look at *Agathis*. Tropical Forestry Papers, no. 11, 60 pp. Commonwealth Forestry Institute, Oxford.
Wilkins, C. P., and Dodds, J. H. (1983). Tissue culture conservation of wood species. *In* "Tissue Culture of Trees" (J. H. Dodds, ed.), pp. 113–136. Croom Helm, London, 147 pp.
Wood, P. J., and Burley, J. (1980). *Ex situ* conservation stands. Paper presented to IUFRO Symposium and Workshop on Genetic Improvement and Productivity of Fast-growing Tree Species. *Silvicultura* **8**, 158–160.
Wood, P. J., Burley, J., and Grainger, A. (1982). "Technologies and Technology Systems for Reforestation of Degraded Tropical Lands." Report 233–5240.0. US Congress Office of Technology Assessment, Washington D.C. 100 pp.
Wright, W. W. (1976). "Introduction to Forest Genetics". Academic Press, New York, 467 pp.
Wyk, G. van (1976). Early growth results in a diallele progeny test of *Eucalyptus grandis* (Hill) Maiden. *Silvae Genet.* **25**, 126–132.
Yap, S. K. (1981). Collection, germination and storage of dipterocarp seeds. *Malay. For.* **44**, 281–300.
Young, P.M., Hutchins, S., and Canfield, M. L. (1984). Use of antibiotics to control bacteria in shoot cultures of woody plants. *Plant Sci. Lett.* **34**, 203–209.

# 13 Selection for Improved Tropical Hardwoods

R. R. B. LEAKEY[1] and D. O. LADIPO[2]

[1]Institute of Terrestrial Ecology, Penicuik, Scotland
[2]Forestry Research Institute, Ibadan, Nigeria

Selection for improvement is one element of the domestication process currently being developed for important indigenous hardwoods of West Africa (Leakey et al., 1982b). *Triplochiton scleroxylon* K. Schum. "Obeche", *Terminalia superba* Engl. & Diels "Afara", *Terminalia ivorensis* A. Chev. "Idigbo" and *Nauclea diderrichii* (De Wild and Dur) Merr "Opepe", together with *Khaya, Entandrophragma* "Mahogany" and *Chlorophora* "Iroko" spp., formed the basis of the West African timber industry, which reached its peak in the 1960s and 1970s. Since the 1930s various efforts have been made to grow these hardwoods commercially, but the susceptibility of the slow-growing mahoganies to shoot-boring insects, other pest problems in *Chlorophora* spp. and the lack of viable seeds in *T. scleroxylon*, made these native species less attractive to foresters than the fast-growing, pulp-producing exotics like *Pinus* and *Gmelina*. In addition, *Tectona grandis* (teak) has been introduced throughout the tropics to substitute for the native quality timbers, but poor stem form has resulted in many of these plantations being rather disappointing, and in West Africa some are coppiced for fuelwood and poles.

Recently, increased demand in West Africa for plywood and construction timbers has again focused attention on native species, particularly those with medium growth rates. Of these, *T. scleroxylon* is the most important. It contributed over 60% of roundwood exports in the 1950s and 60s (Fig. 13.1), until over-exploitation led to declining trade and eventually to a ban on exports from Nigeria in 1975 (Leakey et al., 1983). Consequently, this species was chosen in 1970 as the first species for domestication under the West African Hardwood Improvement Project. This was a joint venture between the Forestry Research Institute of Nigeria (FRIN) and

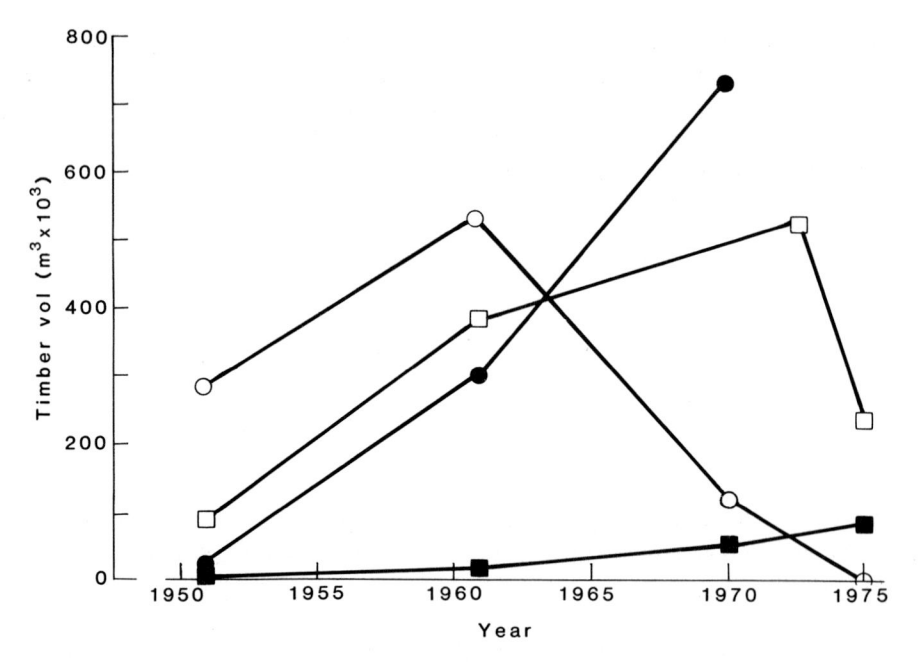

**Figure 13.1.** Annual round- and sawn-wood exports of *Triplochiton scleroxylon* from Cameroon (■), Ghana (□), Ivory-Coast (●) and Nigeria (○).

the Institute of Terrestrial Ecology (ITE), which has been financed by the UK Overseas Development Administration and the Federal Government of Nigeria.

The most important constraint to plantation establishment with *T. scleroxylon* has been the small and irregular supply of seeds (Jones, 1974), as trees flower only once every few years and pests and pathogens reduce the viability of the already short-lived seeds. Vegetative propagation potentially overcomes this problem, first because successive collections of cuttings provide large quantities of planting stock every year and secondly, by selecting and multiplying only high-yielding clones the productivity of the plantations which are established should be considerably greater than either unselected natural populations or stands of selected seed origins.

Although having a reputation as difficult to root, techniques of vegetatively propagating *T. scleroxylon* have been readily developed (Howland, 1975; Leakey *et al.*, 1975, 1982a) using single-node leafy cuttings from juvenile stockplants, set either under mist or enclosed in polythene frames. Clones vary in their auxin requirements for rooting both in optimal concentration and the relative importance of α-naphthalene acetic acid (NAA) and indole-3-butyric acid (IBA), but the rooting of most clones is improved by 40 μg of a 50/50 mixture of NAA and IBA per cutting. The mass propagation of clonal planting stock necessitates reliable rooting of a high percentage of cuttings. This can be achieved provided that the stockplants are managed to maximise the production of easily rooted cuttings. Essentially this is favoured by severe pruning, the restriction of new shoot growth to a few lateral shoots and to growth at low irradiance (Leakey, 1983, 1985a).

## 13.1 CLONAL FIELD TRIALS OF *TRIPLOCHITON SCLEROXYLON*

Using these techniques, clonal trials and gene banks representing the full geographical range of the species have been established at five sites in Nigeria. The main effort has been centred upon Onigambari Forest Reserve, near Ibadan, where detailed investigations have been started to elucidate the criteria for clonal selection of commercial planting stock (Ladipo *et al.*, 1983). Data are presented here from four experiments, all established in 1975. Three (Expts 3/75, 4/75 and 5/75) were planted at a wide spacing (4.9 m between trees) and the fourth (7/75) at a narrow spacing (2.4 m between trees). In each experiment, trees were planted in randomised blocks, with single tree plots, and the clones identified by site of origin and genotype (Table 13.1). In some instances several clones represented the same half-sib, open-pollinated seed collection.

Eighteen months after planting, despite considerable site variation between replicate plots, statistically significant differences, particularly between clones, were evident in almost all parameters measured (Howland *et al.*, 1978). For example, at narrow spacing the mean heights of trees from 14 different seedlots differed significantly, ranging from 2.65 m for seedlot 140, the shortest, to 3.28 m for seedlot 161, the tallest (Fig. 13.2). Seedlot variation in this or other parameters could not, however, be systematically related to the location of source trees, whether considered in terms of latitude, longitude or annual rainfall. However, even at this early age, as has been found in temperate conifers (Kleinschmit, 1974), variation between seedlots was not as great as the variation among the seven clones of the same seedlot. At this stage in the growth of this plantation, selection of the ten tallest clones showed a potential height gain of 16.5% over the mean for all plants (Ladipo *et al.*, 1983). Three and a half years later (5 years after planting), selection of the 33% of this randomly chosen clonal population with both above average mean stem volumes and mean stem scores would have resulted in an overall gain in stem volume of 30.5% (Fig. 13.3), while a 1 in 10 selection pressure

**Table 13.1.** Accession numbers and Nigerian location of individual tree (half-sib) seed collections used for the production of clones planted in four experiments at Onigambari Forest Reserve, near Ibadan, Nigeria in 1975.

| Accession number of seedlots | Origin of seedlots | | | | Accession number of clones (used as subscript to seedlot number) |
|---|---|---|---|---|---|
| | Location | Latitude (°N) | Longitude (°E) | Rainfall zone (mm) | |
| 137 | Olokemeji, Oyo State | 7°21' | 3°32' | 1300–1500 | 2, 4, 5, 9, 10, 11, 12 |
| 139 | Olokemeji, Oyo State | 7°21' | 3°32' | " " | 2, 3, 4, 5, 6, 9, 11, 12 |
| 140 | Onigambari, Oyo State | 7°12' | 3°52' | " " | 1, 2, 3, 4, 6, 9, 10, 11 |
| 142 | Omerelu, Ondo State | 5°17' | 6°55' | 2000–2500 | 1, 3, 4, 6, 7, 10, 11, 12 |
| 144 | Igbo Ora, Ogun State | 7°27' | 3°18' | 1000–1300 | 1, 4, 5, 7, 9, 10, 11 |
| 145 | Bolorunduro, Ondo State | 7°09' | 5°37' | 1300–1500 | 2, 6, 8, 9, 10, 11, 12 |
| 161 | Owo, Ondo State | 7°02' | 5°43' | " " | 1, 3, 4, 5, 7, 9, 12 |
| 166 | Azukala, Anambia State | 7°02' | 6°27' | " " | 1, 2, 3, 6, 8, 9, 11, 12 |
| 175 | Igbado, Ogun State | 6°49' | 4°52' | 1500–2000 | 1, 2, 3, 5, 6, 7, 8, 10 |
| 176 | Ilugun, Oyo State | 7°24' | 3°44' | 1000–1300 | 1, 3, 4, 6, 9, 10, 11 |
| 177 | Ilugun, Oyo State | 7°21' | 3°39' | 1300–1500 | 2, 4, 5, 6, 8, 10, 11 |
| 224 | Ede, Oyo State | 7°42' | 4°26' | " " | 1, 3, 4, 7, 9, 10, 11, 12 |
| 225 | Ede, Oyo State | 7°42' | 4°26' | " " | 1, 2, 3, 4, 8, 9, 10, 12 |
| 226 | Akure, Ondo State | 7°15' | 5°10' | " " | 1, 3, 4, 5, 6, 7, 8 |

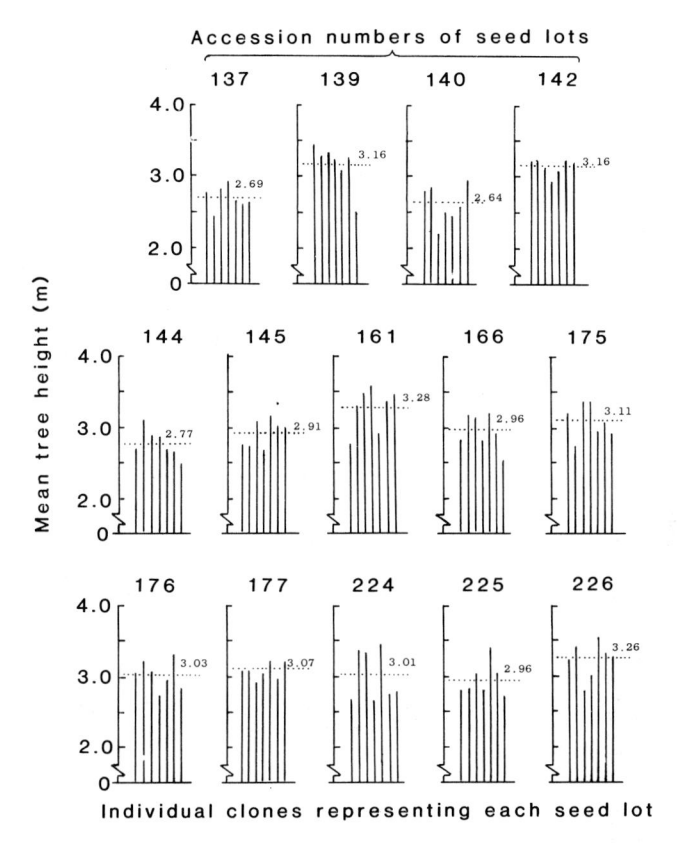

Selection for Improved Tropical Hardwoods 233

**Figure 13.2.** Variation in height, after 18 months, of seven clones of each of 14 seedlots (half-sib) of *T. scleroxylon* collected from different locations, when planted at Onigambari Forest Reserve, Nigeria. Mean height (........) of seedlot; LSD ($P = 0.05$) for individual clones and seedlot means are 0.54 and 0.8, respectively. (From Lapido et al., 1983).

would raise this to 80.5%. The potential yield improvement in plantation-grown *T. scleroxylon* by clonal selection is thus very considerable; extreme performances representing more than an eight-fold difference (Fig. 13.3). At a later date, further tree improvement may be possible by controlled breeding between selected clones, for already some clones have been induced to flower considerably earlier than normal (Leakey *et al.*, 1981) and third-generation progenies have been produced by reciprocal backcrossing in only 7 years.

Despite the success of the field trials reported here, it would be preferable to develop screening techniques which were less laborious,

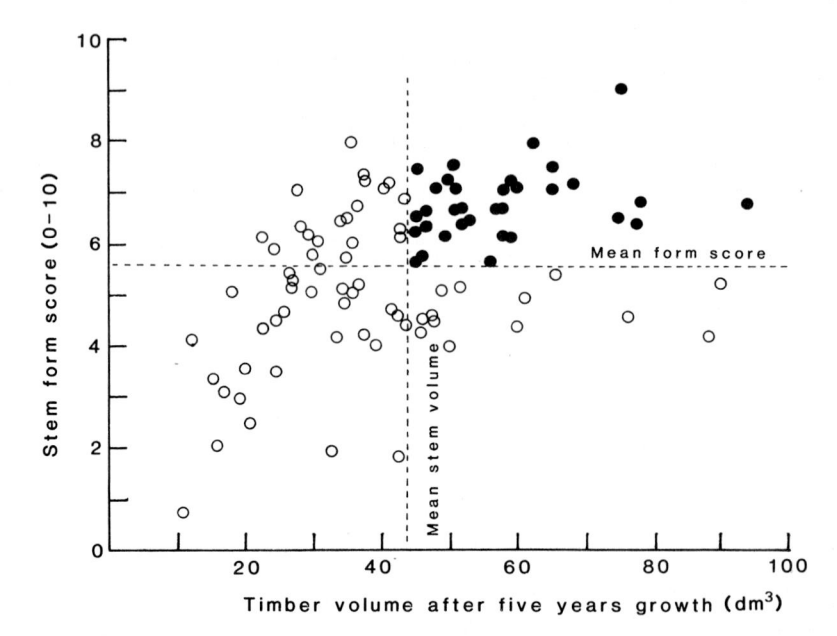

**Figure 13.3.** Relationship, in randomly selected clones of *T. scleroxylon* 5 years after planting, between volume of basal 4 m log and stem form. Selection of clones with above average mean stem volume (○) and mean form score (●) would result in a 31% yield improvement.

space demanding and costly. In addition, it would be desirable to identify superior clones at a very early age, when multiplication by vegetative propagation may be easier than after several years of field growth. However, it should be remembered that, at least in slow-growing temperate hardwoods, juvenile–mature correlations have not been very reliable (Sziklai, 1974), particularly those for yield per unit area (Cannell, 1982). Interestingly however, during the establishment phase of *T. scleroxylon* plantations, a period of almost continuous growth (Howland *et al.*, 1978), there was an apparent relationship in the three experiments at wide spacing, between mean tree height and the mean number of branches per metre of mainstem height. Thus, trees with the fewest branches per unit length of stem were the tallest and vice versa (Table 13.2). Is it possible that branching characteristics such as this are more permanent features retained through the life of a tree and that any advantage possessed by fast-growing clones may thus persist once they are competitively suppressing their neighbours?

Table 13.2. Variations in height, stem diameter and lengths and numbers of primary branches, after 18 months, among clones of *T. scleroxylon* from the same and different seedlots. Experiments planted at a spacing, within and between rows, of 4.9 m.

|  | Accession numbers of clones[a] | Stem diameter 10 cm from base (cm) | Plant height (m) | Number of primary branches | Number of primary branches per metre of mainstem length | Total primary branch length | Mean primary branch length |
|---|---|---|---|---|---|---|---|
| Expt 3/75 | 142/10 | 6.2 | 1.86 | 17.5 | 9.3 | 10.4 | 55.0 |
|  | 166/8 | 5.3 | 2.08† | 10.9 | 5.1 | 6.6 | 55.9 |
|  | 175/1 | 5.4 | 1.90 | 19.7† | 9.9† | 12.6† | 55.8 |
|  | 175/5 | 6.3† | 1.73 | 15.6 | 8.6 | 11.3 | 64.2† |
|  | 177/10 | 4.8 | 1.55 | 13.7 | 9.1 | 6.6 | 46.4 |
| LSD ($p = 0.05$) |  | 1.3 | 0.41 | 5.3 | 1.8 | 5.2 | 16.0 |
| Expt 4/75 | 130/4 | 4.0 | 1.52 | 13.1 | 8.2 | 6.0 | 41.0 |
|  | 140/4 | 4.8 | 1.69 | 12.4 | 7.2 | 7.8 | 58.5 |
|  | 140/9 | 3.7 | 1.37 | 12.1 | 9.3† | 5.3 | 42.6 |
|  | 144/1 | 3.7 | 1.57 | 13.2 | 8.5 | 4.0 | 32.1 |
|  | 224/10 | 5.8† | 2.16† | 13.9† | 6.3 | 9.3† | 61.8† |
|  | 225/3 | 4.3 | 1.49 | 11.1 | 6.9 | 5.6 | 45.9 |
| LSD ($p = 0.05$) |  | 0.8 | 0.24 | 3.3 | 1.6 | 2.6 | 9.3 |
| Expt 5/75 | 139/3 | 5.6 | 1.97 | 12.9 | 6.2 | 8.6 | 60.0 |
|  | 139/9 | 5.8† | 2.25† | 14.5† | 6.1 | 10.1† | 63.1† |
|  | 144/1 | 4.8 | 1.73 | 14.1 | 8.1† | 7.2 | 47.6 |
|  | 224/1 | 5.3 | 1.75 | 11.7 | 6.5 | 7.6 | 61.3 |
| LSD ($p = 0.05$) |  | 0.7 | 0.21 | 2.9 | 1.0 | 2.8 | 8.9 |

Source: Lapipo et al. (1983).
[a] 142/10: 142 refers to seed collection; 10 designates clone within a seed collection.
Values in bold type are highest (with dagger) and lowest in each experiment.

## 13.2 PREDICTION OF BRANCHING HABIT

Clones of *T. scleroxylon* differ markedly in the percentage of buds that make active growth following decapitation, these differences being fairly consistent in plants of different sizes (Leakey and Longman, 1986). This suggests genetic variation in apical dominance, the process controlling the production of branches, and thus perhaps the frequency of branching and the development of clonal differences in branching habit. Could differences in apical dominance between clones be used to predict branching habits? Before testing the hypothesis that differences in bud activity after decapitation could be related to branching habit, it was first necessary to determine the effects of different environmental and physiological factors affecting this expression of apical dominance. Without this, inconsistent clonal responses to decapitation might confound the predictions, as apical dominance is known to be very sensitive both to the environment (McIntyre, 1977), and to the presence of, for example, young leaves (White *et al.*, 1975). In addition, determining and standardising the conditions under which the test optimises genotypic expression would ensure sensitivity to genetic variation.

A series of experiments in tropicalised glasshouses in Edinburgh, and in a nursery in Ibadan, Nigeria have revealed the most critical factors that have to be standardised (Ladipo, 1981; Leakey and Longman, 1986), and has led to the development of a "predictive test" for branching habit (Leakey, 1985c). In particular, it seems that a reduction of the number of leaves per plant to 2–6 enhances bud break in decapitated trees, without deleteriously affecting lateral shoot growth (Table 13.3a). Leaf age also affects sprouting because more buds were active following decapitation when the apical bud and the uppermost node with a fully developed leaf were cut off, than from removal of the top nine nodes (Table 13.3b). Similar within-clone variation occurred when the former decapitation treatment was applied to plants of different age and size (Table 13.3b). Within-clone variation is minimal however, when test plants are of a uniform age and size prior to a uniform decapitation.

Light intensity was the most important environmental factor influencing sprouting. Nearly three times as many buds developed as lateral shoots in full West African sunlight (up to *c.* 2000 $\mu E\ m^{-2}\ s^{-1}$) as under heavy (*ca.* 10% full light) palm frond shading (Table 13.3c). Air temperatures were *ca.* 3°C lower under shade, a difference unlikely to have any significant effect on sprouting when temperatures are near optimal for growth. It is clear therefore that "predictive tests" should be done under uniform light environments in which photon flux is high.

**Table 13.3.** Effects of leaf number, site of decapitation, light intensity, nutrient application and watering regime on peak bud activity of different batches of *T. scleroxylon* clones.

| Treatments | Percentage bud activity at week 4 |
|---|---|
| (a) (i) Five uppermost leaves retained | 78.5 |
| Undefoliated (14 leaves) | 50 |
| (ii) Leafless | 28 |
| Two leaves retained | 54 |
| Four leaves retained | 58 |
| Six leaves retained | 56 |
| (b) Decapitation at node 21 (removal of apical bud and top node) | 70 |
| Decapitation at node 13 (removal of 9 nodes) | 48 |
| Decapitation at node 13 (removal of apical bud and top node) | 54 |
| (c) Full sunlight | 75 |
| Shade (10% full sunlight) | 28 |
| (d) High nutrient concentration (4% solution of 23N : 19.5P : 16K) | 86 |
| Low nutrient concentration (0.004% solution 23N : 19.5P : 16K) | 48 |
| (e) Daily watering (250 ml per pot) | 68 |
| Watering every 12th day (250 ml per pot) | 56 |

*Sources*: Ladipo (1981) and Leakey and Longman (1986).

The role of nutrients were not as great as expected, significantly affecting only maximum bud activity 4 weeks after decapitation when plants had been pre-conditioned for 5 weeks with widely differing applications of a complete NPK liquid fertiliser (Table 13.3d). Less contrasting applications of fertiliser did, however, affect the rate at which the most vigorous (usually the uppermost) lateral shoot suppressed and dominated lower shoots (Leakey and Longman, 1986). It appears therefore that plants for screening by the "predictive test" should be grown and tested in soil with uniform nutrient status, and ideally plants should be re-potted a few weeks before decapitation. Watering regimes are also important, not so much because extreme water stress effects sprouting following decapitation (Table 13.3e), but rather because subsequent lateral shoot growth is very sensitive to relatively mild water stress, making it difficult to determine whether or not decreased bud activity is due to the imposition of dominance by upper lateral shoots. However, some clones are apparently able to regulate sprouting according to the availability of water, although most sprout first and then suffer severe wilting. Obviously therefore, a regular and careful watering regime should be adhered to when doing "predictive tests".

### 13.2.1 Application of the "predictive test"

Very considerable clonal variation in bud activity was identified when 26 clones were screened using the "predictive test" under standardised conditions (Fig. 13.4). A positive relationship ($r = 0.76$, $P = 0.001$) was found when percentage bud activity at week 4 was correlated with numbers

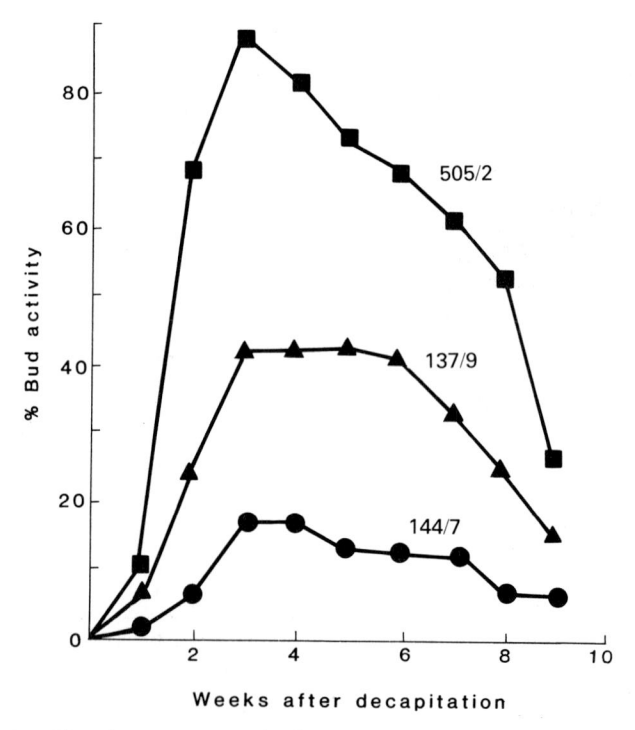

**Figure 13.4.** Clonal variation in bud activity (i.e. percentage of buds growing more than 2 mm per week) following decapitation of small *T. scleroxylon* plants at node 10.

Selection for Improved Tropical Hardwoods

of branches produced in 4 years' growth of the same clones at 2.4 m spacing in Onigambari Forest Reserve, when mean tree height was 7.4 m (Fig. 13.5). From this limited sample it seems clear that the "predictive test" is potentially a very useful method of screening clones for branching habit, at a very early age (i.e. 3–6 months).

### 13.2.2 Possible limitations of the "predictive test"

The wider application of the "predictive test" to different silvicultural regimes, particularly different planting densities which are known to affect branching processes, needs to be tested. In addition, it is important to continue studies relating branching frequency to yield, for although

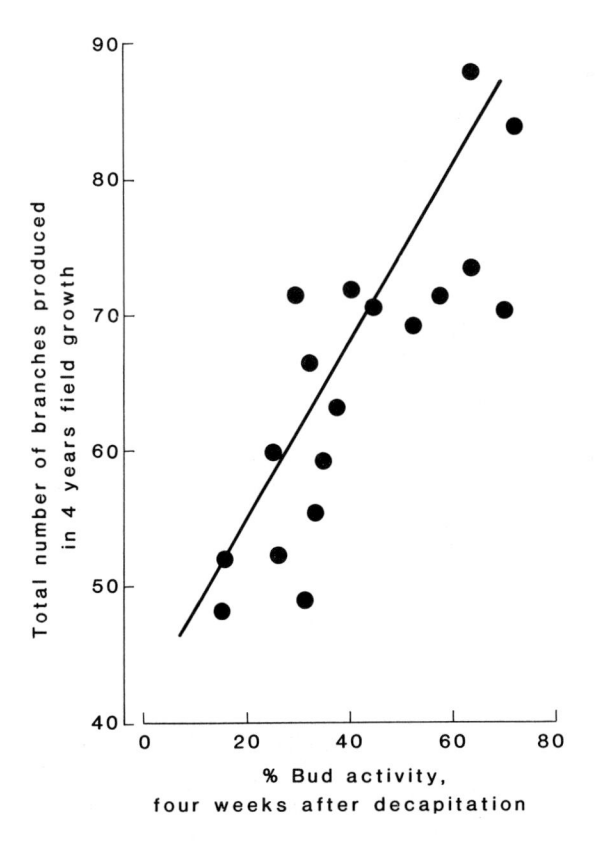

**Figure 13.5.** Relationship between bud activity in small potted plants of *T. scleroxylon* 4 weeks after decapitation and the total number of branches produced during 4 years' growth in plantation.

selection of clones on the basis of branching habit would improve tree form, it is the possibility of being able to select high-yielding clones with good form that is really attractive. Evidence to date suggests that the close relationship found so far between branching habit and yield during the establishment phase of *T. scleroxylon* plantations may decline after canopy closure. Thus, although at wide spacing (4.9 m), tree height (strongly correlated with diameter at breast height; $r = 0.91$ at $P = 0.001$) was still correlated ($r = 0.78$ at $P = 0.01$) 4 years after planting with numbers of branches per metre of mainstem, there was an even stronger relationship ($r = 0.93$ at $P = 0.001$) with the total number of branches. Furthermore, at narrow spacing, any relationship between height (strongly correlated with diameter at breast height; $r = 0.76$ at $P = 0.001$) and branching interval was lost, although total branch number and height were still related ($r = 0.71$ at $P = 0.001$). In the latter instance, in particular, total branch number relates to numbers of branches produced *minus* numbers of lower branches shed. It appears, therefore, that following canopy closure the number of branches constituting the living crown may become critical. Thus in addition to the factors controlling branch production, those factors influencing the ability of lower branches to survive shading and delay self-pruning become important determinants of yield, for which screening methods may have to be developed. In this respect, it is interesting that there is, in *T. scleroxylon*, considerable clonal variation in net photosynthesis, stomatal resistance and mesophyll resistance, and that a good correlation ($r = 0.81$ at $P = 0.05$) between gas exchange parameters and yield can be found if a suitable term is included to allow for respiratory losses from branches, stem and roots (Ladipo *et al.*, 1984). Greater improvements in yield are therefore likely to come from a predictive test in which information on bud activity is combined with information on the $CO_2$ exchange characteristics, and perhaps the apportionment of dry matter to different parts of the tree.

## 13.3 SELECTION FOR WOOD QUALITY

So far, no attempt has been made to examine clonal variation in wood quality in *T. scleroxylon*, although obviously this must be considered before clones are recommended for commercial plantings. However, it seems that, in *Terminalia superba*, the genetic component of variation in wood quality is very considerable, apparently exceeding that which can be attributed to environmental factors (Longman *et al.*, 1979).

## 13.4 IMPROVEMENT OF OTHER TROPICAL HARDWOODS

The vegetative propagation techniques developed for *T. scleroxylon* apply equally well to a wide range of other tropical trees (Leakey *et al.*, 1982b), including many of the other important hardwoods of West Africa. The way is thus open for other trees to be domesticated by selection (Leakey, 1985b). However, the branching patterns of some species like *Terminalia superba*, *Terminalia ivorensis* and *Nauclea diderrichii* are very different from those of *T. scleroxylon*, probably reflecting very different relationships between apical and axillary buds. Nevertheless, apical dominance, as determined by decapitation, is very variable between clones and it remains to be seen how these differences are expressed in the growing tree, and whether it will be necessary to develop very different early selection procedures.

## REFERENCES

Cannell, M. G. R. (1982). 'Crop' and 'Isolation' ideotypes: evidence for progeny differences in nursery-grown *Picea sitchensis*. *Silvae Genet.* **31**, 60–66.

Howland, P. (1975). Vegetative propagation methods of *Triplochiton scleroxylon* K. Schum. *Proc. Symp. on Variation and Breeding Systems of* Triplochiton scleroxylon *K. Schum.*, pp. 99–109. Ibadan, Nigeria.

Howland, P., Bowen, M. R., Ladipo, D. O., and Oke, J. B. (1978). The study of clonal variation in *Triplochiton scleroxylon K. Schum.* as a basis for selection and improvement. *Proc. Joint Workshop IUFRO Working Parties S2.02–08 and S2.03–1*, Brisbane, 1977 (D. G. Nikles, J. Burley and R. D. Barnes, eds), pp. 898–904. Commonwealth Forestry Institute, Oxford.

Jones, N. (1974). Records and comments regarding the flowering of *Triplochiton scleroxylon* K. Schum. *Commonw. For. Rev.* **53**, 53–56.

Kleinschmit, J. (1974). A programme for large-scale cutting propagation of Norway Spruce. *NZ J. For. Sci.* **4**, 359–366.

Ladipo, D. O. (1981). Branching patterns of the tropical hardwood *Triplochiton scleroxylon* K. Schum. with special reference to the selection of superior clones at an early age. Ph.D. thesis, University of Edinburgh, 248 pp.

Ladipo, D. O., Grace, J., Sandford, A., and Leakey, R. R. B. (1984). Clonal variation in photosynthesis, respiration and diffusion resistances in the tropical hardwood *Triplochiton scleroxylon* K. Schum. *Photosynthetica* **18**, 20–27.

Ladipo, D. O., Leakey, R. R. B., Longman, K. A., and Last, F. T. (1983). A study of variation in *Triplochiton scleroxylon* K. Schum.: some criteria for clonal selection, *Silvicultura* **8**, 333–336.

Leakey, R. R. B. (1983). Stockplant factors affecting root initiation in cuttings of *Triplochiton scleroxylon* K. Schum., an indigenous hardwood of West Africa. *J. hort. Sci.* **58**, 277–290.

Leakey, R. R. B. (1985a). The capacity for vegetative propagation in trees. *In* "Attributes of Trees as Crop Plants", (M. G. R. Cannell and J. E. Jackson, eds), pp. 110–133. Institute of Terrestrial Ecology, Abbots Ripton, Huntingdon, England.

Leakey, R. R. B. (1985b). Cloned tropical hardwoods—quicker genetic gain. *Span* **29**, 35–37.

Leakey, R. R. B. (1985c). Prediction of branching habit in clonal *Triplochiton scleroxylon*. In "Crop Physiology of Forest Trees", (P. M. A. Tigerstedt, P. Puttonen and V. Koski, eds), pp. 71–80. University of Helsinki Viikki, Finland.

Leakey, R. R. B., and Longman, K. A. (1986). Physiological, environmental and genetic variation in apical dominance as determined by decapitation in Triplochiton scleroxylon. *Tree Physiol.* **1**, 193–207.

Leakey, R. R. B., Chapman, V. R., and Longman, K. A. (1975). Studies on root initiation and bud outgrowth in nine clones of *Triplochiton scleroxylon* K. Schum. *Proc. Symp. Variation and Breeding Systems of* Triplochiton scleroxylon K. Schum., pp. 86–92. Ibadan, Nigeria.

Leakey, R. R. B., Chapman, V. R., and Longman, K. A. (1982a). Physiological studies for tropical tree improvement and conservation. Some factors affecting root initiation in cuttings of *Triplochiton scleroxylon* K. Schum. *For. Ecol. Mgmt* **4**, 53–66.

Leakey, R. R. B., Ferguson, N. R., and Longman, K. A. (1981). Precocious flowering and reproductive biology of *Triplochiton scleroxylon* K. Schum. *Commonw. For. Rev.* **60**, 117–126.

Leakey, R. R. B., Longman, K. A., and Last, F. T. (1982b). Domestication of forest trees: a process to secure the productivity and future diversity of threatened tropical ecosystems. *Commonw. For. Rev.* **61**, 33–42.

Leakey, R. R. B., Last, F. T., Longman, K. A., Ojo, G. O. A., Oji, N. O., and Ladipo, D. O. (1983). *Triplochiton scleroxylon*: a tropical hardwood for plantation forestry. *Silvicultura* **8**, 346–348.

Longman, K. A., Leakey, R. R. B., and Denne, M. P. (1979). Genetic and environmental effects on shoot growth and xylem formation in a tropical tree. *Ann. Bot.* **44**, 377–380.

McIntyre, G. I. (1977). The role of nutrition in apical dominance. In "Integration of Activity in the Higher Plant" (D. H. Jennings, ed.), pp. 251–273. Cambridge University Press.

Sziklai, O. (1974). Juvenile–mature correlation. *Proc. IUFRO Meeting*, Stockholm, pp. 217–235.

White, J. C., Midlow, G. C., Hillman, J. R., and Wilkins, M. B. (1975). Correlative inhibition of lateral bud growth in *Phaseolus vulgaris* L. Isolation of indoleacetic acid from the inhibitory region. *J. exp. Bot.* **26**, 419–424.

# Part IV Sources and Exploitation of Genetic Variation

# 14 Genetic Variation in Temperate Forest Trees

J. KLEINSCHMIT

*Lower Saxony Forest Research Institute, Escherode, FRG*

The study of genetic variation in forest tree species has a long tradition (Langlet, 1964). The starting point was the practical interest of foresters in exploiting natural variation for the improvement of stands. Differences in survival, quality and growth rate between different geographical sources of the same species were obvious and this led, as early as 1745, to the establishment of the first comparative test with pine by Duhamel du Monceau. From this time, a steadily increasing number of provenance and progeny tests for the different temperate forest tree species has been established, giving information on the variation between different geographical sources of the same species and between individuals within populations.

The results of these studies had implications for taxonomic considerations and population genetics.

One of the problems in provenance or progeny tests is the long lifespan of forest trees, which makes a long follow-up of the experiments necessary before results can be transferred to practical forestry. The number of trees in a forest is reduced by about a factor of ten between planting and the final harvest; during the development period of up to 100–300 years, natural selection acts on the material, changing the genetic composition in the direction of the best-adapted trees for the particular site. This means that it is not possible to judge clearly the suitability of a certain source at a more advanced age, especially if the selection pressure changes with increasing age, thereby affording different adaptations at different developmental stages (Barber, 1958). The total genetic variation of a specific material is only available at the beginning of the test period. This total variability is necessary for assessment of the adaptational potential of the source. This shows one of the principal problems in studying genetic variation in forest tree species in long-term experiments.

Conclusions about adaptation at a young age have often been shown to be wrong. Most of the studies are based on a juvenile stage of development; in the light of the problems this has some justification as far as the study of the total genetic variation is concerned. It is more questionable, however, if the objective is to provide recommendations for forestry, especially if little or no information on the long-term adaptational behaviour is available. One of the consequences of natural selection in populations during testing is that the genetic differentiation between populations decreases as the experiments become older. Inevitably, those individuals best adapted to the local site are selected. If the initial sources have a broad genetic base, there can be a considerable overlap, which leads, with time, to an increasing similarity of the different sources (Fig. 14.1). Fischer (1949) has demonstrated that Norway spruce from a low-elevation source planted at high elevations was eventually more like that from high-elevation sources in phenological characters than from a low-elevation source. We find similar situations during the development of "land races"; for example, in Douglas fir, where by natural selection the succeeding generation may be much better adapted, at least in phenological characteristics, than the original source (Pellati, 1969).

## 14.1 LEVELS OF VARIATION

As in other plant species, we find three levels of variation within forest tree species;

(a) variation between populations within species;
(b) variation between individuals within populations; and
(c) variation within individuals (degree of heterozygosity).

I shall not consider variation between species here. There is much experimental information available for all three levels of variation within species. Variation between populations has been described in the classical work of Langlet (1963, 1964), which gives most of the basic information. This has been extended mainly by biochemical studies during recent years. Variation between individuals is seen in most forest tree breeding programmes.

The total individual variation can be subdivided into environmental and genetic components, especially with vegetatively propagated species. Subdivision into additive, dominant and sometimes epistatic gene effects is evident from the results of different crossing programmes, thus giving

Genetic Variation in Temperate Forest Trees

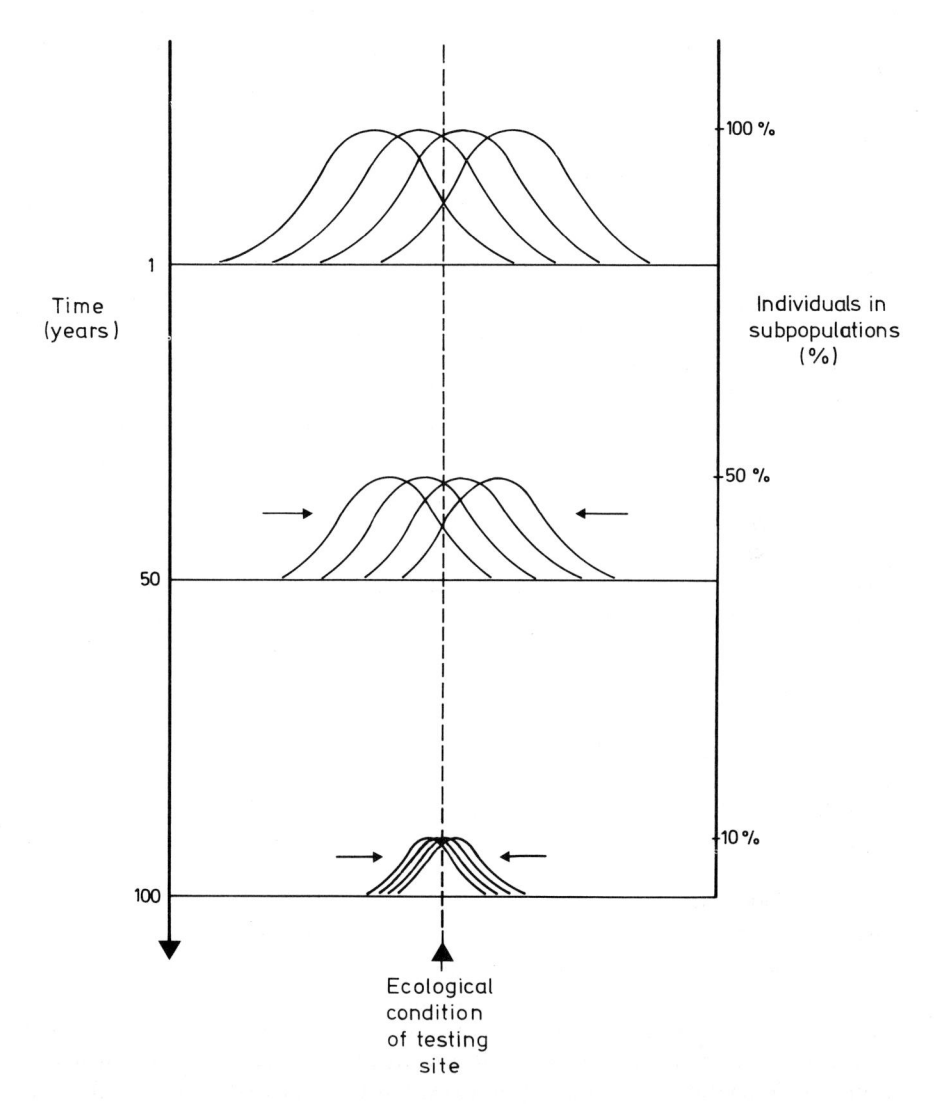

**Figure 14.1.** Result of natural selection on subpopulation differences.

the breeders an idea of how far this level of variation can be used in seed production breeding programmes. Biochemical studies of genetic variation within single coniferous trees are possible, using the haploid endosperm of the seed (Bartels, 1971).

This chapter summarises the most important results and considers a specific case study, using the Norway spruce. This is the most economically important temperate forest tree species in Europe, and one for which vegetative propagation is a good tool, the use of which should enable a better understanding of the variation in respect of practical forestry.

### 14.1.1. Variation between populations (provenances)

Forest tree populations have been adapted to the ecological conditions of their habitat by natural selection. Different species with a similar natural range often show parallel evolution (Kung and Wright, 1972). Climatic factors are by far the most important influences. It is still not completely known how far edaphic conditions have a significant influence on the selection of populations, as in annual plants.

Such influences may exist, as shown for beech by Le Poutre and Teissier du Cros (1979), and for white spruce by Farrar and Nicholson (1966) and Teich and Holst (1974). Giertych (1969) suspects that often they have not been detected due to the lack of precise data. Weiser (1964a, b, 1965, 1974) and von Schönborn (1967) did not find "soil ecotypes" for *Fraxinus*. The situation is probably different in different tree species and according to the characteristics observed. At least the size of edaphic influences on the variation between populations of tree species is minor compared to that of climatic influences.

The different characters may have very different correlations with ecological variables, due to their heritability and to adaptive importance (e.g. Holzer, 1978). For most characters a strong clinal component of variation was found. But in the same species, clinal and ecotypic variation may occur, depending on the characters observed. This has been demonstrated for different species (Stern, 1964; Burley, 1965).

Between species there are pronounced differences due to their natural range. Species with a large continuous range and steady gene-exchange between populations have more clinal variation (Koski, 1973, 1974). Species with discontinuous patterns, in mixed stands and specific ecological niches, have more ecotypic variation, even within related groups of trees, and a certain degree of inbreeding (Sakai *et al.*, 1972). This occurs in spite of the strong selection against inbreeding in most forest tree species (Sorensen, 1971; Libby *et al.*, 1981).

Colonising species are less closely adapted to clines and ecological conditions; they often reproduce from one or few individuals and therefore often show larger differences between populations than climax species (Stern, 1964; Rehfeldt and Lester, 1969). Studies of biochemical variation, using isozymes, have been summarised by Feret and Bergmann (1976).

In many tree species, introgressive hybridisation creates new variability in the contact zones of populations (Bobrov, 1965; Morgenstern and

Farrar, 1965; Boden, 1966; Sutilov, 1968; Dancik and Barnes, 1972). The improved understanding derived from biochemical studies in forest trees has helped to clarify instances of natural hybridisation (Hunt and von Rudloff, 1974, 1979; Bonnet-Masimbert and Bikay-Bikay, 1978; Zavarin et al., 1980; Forrest, 1981).

As a rule, forest trees seem to be much less adapted to a "fine grained pattern" of environment than annual plants. This can be explained by random events during the historical development of the species, by the longevity of the forest trees, the late time of reproduction and the changes of their environment in time and space. Compared with annual plants, forest trees need 100–1000 times the duration to reach the same degree of specialisation. This is further complicated by regularly occurring successions and the changes in environment as trees grow. They may germinate in the litter under reduced light conditions, and in competition with other species. Later on, they reach different soil horizons and open light which induce different adaptational behaviour. In addition, microclimate and sometimes macroclimate, e.g. by pollution, may change. Therefore, as a rule, very specific adaptations would not be too useful for most tree species. The results of testing different provenances of forest tree species have been used for provenance selection and for the establishment of general rules for seed transfer (see Langlet, 1936; Campbell, 1974; Campbell and Sorensen, 1978).

### 14.1.2 Variation between individuals within populations

All studies show that there is much variation within populations. This is true for all characteristics observed. Exceptions are species, such as aspen, which also reproduce vegetatively in nature. Biochemical studies have helped to describe this variation, but they do not always correlate well with morphological or physiological variation. Some examples demonstrate the adaptational importance of such traits, while others do not.

Variation in provenances is invariably greater than variation between provenances, and for some biochemical characteristics this difference is 90 times greater (Hunt and von Rudloff, 1977; Yeh and El-Kassaby, 1980; Yeh and Malley, 1980). The variation within subpopulations decreases in marginal populations due to increasing selection pressure (Matsura and Sakai, 1972; Bergmann and Gregorius, 1978). Declining populations may have similar characteristics to those in marginal populations (Langner, 1959). Therefore, the potential for selection may be quite different from one population to another.

However, the evidence that subpopulations of forest trees regularly maintain a large amount of variation is a striking fact. This is also true

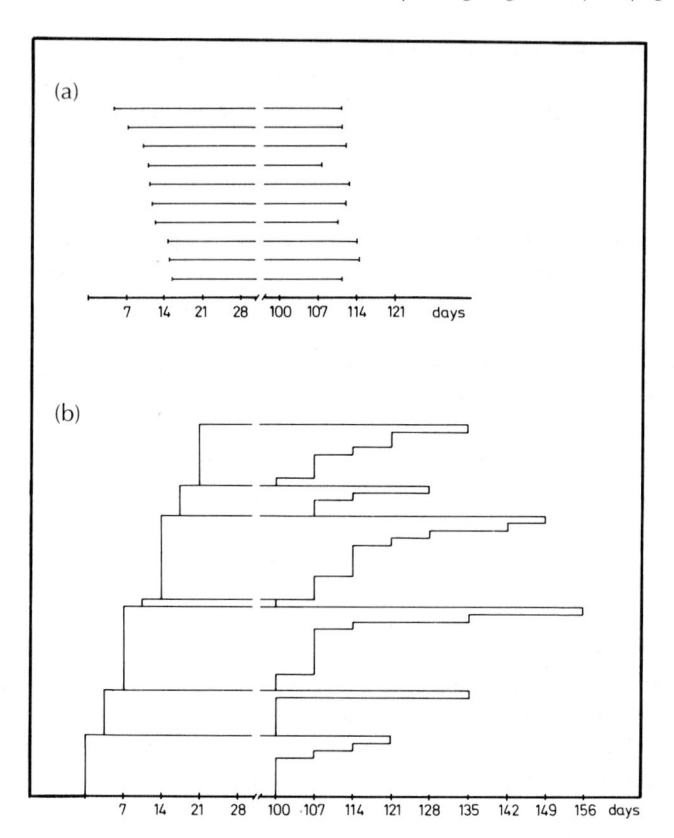

**Figure 14.2.** Variation in time of flushing and bud set of Norway spruce. (a) Variation between subpopulations (n=10); (b) variation within subpopulations (n=50)

for characteristics such as flushing and bud set (Fig. 14.2) that respond quickly to selection due to their high heritability and adaptive importance. This can be explained by the changing environment in space and time and/or by superiority of heterozygous individuals.

Studies using clonal material have helped to assess the size of the total genetic component of this variation in Norway spruce (see below).

### 14.1.3 Degree of heterozygosity

Most temperate forest tree species seem to maintain a high degree of heterozygosity and a high genetic diversity (Gregorius, 1986). A degree of selfing of between 0–26% was estimated by Lundkvist (1978) for Norway spruce. The selfing rate probably differs with different species, and is especially low in species with a large continuous natural area (Sorensen, 1971; Koski, 1973, 1974); but higher in relict stands (Langner, 1959) or small isolated populations. However, seed set is lower in selfed flowers (Cram, 1984), and the survival of seedlings from selfed-seed is low, which reduces this rate considerably. Moreover, plants from selfed-seed have reduced growth rates and are thus eliminated by natural selection, as discussed earlier. Most of the species are mainly cross-fertile and heterozygous individuals often have a selection advantage. This leads to an increase in heterozygosity in the surviving part of the population with time. The high degree of heterozygosity seems to be necessary for the trees to survive in the heterogeneous environment that faces them during their ontogenetic development. With the help of isozyme techniques, it was possible to trace selection processes in tree species and to show that heterozygous individuals have a higher survival rate under stress conditions (Scholz and Bergmann, 1984; Bergmann and Scholz, 1985; Müller-Starck, 1985).

## 14.2 STUDIES IN NORWAY SPRUCE (*Picea abies* Karst.)

Norway spruce has been intensively studied for a long time (see review by Schmidt-Vogt, 1978). The following account concentrates on our own work with Norway spruce, using vegetative propagation as a method for studying genetic variation, and the use of this information in our breeding programme. This may serve as an example for other temperate tree species, for which we are developing methods of vegetative propagation with the intention of integrating them into the breeding programmes (Kleinschmit *et al.*, 1975; Vieitez *et al.*, 1977; Spethmann, 1985a, b, 1986).

There is considerable variation between provenances in Norway spruce (Weisberger *et al.*, 1976, 1977). The most intensive study by Langlet (1964) includes 1100 provenances representing the whole natural range. Dietrichson *et al.* (1976) published the results from 38 9–12-year-old field experiments in eight countries. More recent results of the same set of experiments were presented at an IUFRO meeting in Vienna, 1985. One of the interesting results of this study is that some provenances perform well in a wide range of sites and climates. This supports the hypothesis that specialisation is comparatively low in many temperate forest tree species. Some of the generally good provenances originated from White Russia, Rumania and Czechoslovakia.

We followed the performance of 500 individuals from ten German provenances for a study of the genetic variation within provenances and at the same time, started a broad selection programme which includes

55 000 clones. Roughly two-thirds of these have already been rejected through testing; the rest are still being propagated and are included in clonal tests.

In the course of this programme we started species hybridisation with all the spruce species we had at hand. More than 200 different hybrid progenies have resulted from this work so far, and many superior genotypes are included in the clonal selection scheme (Hoffmann and Kleinschmit, 1979).

### 14.2.1 Genetic variation in Norway spruce

The total variation in height at age 10 years for 1100 provenances in the Langlet experiments ranged from 50 to 150% of the overall mean, including all extremes, with heritability estimates of $h^2_b = 0.25-0.35$. Since we took only German provenances, which cover roughly 40% of the total range of all growing provenances, for our study, this explains the more limited range between provenance means. The clones had passed only a first phenotypical selection which may have reduced the total variation for height growth to 70% (Fig. 14.3). This leads to an overestimation of clonal influences as compared with the total species base. However, it is a realistic estimate, since it includes the material which would regularly enter breeding programmes. There were three absolute replicates of each clone. On this basis we calculated the genetic variation for different characters of the provenances and individuals (Table 14.1), and the possible genetic gain by using either provenance selection or clonal selection only.

Clones account for far more of the total variation than provenances and, depending on the character, exceed provenance variation by a factor of 3–50. This is reflected in the much better scope for genetic improvement by clonal selection following provenance selection (Sauer-Stegmann et al., 1978). Thus, height gain by clonal selection is roughly three times that obtained by provenance selection. Clonal selection is especially effective for phenological characters, though it seems better to maintain genetic variation to make use of the heterogenous environment and the annual weather fluctuations.

### 14.2.2 Correlations and grouping

If one compares the correlations between height and the different characters observed, this can be interpreted as a measure of co-adaptation. In this study we calculated genetic correlations for clones and provenances and phenotypic correlations for single plants (Table 14.2).

(a)

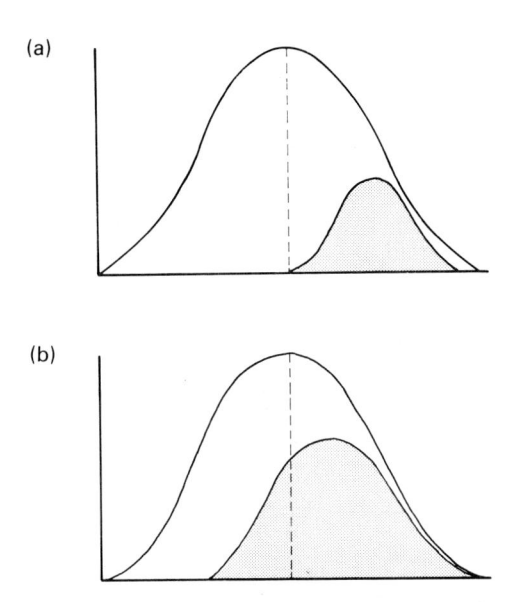

(b)

**Figure 14.3.** Situation of tested material in the total range of variation: (a) provenances; (b) individuals.

**Table 14.1.** Results of analysis of variance (hierarchical, orthogonal and random effects).

|  | F-values |  | Components of variance (%) |  |  |
|---|---|---|---|---|---|
| Characters | Provenances | Clones | Provenances | Clones | Residual |
| Flushing | 8.88*** |  | 13.8 | 86.2 | 0 |
| Bud set | 7.99*** | 0 | 12.2 | 87.8 | 0 |
| Lamma shoot formation | 1.81 n.s. | 0 | 1.6 | 98.4 | 0 |
| Height (1972) | 12.17*** | 4.71*** | 13.7 | 45.2 | 41.1 |
| Height (1973) | 15.9*** | 5.08*** | 16.9 | 48.6 | 34.5 |
| Root collar diameter | 8.22*** | 2.00*** | 7.9 | 32.6 | 59.6 |
| Branch angle | 4.34*** | 8.31*** | 5.8 | 66.7 | 27.5 |
| Branch length | 15.42*** | 5.43*** | 17.4 | 47.3 | 35.3 |
| Needle length | 3.29*** | 4.02*** | 3.5 | 49.7 | 46.8 |
| Needle width | 12.79*** | 6.25*** | 14.7 | 55.4 | 29.9 |
| Nitrogen uptake | 11.05*** | 9.72*** | 14.1 | 62.6 | 23.2 |
| Phosphorus uptake | 5.83*** | 11.77*** | 6.5 | 72.8 | 20.7 |
| Potassium uptake | 5.61*** | 10.32*** | 6.4 | 71.8 | 21.8 |
| Calcium uptake | 3.76*** | 11.33*** | 4.1 | 74.5 | 21.4 |
| Magnesium uptake | 17.96*** | 17.21*** | 22.3 | 65.4 | 12.3 |

***Significant on 0.01% level.

**Table 14.2.** Simple linear correlations of height with different characters in Norway spruce.

|  | Provenance level | Clonal level | Single plant level (phenotypic) |
|---|---|---|---|
| Diameter | 0.87 | 0.75 | 0.74 |
| Branch length | 0.81 | 0.73 | 0.70 |
| Needle length | −0.01 | 0.23 | 0.18 |
| Needle width | 0.50 | 0.25 | 0.27 |
| Branch angle | −0.18 | −0.03 | 0.04 |
| Tropism | −0.10 | −0.15 | −0.10 |
| Shape | 0.12 | 0.34 | 0.28 |
| Number of needles | −0.52 | −0.30 | −0.23 |
| Needle colour | 0.58 | 0.16 | 0.16 |
| Needle position | −0.31 | −0.14 | −0.13 |
| Needle tip | 0.51 | 0.03 | 0.01 |
| Needle barbs | −0.36 | −0.24 | −0.25 |
| Flushing | 0.69 | 0.19 | 0.17 |
| Bud set | 0.06 | 0.17 | 0.15 |
| Lamma shoot | −0.69 | 0.33 | 0.29 |
| Nitrogen content | −0.34 | −0.24 | −0.17 |
| Potassium content | −0.23 | −0.07 | −0.05 |
| Phosphorus content | 0.04 | 0.09 | 0.06 |
| Calcium content | 0.55 | 0 | −0.01 |
| Magnesium content | 0.57 | 0.08 | 0.04 |

These results show that the correlations based on provenance are generally higher than on a clonal level, and these differences are significant. The same is true for the comparison of correlations between clones and single plants.

On the higher integrated level of the population, co-adaptation is more obvious than on the individual level. The individuals are more variable than the provenances and retain greater flexibility for the whole system.

There is no clear general trend for the different relationships within the provenances. Flushing, a characteristic of economic importance in Norway spruce, has been correlated with height in various studies, but the results have often been contradictory. This can be partly explained by differences in climate of the growing sites. However, this does not seem to be the only reason. As demonstrated in Fig. 14.4, the individuals of different provenances may behave quite differently. Positive and negative correlations occur on the same growing site. At the clonal level, even the length of growing season shows only weak positive correlation with height ($r = 0.32$, $P = 0.01$). There are clones with a short vegetative period and excellent growth and vice versa (Fig. 14.5).

Genetic Variation in Temperate Forest Trees

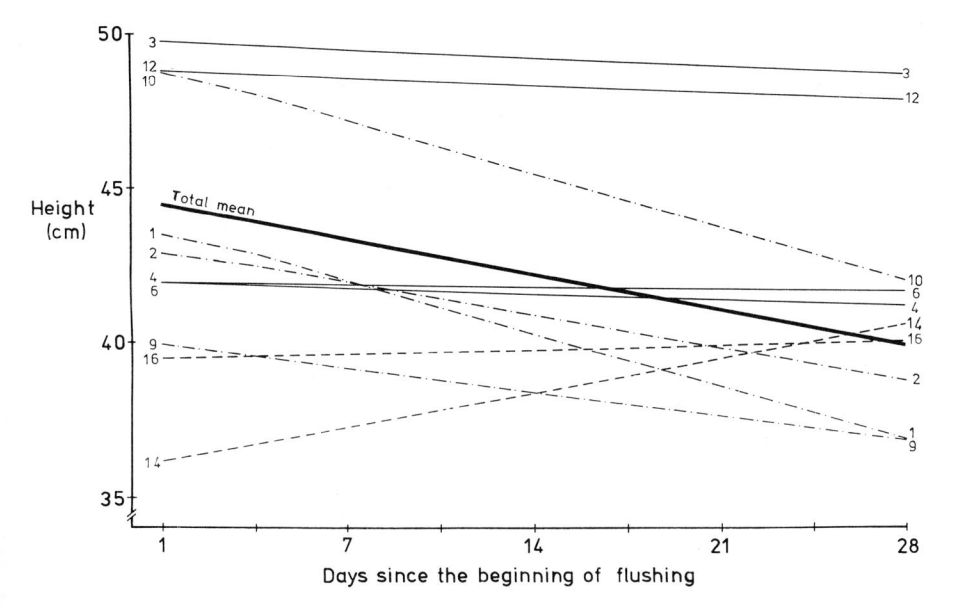

**Figure 14.4.** Regressions of height against flushing time for Norway spruce from ten provenances (each represented by 50 clones) in 1973.

The single plants of a provenance react far less predictably than the provenances themselves. This is reflected in the greater production stability of subpopulations compared to individuals, since the single components in an integrated system of a subpopulation are able to make use of a heterogeneous environment much better than each separate unit. This can be due to natural selection in the mixture as discussed earlier or to compensatory effects, e.g. following annual weather fluctuations. The single components may then support total production differently (Kleinschmit et al., 1981a,b).

In a cluster analysis, using "Ward's Error sum of squares method" and the euclidic distance, we made an automatic grouping of the single plants according to their similarity. Altogether there were 99 characteristics for every clone, including analyses of the polyphenol and terpene content. There were 1500 candidates to be grouped according to the variation in these 99 characteristics.

The results showed some interesting features:

(a) Single plants of each clone were allocated to the same subgroup. This means that a clonal identification on the basis of 99 characters is possible.

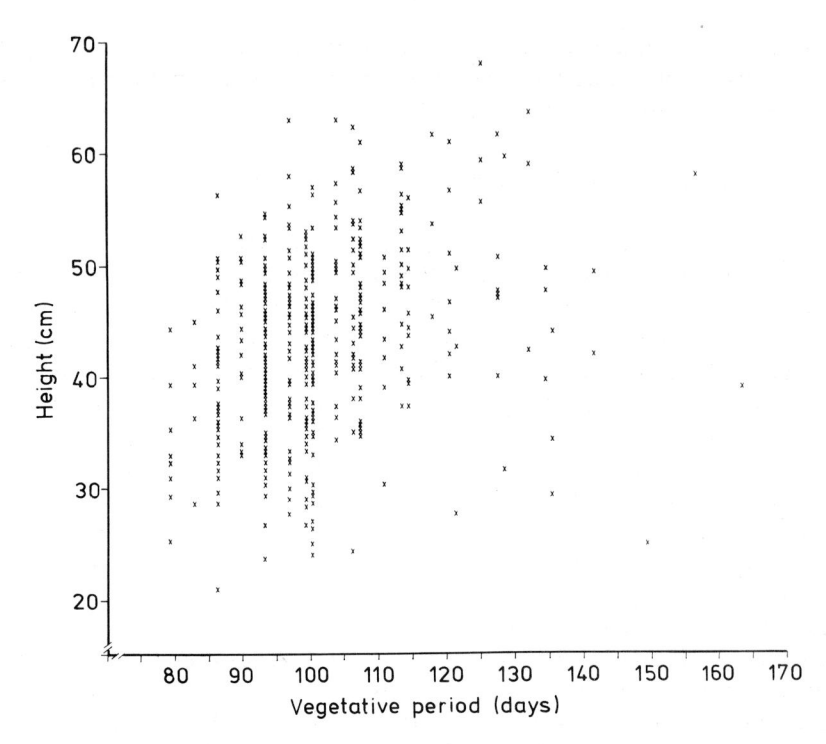

**Figure 14.5.** Correlation between tree height and length of vegetative period of Norway spruce in 1973.

(b) Also, clones of different provenances can be allocated to the same group, while clones of the same provenance may be placed in quite different groups.
(c) Some clones of the same population can vary as much as clones from different populations. Only a few clusters of clones of the same provenance exist.

This shows that on a clonal level even in a multidimensional space there is little co-adaptation. Heterogeneity is more common at the clonal level than are the common characteristics within one population. These results could be partly influenced by the fact that not all provenances were autochthonous and that during stand development there was cross-pollination, and in some cases a mixing of sources may have happened. But these findings are true also for those provenances which we believe to be autochthonous and more-or-less unchanged in their natural range.

The high amount of genetic variation within populations is one characteristic of temperate forest tree species and gives an excellent base for selection and tree improvement.

### 14.2.3 Selection and testing programme

In the course of our breeding programme we try to make use of this variation by selection of outstanding clones and, at the same time, creating new variation by hybridisation. One consequence of the results is that we try to maintain genetic variation in our clonal material in all but production characteristics, since the high variation seems to have selective advantages. But other considerations show that a strategy of stepwise reduction of genetic variation is wise (Kleinschmit, 1979). Theoretically, it is an excellent procedure to use cuttings for mass propagation and in this way transmit genetic gain quickly into practice, as stressed repeatedly by Libby (1983).

However, there are some restraints. The most serious of all is probably ageing, especially in conifers, which can limit progress considerably. The restriction of genetic diversity may also cause trouble, especially if propagation is based only on early testing. Controlling nomenclature becomes a problem, if mixtures of cuttings are used commercially. More information is needed, also, on clonal stability over long periods of time and in different environments. Further research in this field is urgently needed.

## REFERENCES

Barber, H. N. (1958). The process of natural selection. *Proc. 10th Int. Congr. Gen., Montreal*, pp. 13–14.
Bartels, H. (1971). Genetic control of multiple esterases from needles and macrogametophytes of *Picea abies*. *Planta* **99**, 283–289.
Bergmann, F., and Gregorius, H. R. (1978). Comparison of genetic diversities of various populations of Norway Spruce (*Picea abies* K.). *Proc. Conf. Biochem. Genet. For. Trees*, Report 1.
Bergmann, F., and Scholz, F. (1985). Effects of selection pressure by $SO_2$ pollution on genetic structures of Norway spruce (*Picea abies*). *In* "Population Genetics in Forestry", (H. R. Gregorius, ed.) pp. 267–275. Springer-Verlag, Berlin.
Bobrov, E. G. (1965). Le rôle de l'hybridization introgressive dans la flore de la Sibérie et de l'Europe orientale. [The role of introgressive hybridization in the flora of Siberia and eastern Europe.] *Acta bot. hung.* **12**, 1–8.
Boden, R. W. (1966). Hybridisation in *Eucalyptus*. *Ind. For.* **90**, 581–586.
Bonnet-Masimbert, M., and Bikay-Bikay, V. (1978). Variabilité intraspécifique des isozymes de la glutamate–oxaloacetate transaminase chez *Pinus nigra* Arnold Intérêt pour la taxonomie des sous espèces. *Silvae Genet.* **27**, 71–79.

Burley, J. (1965). Genetic variation in *Picea sitchensis* (Bong.) Carr.—a literature review. *Commonw. For. Rev.* **44**, 47–59. [Ref. in *Silvae Genet.* **15**, 1–192, 194.]

Campbell, R. K. (1974). A provenance-transfer model for boreal regions. *Norsk. Inst. Skogsforskning* **1**, 544–564.

Campbell, R. K., and Sorensen, F. C. (1978). Effect of test environment on expression of clines and on delineation of seed zones in Douglas fir. *Theoret. appl. Genet.* **51**, 233–246.

Cram, W. H. (1984). Some effects of self-, cross- and open pollinations in *Picea pungens*. *Can. J. Bot.* **62**, 292–296.

Dancik, B. P., and Barnes, B. V. (1972). Natural variation and hybridization of Yellow Birch and Bog Birch in Southeastern Michigan. *Silvae Genet.* **21**, 1–9.

Dietrichson, J., Christophe, C., Coles, J. F., De Jamblinne, A., Krutzsch, P., König, A., Lines, R., Magnesen, S., Nanson, A., and Vins, B. (1976). The IUFRO Provenance Experiment of 1964/68 on Norway spruce. *IUFRO–Norwegian For. Res. Inst.*, pp. 1–14.

Duhamel du Monceau (1745). *In* Langlet, O. (1971). Two hundred years' genecology. *Taxon* **20**, 653–722.

Farrar, J. L., and Nicholson, I. I. (1966). Tree breeding work at the Faculty of Forestry, University of Toronto. *Proc. 10th Meeting Comm. For. Tree Breed. Can.* **2**, 35.

Feret, P. P. and Bergmann, F. (1976). Gel electrophoresis of proteins and enzymes. *In* "Modern Methods in Forest Genetics", (J. P. Miksche, ed.), pp. 49–77. Springer-Verlag, Berlin.

Fischer, F. (1949). Ergebnisse von Anbauversuchen mit verschiedenen Fichtenherkünften (*Picea abies* (L. J. Karst.). *Mitt. schweiz. Anst. forstl. VersWes.* **26**, 153–204.

Forrest, G. J. (1981). Geographical variation in oleoresin monoterpene composition of *Pinus contorta* from natural stands and planted seed collections. *Biochem. Syst. Ecol.* **9**, 97–103.

Giertych, M. M. (1969). Growth as related to nutrition and competition. *Second Wld, Consult. For. Tree Breed, Wash.* **I**, 37–59.

Gregorius, H. R. (1986). The importance of genetic multiplicity for tolerance of atmospheric pollution. *Proc. 18th IUFRO Wld Congr. Div. 2* **I**, 295–305.

Hoffmann, D., and Kleinschmit, J. (1979). An utilization program for provenance and species hybrids. *IUFRO Norway Spruce Meeting, Bucarest*, pp. 216–236.

Holzer, K. (1978). Die Kulturkammertestung zur Erkennung des Erbwertes bei Fichte (*Picea abies* (L.) Karst.): 2. Merkmale des Vegetationsablaufes. *Centralblatt für das Gesamte Forstwesen* **95**, 30–51.

Hunt, R.S., and Von Rudloff, E. (1974). Chemosystematic studies in the genus *Abies*: I. Leaf and twig oil analysis of alpine and balsam firs. *Can. J. Bot.* **52**, 477–487.

Hunt, R. S., and Von Rudloff, E. (1977). Leaf–oil–terpene variation in Western White Pine populations of the Pacific Northwest. *For. Sci.* **23**, 507–513.

Hunt, R. S., and Von Rudloff, E. (1979). Chemosystematic studies in the genus *Abies*: IV. Introgression in *Abies lasiocarpa* and *Abies bifolia*. *Taxon* **28**, 297–309.

Kleinschmit, J. (1979). Limitations for restrictions of the genetic variation. *Silvae Genet.* **28**, 61–67.

Kleinschmit, J., Lunderstadt, J., and Svolba, J. (1981b). Charakterisierung von Fichtenklonen (*Picea abies* Karst.) mit Hilfe morphologischer, physiologischer and biochemischer Merkmale (III). *Silvae Genet.* **30**, 122–127.

Kleinschmit, J., Witte, R., and Sauer, A. (1975). Möglichkeiten der züchterischen Verbesserung von Stiel- und Traubeneiche: II. Versuche zur Stecklingsvermehrung von Eiche. *Allg. Forst-u. Jagdztg* **146**, 179–186.
Kleinschmit, J., Sauer-Stegmann, A., Lunderstadt, J., and Svolba, J. (1981a). Charakterisierung von Fichtenklonen (*Picea abies* Karst.): II. Korrelation der Merkmale, *Silvae Genet.* **30**, 74–82.
Koski, V. (1973). On self-pollination, genetic load, and subsequent inbreeding in some conifers. *Communs Inst. Forest. Fenn.* **78**, 1–42.
Koski, V. (1974). On the effective population size in an boreal continuous forest. *IUFRO Meeting of Population Genetics, Breeding Theory and Progeny Testing in Sweden*, pp. 1–18.
Kung, F. H., and Wright, J. W. (1972). Parallel and divergent evolution in Rocky Mountain trees. *Silvae Genet.* **21**, 77–85.
Langlet, O. (1936). Studier över tallens fysiologiska variabilitet och dess samband med climatet. *Medd. Statens Skogsf. Anst.* **29**, 1–4.
Langlet, O. (1963). Patterns and terms of intraspecific ecological variability. *Nature* **200**, 347–348.
Langlet, O. (1964). Proveniensvalets betydelse för produktion och skogsträdsförädling av gran. *Sv. Skogsv. Fören. Tidskr.* 1964, 145–155. [Ref. in *Silvae Genet.* **16**, 117.]
Langner, W. (1959). Selbstfertilität und Inzucht bei *Picea omorika*. *Silvae Genet.* **8**, 69–104.
Lepoutre, B., and Teissier du Cros, E. (1979). Increment and nutrition of one year old beech seedlings (*Fagus silvatica* L.) from different provenances, on natural acid substrate and the same substrate added with lime. *Annls Sci. For.* **36**, 239–262.
Libby, W. J. (1983). Potential of clonal forestry. *Proc. 19th Meeting Can. Tree Improvement Assoc. Toronto*, pp. 1–11.
Libby, W. J., McCutchan, B. G., and Miller, C. J. (1981). Inbreeding depression in selfs of redwood. *Silvae Genet.* **30**, 15–25.
Lundkvist, K. (1978). Allozymes in Population Genetic Studies of Norway Spruce (*Picea abies* K.). Dissertation, Universität Umea, pp. 42–70.
Matsura, T., and Sakai, K. (1972). Geographical variation on an isoenzyme level in *Abies sachalinensis*. *IUFRO Genetics — Sabrao Joint Symposium*, Tokyo, pp. 1–11.
Morgenstern, E. K., and Farrar, J. L. (1965). Introgressive hybridization in red spruce and black spruce. *Univ. Toronto, Fac. Forestry, Tech. Rep.* **4**, 46 pp. (1964). [Ref. in *Silvae Genet.* **14**, 136.]
Müller-Starck, G. (1985). Genetic differences between "tolerant" and "sensitive" beeches (*Fagus sylvatica* L.) in an environmentally stressed adult forest stand. *Silvae Genet.* **34**, 241–247.
Pellati, E. de Vecchi (1969). Evolution and importance of land races in breeding. *Second World Consult. For. Tree Breed., Wash.* **II** (1970), 1263–1278.
Rehfeldt, G. E., and Lester, D. T. (1969). Specialization and flexibility in genetic systems of forest trees. *Silvae Genet.* **18**, 118–123.
Sakai, K. I., Iyama, S., Miyazaki, Y., and Iwagami, S. (1972). Genetic studies in natural populations of forest trees. *IUFRO Genetics — Sabrao Joint Symposium, Tokyo*, pp. 1–13.
Sauer-Stegmann, A., Kleinschmit, J., and Lunderstädt, J. (1978). Methoden zur Charakterisierung von Fichtenklonen (*Picea abies* Karst.). *Silvae Genet.* **27**, 109–117.
Schmidt-Vogt, H. (1978). Monographie der *Picea abies* (L.) Karst. unter Berücksichti-

gung genetischer and züchterischer Aspekte. *Forstwiss. Centralblatt* **97**, 281–302.
Scholz, F., and Bergmann, F. (1984). Selection pressure by air pollution as studied by isoenzyme–gene-systems in Norway Spruce exposed to sulphur dioxide. *Silvae Genet.* **33**, 238–241.
Sorensen, F. (1971). Estimate of self-fertility in coastal Douglas fir from inbreeding studies. *Silvae Genet.* **20**, 115–120.
Spethmann, W. (1985a). Stecklingsvermehrung von Traubeneiche. *Allg. Forstz.* **41**, 693–695.
Spethmann, W. (1985b). Propagation of *Prunus avium* by summer and winter cuttings and survival after different storage treatments. *Acta Hort.* **169**, 353–362.
Spethmann, W. (1986). Stecklingsvermehrung von Stiel- und Traubeneiche (*Quercus robur* L. und *Quercus petraea* (Matt.) Liebl.). *Schriften aus der Forstlichen Fakultät der Universität Göttingen und der Niedersächsischen Forstlichen Versuchsanstalt* **86**, 1–99.
Stern, K. (1964). Herkunftsversuche für Zwecke der Forstpflanzenzüchtung, erläutert am Beispiel zweier Modellversuche. *Züchter* **34**, 181–219.
Sutilov, V. A. (1968). Introgressive Hybridization und Variabilität von Kaukasischen Eichen-Arten. *Botan. Zurn* **53**, 243–253. [Ref. *Silvae Genet.* **17**, 198.]
Teich, A. H., and Holst, M. J. (1974). White spruce limestone ecotypes. *For. Chronicle* **50**, 1–2.
Vieitez, A. N., Ballester, A., and Kleinschmit, J. (1977). Einfluß von Alter und Wachstumsbedingungen auf die Entwicklung und Form von Douglasienstecklingen. *Forstarchiv* **48**, 74–79.
Von Schönborn, A. (1967). Gibt es Bodenrassen bei Waldbäumen? *Allg. Forstztg* **22**, 294–296.
Weiser, F. (1964a). Beitrag zum Problem der sog. Bodenrassen bei unseren Waldbaumarten, unter besonderer Berücksichtigung der Esche, *Fraxinus excelsior* L. *Forstwiss. Centralblatt* **83**, 23–33.
Weiser, F. (1964b). Anlage und erste Ergebnisse vergleichender Anbauversuche mit generativen Nachkommenschaften von Eschen (*Fraxinus excelsior* L.) trockener Kalkstandorte und grundwasserbeeinflußter Standorte. *Forstwiss. Centralblatt* **83**, 193–211.
Weiser, F. (1965). Untersuchungen generativer Nachkommenschaften von Eschen (*Fraxinus excelsior* L.) trockener Kalkstandorte und grundwasserbeeinflußter Standorte im Gefäßversuch bei differenzierten Wasser- und Kalkgaben. *Forstwiss. Centralblatt* **84**, 44–64.
Weiser, F. (1974). Ergebnisse zehnjähriger vergleichender Anbauversuche mit generativen Nachkommenschaften von Eschen (*Fraxinus excelsior* L.) trockener Kalkstandorte und grundwasserbeeinflußter Standorte. *Beitr. Forstwirtsch.* **1**, 11–16.
Weisgerber, H., Dietze, W., Kleinschmit, J., Racz, J., Dietrich, H., and Dimpflmeier, R. (1976). Ergebnisse des Internationalen Fichten-Provenienzversuches 1962. Teil I: Phänologische Beobachtungen und Höhenwachstum bis zur ersten Freilandaufnahme. *Allg. Forst- u. Jagdztg* **147**, 227–235.
Weisgerber, H., Dietze, W., Kleinschmit, J., Racz, J., Dietrich, H., and Dimpflmeier, R. (1977). Ergebnisse des Internationalen Fichten-Provenienzversuches 1962. Teil II: Weitere Entwicklung bis zum Alter 13. *Allg. Forst- u. Jagdztg* **148**. 217–227.
Yeh, F. C., and El-Kassaby, Y. A. (1980). Enzyme variation in natural populations of Sitka spruce (*Picea sitchensis*): 1. Genetic variation patterns among trees from 10 IUFRO provenances. *Can. J. For. Res.* **10**, 415–422.

Yeh, F. C., and Malley, D. O. (1980). Enzyme variations in natural populations of Douglas fir, *Pseudotsuga menziesii* (Mirb.) Franco, from British Columbia: 1. Genetic variation patterns in coastal populations. *Silvae Genet.* **29**, 83–92.

Zavarin, E., Snajberg, K., and Debry, R. (1980). Terpenoid and morphological variability of *Pinus quadrifolia* and its natural hybridization with *Pinus monophylla* in Northern Baja California and adjoining United States. *Biochem. Syst. Ecol.* **8**, 225–235.

# 15 Phase Change and Vegetative Propagation

P. F. WAREING

*Department of Botany and Microbiology, University College of Wales*

It is well known to horticulturists and foresters that, as a tree grows and the branch system increases in size and complexity, there is a gradual decline in the vigour of the annual shoots as measured by the annual increment in extension growth. In parallel with this decline in vigour of extension growth are other changes, including less pronounced apical dominance and reduced geotropic responses, so that the twigs are not strongly negatively geotropic but appear to be more or less ageotropic in that they frequently appear to be randomly orientated.

The decrease in the vigour of vegetative growth can usually be counteracted by pruning, and this is indeed one of the objects of pruning fruit trees.

If a scion is taken from the branch system of a mature tree and grafted on to a seedling stock, there is usually a rapid increase in the vigour of the scion; alternatively, if the tree is of a species in which cuttings from the mature part of the tree can be rooted, there is a similar renewal of vegetative vigour. Thus, the decline in the vigour of growth observed as the tree increases in size can usually be reversed to a large extent by various treatments, and presumably the decline reflects progressive deterioration in the nutritional status of the branch system as it increases in size and complexity. I have called these readily reversible characteristics of older shoots "ageing" (Wareing, 1959).

However, where trees have been grown from seed other changes may be observed during their development, which are not so easily reversible. Thus, it is well known that flowering is delayed for several years—in some species for as long as 20–30 years—in trees which have been grown from seed, and the transition from the juvenile to the adult (or "mature") flowering state may be accompanied by other changes affecting leaf characters, phyllotaxy, thorn development and so on. Many of these changes are *not readily reversible* if cuttings or scions are taken from the mature parts of the tree. This phenomenon has been called "phase change" (Brink, 1962).

Thus, it is important to make a distinction between the easily reversible changes, referred to as "ageing", and the other changes which are not so easily reversible, which may be called "maturation" (Wareing, 1959). Unfortunately, this distinction is frequently not recognised and although the terms "juvenile" and "adult" are often applied to the states associated with phase change, the term "rejuvenation" is frequently applied to the increased vigour resulting from pruning or grafting, but is really the reversal of "ageing" and would best be called "re-invigoration". A true reversion to the juvenile phase is best called "phase-reversal".

## 15.1 THE NATURE OF PHASE CHANGE

A very striking example of phase change is seen in ivy (*Hedera helix*), which has been the subject of much experimental work (Hackett, 1985; Zimmerman *et al.*, 1985).

The juvenile and adult phases of ivy differ in a number of vegetative characters, including leaf shape, phyllotaxy, pigmentation, pubescence, growth habit and production of adventitious roots, as well as in flowering. If cuttings are taken from the juvenile and adult parts of the ivy vine, they continue to show for many years the characteristics of the shoots from which they were derived. Normally, the vine remains juvenile at the base, even after the upper part has attained the adult phase. Inspection of a relatively young vine attached to a tree shows that at a certain height the leaves show a change in shape and become intermediate in character between fully juvenile and fully adult type leaves. This is the zone of transition between the juvenile and adult regions of the vine.

To ascertain whether the intermediate forms in the transitional zone are stable or unstable, vines 4 m in height were cut into 15 cm sections and rooted, and observations were made on the morphology and development of the lateral shoots which grew out from them, over a period of 3 years (Robinson, 1962). Cuttings taken from the base of the vines gave plants showing all juvenile characters, while all those from the top of the vine showed adult characters. Cuttings taken from the transitional zone produced shoots which showed mixtures of both juvenile and adult characters. For example, shoots were observed which showed adult characters with respect to leaf shape and lack of anthocyanin, but these were associated with a horizontal growth habit and opposite phyllotaxis, both juvenile characters. The majority of cuttings showing intermediate characters changed to either the fully juvenile or the fully adult condition in the course of time, but a smaller number of shoots retained their transitional character for up to 3 years.

The occurrence of "mixed" shoots showing both juvenile and adult characters indicates that phase change does not occur simultaneously in all the characters distinguishing the two phases, a conclusion to which I shall refer later.

Although the adult phase of ivy is remarkably stable, reversion to the juvenile state occurs if rooted adult cuttings are kept in a warm glasshouse, or if they are treated with relatively high doses of gibberellic acid (Robbins, 1957; Frydman and Wareing, 1974).

What is the molecular basis of phase change? Clearly, the transition from the juvenile to the adult condition cannot involve a permanent change in the genome since it is the adult phase which produces the seeds which, in turn, give rise to seedlings with juvenile characters. The characters of juvenile and adult shoots are the products of their respective apical meristems, and it seems likely that individual cells of these meristems are "programmed" as juvenile or adult. Navarro *et al.* (1975) successfully grafted apical meristems, consisting of the apical dome and only three small leaf primordia from adult shoots of *Citrus*, on to the hypocotyls of seedlings. These grew into small plants which flowered and produced fruits, indicating that the grafted apical meristems were in an adult condition.

That there are stable differences between the cells of juvenile and adult ivy is also indicated by the differences shown by callus tissues derived from juvenile and adult shoots grown aseptically. Juvenile callus grows more rapidly and shows a greater capacity to regenerate roots than adult callus grown on the same culture medium (Stoutemyer and Britt, 1965, 1969). Stable differences between callus cultures are also shown in the phenomenon known as "habituation". Many tissue cultures require exogenous auxin and cytokinin when first established, but later acquire the ability to grow on media without these hormones and are then said to be "habituated". Meins and his co-workers (Meins and Binns, 1978, 1979) have shown that habituation is a property of individual cells, and that the habituated state is reversed if whole plants are regenerated from habituated tissue, since new callus cultures established from these latter plants are again found to require exogenous auxin and cytokinin. Thus, the changes involved in habituation cannot involve mutation, and Meins suggests that they are "epigenetic", i.e. they involve stable differences in *gene expression* but not in the genome itself.

Further evidence that even actively dividing cells can retain stable differences reflecting the type of tissue from which they were derived is provided by the work of Raff et al. (1979). Stems, leaves, pistils and anthers of *Prunus avium* examined by immunological methods were found to share some antigenic determinants, while others were unique to a particular organ. Callus cells derived from different organs shared some determinants, while others were specific; moreover, the parental organ antigens were still expressed in callus cells after four subcultures. In more recent work (Raff and Clarke, 1981), the surfaces of protoplasts of cells grown in suspension cultures were examined for their capacity to bind antibodies raised to extracts of cell homogenates. Variation was detected in the binding to the surface of protoplasts derived from different tissues.

It thus seems well established that actively dividing cells can show stable differences and can be said to be "determined" (Wareing, 1978, 1982). The occurrence of persistent differences between cells in culture and the transition from one stable form to another seen in both habituation and phase change suggest that both phenomena have a similar molecular basis, and that the differences between the juvenile and adult states are probably of an epigenetic nature. (For a discussion of cell determination in relation to phase change, see Wareing, 1987.)

How can such differences in gene expression be maintained through cell and nuclear division? If phase change does not involve permanent irreversible changes in the genome, it would be expected that the DNA content of juvenile and adult cells would not be different. However, there have been several reports of differences in the DNA content of juvenile and adult tissues of ivy. Thus, Millikan and Ghosh (1971) and Kessler and Reches (1977) found that the DNA content of adult leaf tissue was lower than that of juvenile leaf tissue. Schaeffner and Nagl (1979) analysed cells of whole buds and leaves and reported that nuclei of cells of adult tissue contained 71% more DNA than those of juvenile tissues. Nagl (1979) has argued that these differences were due to differential DNA replication of parts of the genome of adult nuclei.

In contrast to these results, analysis of the nuclei of apical meristem cells of ivy show no differences in DNA content between juvenile and adult shoots (Wareing and Frydman, 1976). Similarly, calli established from juvenile and adult apical meristems showed no differences in DNA content (Polito and Alliata, 1981).

The discrepancy between these various reports may lie in the fact that the investigations in which differences in DNA content were reported were made on mature leaf tissue, whereas studies made on meristem cells or callus derived from them yielded no differences. Moreover, Nagl (1979) reported that the DNA content of cells of the flowers of ivy (borne on adult shoots) corresponded to that of juvenile leaf tissue. Thus, there is no conclusive evidence to support the view that the difference between juvenile and adult meristem cells is primarily due to differential DNA replication.

On the other hand, it has been suggested that cell differentiation may involve modification of sections of the DNA by methylation of specific bases (Holliday and Pugh, 1975). Such base modification, once effected, could be passed on from cell to cell through successive generations. It is postulated that the susceptible bases are located at protein-binding sites such as operators and promoters, and that modification alters the affinity of such sites for their binding proteins. Enzymes capable of effecting DNA modification are known to occur in bacteria, and evidence is accumulating that, in eukaryotes, DNA sequences that are being transcribed are very low in 5-methyl-cytosine, with a corresponding excess of 5-methyl-cytosine in non-transcribed sequences. Thus, it is possible that stable epigenetic differences may be due to methylation or other modification of specific DNA sequences.

## 15.2 PHASE CHANGE AND VEGETATIVE PROPAGATION

Since this volume is concerned with vegetatively propagated plants, it is appropriate to consider the implications of phase change for vegetative propagation both by conventional cuttings and by micropropagation.

With the majority of woody plants, it is generally found that cuttings from mature trees are more difficult to root than those from young trees. With this diminished rooting ability, are we dealing with an ageing phenomenon, which can easily be reversed, or is the reduced rooting ability an aspect of phase change and therefore much more difficult to reverse? In many cases it appears that this reduced rooting ability is an ageing phenomenon and can readily be reversed by treatments which result in increased vigour of growth, e.g. by pruning or growing as hedges which are repeatedly cut back (Franclet, 1979). It is important to note that such treatments to increase vigour do not normally cause a reversion to the juvenile phase so far as flowering is concerned; flowering may be delayed for a few years in vigorously growing ramets, but it recommences much earlier than would occur in a seedling of the same species. Hence, re-invigoration does not normally lead to phase-reversal with respect to flowering.

Although, in most species, reduced rooting ability of cuttings taken from older trees appears to be an ageing effect, in some species it appears that it is also partly a phase-change effect. Thus, juvenile shoots of ivy produce copious adventitious roots which are not normally produced on adult shoots, and rooting can only be induced in cuttings of adult ivy under favourable conditions. Striking evidence that phase change affects rooting in redwood (*Sequoia sempervirens*) is also provided by a long-term experiment carried out by Professor Libby in California (pers. comm.). Cuttings were taken at various distances from the bases of three large (100 m) redwood trees which had been felled. Over 90% of cuttings taken from the basal region rooted, whereas only 16–17% of those from the tops rooted successfully; cuttings from various intermediate heights gave a gradient in percentage rooting. The young trees from the cuttings were grown for 5 years in the nursery, when cuttings were again taken from them; again, the same rates of rooting were achieved as with the original cuttings from which the young trees were derived, i.e. high rooting from young trees derived from basal cuttings and low rooting from those derived from cuttings taken from the crowns of the original trees. Thus, differences in rooting ability persisted over a period of 5 years, strongly suggesting a phase-change effect.

Franclet (1979) has reported that with some woody species which do not respond to treatments such as pruning or growing in hedges, rooting ability can gradually be improved by successive grafting of scions of adult shoots on to seedlings, e.g. in *Eucalyptus camaldulensis*. Similarly, in the micropropagation of apple trees from shoot tips, it is difficult to obtain rooting with the initial cultures, but with successive transfers rooting ability improves (O. P. Jones, pers. comm.). In such cases it is difficult to say whether the improved rooting is due to re-invigoration or to rejuvenation (phase-reversal). If the changes were due solely to re-invigoration, it is difficult to understand why it is a slow process requiring successive graftings or successive transfers in aseptic culture. However, in the case of apples rooted by micropropagation, it is thought that they still retain the potential for flower initiation (O. P. Jones, pers. comm.) so that evidently the treatment does not cause complete phase-reversal.

The "transitional" forms of ivy described above indicate that phase change and its reversal can take place by stages, so that plants showing a mixture of juvenile and adult characters can be obtained. Thus, it is possible that successive grafting or repeated transfer to fresh medium in micropropagation may result in phase-reversal with respect to rooting ability but not flowering potential. If this interpretation is correct, it would appear possible to obtain partial phase-reversal. A better understanding of these phenomena is clearly of vital importance as new techniques for vegetative propagation are developed.

## REFERENCES

Brink, R. A. (1962). Phase change in higher plants and somatic cell heredity. *Q. Rev. Biol.* **37**, 1–22.

Franclet, A. (1979). Rejeunissement des arbres adultes en vue de leur propagation végétative. *In* "Micropropagation D'Arbres Forestiers", pp. 3–18. AFOCEL Etudes et Recherches, No. 12.

Frydman, V. M., and Wareing, P. F. (1974). Phase change in *Hedera helix* L. III. The effects of gibberellins, abscisic acid and growth retardants on juvenile and adult ivy. *J. exp. Bot.* **25**, 420–429.

Hackett, W. P. (1985). Juvenility, maturation and rejuvenation in woody plants. *Hort. Rev.* **7**, 109–155.

Holliday, R., and Pugh, J. E. (1975). DNA modification mechanisms and gene activity during development. *Science* **187**, 226–232.

Kessler, B., and Reches, S. (1977). Structural and functional changes of chromosomal DNA during ageing and phase changes in plants. *Chromosomes Today* **6**, 237–246.

Meins, F., and Binns, A. N. (1978). Epigenetic clonal variation in the requirement of plant cells for cytokinins. *In* "The Clonal Basis for Development" (S. Subtelny and I. M. Sussex, eds), pp. 185–201. Academic Press, New York.

Meins, F., and Binns, A. N. (1979). Cell determination in plant development. *Bioscience* **29**, 221–225.

Millikan, D. F., and Ghosh, B. N. (1971). Changes in nucleic acids associated with maturation and senescence in *Hedera helix*. *Physiol. Plant.* **24**, 10–13.

Nagl, W. (1979). Differential DNA replication in plants: a critical review. *Z. Pflanzenphysiol.* **95**, 283–314.

Navarro, L., Roistacher, C. N., and Murashige, T. (1975). Improvement of shoot tip grafting *in vitro* for virus-free *Citrus*. *J. Am. Soc. hort. Sci.* **100**, 471–479.

Polito, V. S., and Alliata, V. (1981). Growth of calluses derived from shoot apical meristems of adult and juvenile English ivy (*Hedera helix* L.). *Pl. Sci. Lett.* **22**, 387–393.

Raff, J. W., and Clarke, A. E. (1981). Tissue-specific antigens secreted by suspension cultured callus cells of *Prunus avium* L. *Planta* **153**, 115–124.

Raff, J. W., Hutchinson, J. F., Knox, R. B., and Clarke, A. E. (1979). Cell recognition: antigenic determinants of plant organs and their cultured callus cells. *Differentiation* **12**, 179–186.

Robbins, W. J. (1957). Gibberellic acid and the reversal of adult *Hedera* to a juvenile state. *Am. J. Bot.* **44**, 743–746.

Robinson, L. W. (1962). A Study of Juvenility in Plants. Ph.D. thesis, University of Wales.

Schaeffner, K.-H., and Nagl, W. (1979). Differential DNA replication involved in transition from juvenile to adult phase in *Hedera helix*. *Plant. Syst. Evol. Supp.* **2**, 105–110.

Stoutemyer, V. T., and Britt, O. K. (1965). The behaviour of tissue culture from English and Algerian ivy in different growth phases. *Am. J. Bot.* **52**, 805–810.

Stoutemyer, V. T., and Britt, O. K. (1969). Growth and habituation in tissues cultures of English ivy *Hedera helix*. *Am. J. Bot.* **50**, 222–226.

Wareing, P. F. (1959). Problems of juvenility and flowering in trees. *J. Linn. Soc. Lond. (Bot.)* **56**, 282–289.

Wareing, P. F. (1978). Determination in plant development. *Bot. Mag., Tokyo*, Special Issue 1, 3–18.

Wareing, P. F. (1982). Determination and related aspects of plant development. *In* "The Molecular Biology of Plant Development" (H. Smith and D. Grierson, eds), pp. 517–541. Blackwell Scientific, Oxford.

Wareing, P. F. (1987). Juvenility and cell determination. *In* "Manipulation of Flowering" (J. G. Atherton, ed.), pp. 83–92. Butterworth, London.

Wareing, P. F., and Frydman, V. M. (1976). General aspects of phase change, with specific reference to *Hedera helix* L. *Acta Hort.* **56**, 57–69.

Zimmerman, R. H., Hackett, W. P., and Pharis, R. P. (1985). Hormonal aspects of phase change and precocious flowering. *Encycl. Plant Physiol.* **11**, 79–115.

# 16 The Chimeral Problem

R. A. E. TILNEY-BASSETT

*School of Biological Sciences, University College of Swansea, Swansea*

## 16.1 CHIMERA INDUCTION

The perpetual demand for new plants is only satisfied by the constant quest for new sources of variation, which is achieved through the processes of mutation, hybridisation, recombination and selection. But hybridisation and recombination depend upon successful breeding, and in plants restricted to vegetative propagation this is not an option. Consequently, the improvement of such plants is limited to the slow increase in variation following mutation coupled with the selection of the more favourable new genotypes that arise. Fortunately, this is not always a disadvantage. Quite often hybridisation and recombination break up the highly successful and unique genotype of a commercially valuable stock so that only through individual gene mutations are small, beneficial changes possible. This lesson has given rise to the practice of "mutation breeding", a technique useful for a wide range of plants and not just vegetatively propagated ones (Broertjes and Van Harten, 1978), especially with respect to the improvement of the commercially important characters such as flower colour.

Useful spontaneous mutations occur too infrequently to meet fully the demand for new variants and so the artificial induction of mutation is widely practised. The physical mutagens used include X-rays, γ-rays and thermal neutrons, while the chemical mutagens used include a wide range of compounds, although most workers prefer to use well tried and tested ones such as ethyl methane sulphonate and sodium azide (Gottschalk and Wolff, 1983). Usually, the intention is to induce a spread of nuclear gene mutations expressing a wide range of phenotypes, but ethyl methane sulphonate and nitrosomethyl urea also induce plastid mutants leading to variegated shoots (Tilney-Bassett, 1986).

The vast majority of nuclear mutants are recessive and so their mutant lineages are not discernible in the $M_1$ generation. In sexually reproducing plants these are identified by their segregation in the $M_2$ after selfing or intercrossing the $M_1$ plants containing the newly induced heterozygous lineages. Where breeding is not possible, these heterozygous lineages are unlikely to be revealed. Hence, with vegetatively propagated plants, the mutation breeder relies upon the occurrence of dominant mutations expressed directly in the $M_1$ generation, or he depends upon the existence of stocks that are already heterozygous for genes controlling important characters to create a homozygous recessive lineage directly.

Mutations occur within the nucleus or cytoplasmic organelle of a single cell, which in the apical meristem of a shoot is going to be surrounded by normal cells. If the mutant cell is favourably located as a growing-point initial cell, its descendants give rise to a mutant cell lineage forming a mutant sector within the larger body of the normal cells. In the case of plastid mutation, the mutant sector becomes visible as a white or variegated area within the shoot. By careful selection through pruning, the variegation may be encouraged to spread into side shoots and the developing mixture of green and white tissues allowed to become sorted out into a stable periclinal chimera, and propagated further by cuttings. With many nuclear mutants there is no sign of the newly arisen mutant lineage; after all, a mutation affecting flower colour is not going to be revealed until the flower opens. In such cases, the experimenter needs to adapt the best strategy for maximising his chances of recovering any hidden mutations that arise, by encouraging the growth and subsequent propagation of laterals from all around the circumference of the treated shoots. In such a way, from one treated shoot he has many to test for signs of mutation and, if there were any mutations, he has greatly increased his chances of finding them.

Occasionally, if the experimenter is fortunate, he is able to isolate, and propagate further, a pure mutant shoot—a solid mutant. More often, he obtains a chimera initially sectorial or mericlinal but eventually stabilising as a periclinal chimera. This is because many, perhaps even most, flowering plants have layered apices in which the cells are situated in two zones. The outer zone consists of one or two, rarely three, peripheral layers of cells called tunicas (LI, LII), in which divisions are predominantly anticlinal (at right angles to the surface). The inner zone, called the corpus (LIII), consists of a central core of cells in which divisions are anticlinal and periclinal (parallel to the surface), and in other planes too. Within this apical meristem, the tunica layers and the corpus generally remain discrete and very constant. Consequently, when a mutation occurs within a single cell of a tunica layer, or the corpus, it may eventually come to replace all the normal cells within that layer, but does not usually displace the cells of the adjoining layers. Whenever one layer surrounds, or is surrounded by, a layer of different genotype it becomes known as a periclinal chimera. Some plants have a single tunica over the corpus; in others there are commonly two tunicas so the mutant layer may occupy any one of two or three layers. Hence we are able to distinguish between periclinal chimeras with a thin skin (MNN or NMM), a thick skin, (MMN or NNM), or a sandwich structure (NMN or MNM), where N represents a normal and M a mutant layer (Tilney-Bassett, 1986).

The tendency to form periclinal shoots after mutant induction is a mixed blessing. The chimera itself is often highly desirable: on other occasions the experimenter is more concerned to obtain a solid mutant, which he then has to try and isolate from the chimera. As we shall see below, methods are available for separating the layers of a chimera, but these may prove too irksome so that it is preferable to use methods of mutant induction that avoid the inevitable formation of chimeras.

### 16.1.1 Advantageous Chimeras

A wide range of chimeras are commonly grown. A most obvious and very important group are those with white-margined, green-margined and striped variegated leaves widely used in parks and gardens, office blocks and public buildings for their decorative foliage; it is the contrast between the white and green colours that makes these plants so attractive. They make a significant contribution to the ornamental horticultural industry of many countries. Some have originated by plastid mutation, followed by the sorting-out of plastids first into pure green or white cells and later into pure green or white tissues, and thence into periclinal chimeras. The mutant white tissues, sometimes cream or yellow, would not survive without the support of the green, as the white tissues cannot photosynthesise, and so the chimeral structure is essential for the maintenance of the mutation.

Often the user is quite unaware of the chimeral nature of his commercial stocks. For example, I have listed many potato tubers or apple fruit that are chimeral for a coloured or russet skin (Tilney-Bassett, 1986). Such cultivars are maintained because of their many desirable qualities of growth, appearance, taste, disease resistance and so forth. At the time of origin as new variants, they must have compared reasonably well with the existing cultivars of the time, and so were propagated further. In most cases it was incidental that the cultivar was actually a chimera; many desirable qualities were probably unrelated to the chimeral structure. It is likely that the character for which the plant was chimeral was determined by a single allelic difference in one gene controlling a trivial, although desirable, aspect of the cultivar such as the skin colour. But once this trivial character had become a property of the cultivar, the constancy of the cultivar would have to be vigilantly maintained at the time of propagation by roguing out any variants arising through alterations in the chimeral structure. Evidently, with very stable chimeras, roguing might be a minor, insignificant problem, whereas with less stable chimeras the need for constant roguing might prove to be a most undesirable property. The borderline between whether a chimera is sufficiently stable to be worth keeping, or whether its instability makes it too much of a nuisance, is likely to depend on the aesthetic and commercial considerations of the grower. These problems are equally applicable to the cultivation of variegated-leaf chimeras; not all are worth the effort.

A few chimeras are grown as curiosities. The sight of the yellow flowers of laburnum (*Laburnum anagyroides*) growing alongside the purple sprays of broom (*Cystisus purpureus*), and their corresponding differences in leaf, on the graft chimera *Laburnocytisus adamii* is always fascinating. Another delightful graft chimera is *Crataegomespilus potsdamiensis*, which exists as both thin-skinned and thick-skinned forms of medlar (*Mespilus germanica*) over a core of hawthorn (*Crataegus monogyna*) (Bergann and Bergann, 1984). Unlike the classical graft chimeras between these species, these forms were derived by grafting medlar to the double red-flowered hawthorn, Pauli, which regularly reappears as a vivid sporting branch. Such curious plants attract people into botanic gardens and provide a talking point for visitors.

The majority of flowering plants appear to have flowers with anthocyanin pigments localised in the epidermal cells of the petals (Kay *et al.*, 1981), but a few flowers have pigment in the underlying mesophyll cells too. Moreover, in some of these the epidermis divides periclinally, as well as anticlinally, to produce any width from a wide margin to a narrow edge to each petal. Consequently, when the genotypes of the two outer layers differ and both are expressed, a chimeral pattern results. The forget-me-nots, *Myosotis arvensis* Star of Zurich and Weirleigh Surprise had, respectively, blue flowers with a central white band, or the reverse pattern, on each petal (Chittenden, 1927, 1928), indicating that most of the petal developed from LI but that there was a central area derived from LII, with gene expression in both layers. White-edged flowers with fringes of varying width were reported in the zonal pelargonium *Pelargonium* × *hortorum* Mr Wren (Cassells and Minas, 1983), *Rhododendron obtusum* (Imai, 1937), *R. indicum, R. lateritium, Punica granatum, Chaenomeles lagenaria* and *Camellia japonica* (Imai, 1935) and *R. simsii* (de Loose, 1979). These attractive flowers are uniquely dependent on the combination of a periclinal chimera structure, and gene expression for anthocyanin synthesis in both LI and LII. They would not exist as solid mutants.

A unique effect of a periclinal structure was observed by Péreau-Leroy (1969, 1974). He discovered that the carnations, *Dianthus caryophyllus* Jacqueline and Jacky, which had orange flowers, were thin-skinned chimeras with a yellow LI over a genotypically red, but phenotypically colourless, LII. It appears that the skin was orange and not yellow in the chimera, because of a physiological interaction between the genetically distinct tissues in which the yellow LI was modified by the red LII even though LII was not itself coloured. A similar example occurred with the carnation Dusty (Johnson, 1980), in which flowers developed from LI alone were pale pink with red flecks, and flowers developed from LII alone were red, but the chimeral flowers were salmon pink with red flecks, together with a red sinus blotch, where the LII tissue had pushed through the pale skin.

Physiological advantages of chimeras have been demonstrated on several occasions. Krenke (1933) found that the graft chimera with a thin skin of *Solanum memphiticum* provided frost protection to the sensitive core of tomato (*Lycopersicum esculentum*). Klebahn (1918) infected four of Winkler's tomato–nightshade (*Solanum nigrum*) graft chimeras with the fungi *Septoria lycopersicii* and *Cladosporium fulvum*, to which tomato is susceptible and nightshade is resistant. He found that a sub-epidermal layer of nightshade resisted infection, whereas a sub-epidermal layer of tomato did not. Another graft chimera with a thin skin of *S. pennelli* protected the underlying and highly susceptible tomato core against the greenhouse white fly, but a similar resistance against the potato aphid was not achieved (Clayberg, 1975). These advantages of chimeras are mainly of academic interest as no effort has been made to exploit them, and it is unlikely that it would be a practical proposition to do so. In other ways, too, chimeras are interesting objects of study. Synthetic graft chimeras have been used to study problems of incompatibility (Günther, 1961), the role of the epidermis in the perception of and response to light (Mayer *et al.*, 1973; Junker and Mayer, 1974) and to assess the stomatal response to light (Heichel and Anagnostakis, 1978). Cytochimeras of apple (*Malus pumila domestica*), cranberry (*Vaccinium oxycoccus*), peach (*Prunus persica*) and thornapple (*Datura stramonium*) have been particularly valuable in studies of histogenesis. A variety of different types of chimeras have helped in our growing comprehension of the behaviour of apical meristems, particularly in regard to the problem of the numbers of apical initial cells in a tunica layer, or in the initiation of an adventitious bud (Moh, 1961; Steffensen, 1968; Verkerk, 1971; Broertjes and Keen, 1980; Smith and Norris, 1983; Broertjes and Van Harten, 1985; Tilney-Bassett, 1986). The exploitation of chimeras has also found uses in the analysis of cells in tissue culture (Bush *et al.*, 1976, Sree Ramulu *et al.*,

1976; Cassells and Minas, 1983). In all these research activities the existence of two or three genetically distinct tissues within the periclinal chimera has provided an excellent marker system with which to follow the behaviour of cells during development and differentiation.

### 16.1.2 Disadvantageous chimeras

Plant breeders often use a spindle inhibitor, such as colchicine solution, to double the chromosome complement of cells by allowing the chromosomes to divide once without cell division. The breeder is interested in creating a pure tetraploid line from the diploid parent, but tetraploidy is frequently achieved in just one layer, so producing a diploid–tetraploid cytochimera, as was found in about 20 species tested (Tilney-Bassett, 1986), of which the most common had either a thin diploid skin and tetraploid core (244) or a thin tetraploid skin and diploid core (422). In the first instance, the breeder needs to recognise that he has induced a cytochimera, which may prove somewhat confusing. The first indications of a 244 chimera are such features as greater thickness of the leaves, their broader shape, their increased brittleness and so on; yet, upon examination of the leaf epidermis, the stomatal guard cells and number of chloroplasts within them may be no different from the diploid, raising doubt as to the significance of the other growth changes. Conversely, the 422 chimera may reveal larger guard cells with more chloroplasts but no significant increase in growth of the leaves as a whole. Thus, the chimera structures produce contrary indications of the existence of tetraploid tissue and, without further analysis by chromosome counts, the preliminary indications are confused and the successful induction of tetraploidy easily overlooked. When the cytochimera is recognised, the experimenter may still have to face the problem of isolating a solid tetraploid shoot from the chimeral one. In the case of the 244 chimera, the tetraploid layer may be extracted from the chimera by the formation of adventitious buds on roots, which develop from the central core tissue derived from LIII. Alternatively, if fertility is not completely impaired, seedlings may be obtained which trace back to the origin of the germ cells from LII. In the case of 422 chimeras, both techniques produce the parental diploid tissue again. Hence it becomes necessary to stimulate the epidermal layer to divide periclinally and duplicate, so displacing the inner layers. This has been achieved in apple cytochimeras by treating the cytochimeral shoots with X-ray irradiation or thermal neutrons (Pratt, 1960, 1963), although the effect is not specific and induces other layer changes as well. So, in many ways, the cytochimeras are intermediate between the diploid parents and the sought-after tetraploids; they have their uses but it would be better if solid tetraploid shoots could be obtained directly.

The cytochimera may be an awkward intermediate between ploidy levels, but at least its effects are visible on close examination. In contrast, because of the absence of gene expression in the altered layer, some chimeras have little or no visible effects. This is particularly true in respect of the surface colouring of tubers, fruits or flowers, in which the colour depends wholly, or mostly, on gene expression in the epidermis. In such cases a mutation leading to an alternative genotype in LII or LIII remains invisible, and so there is no outward sign of the existence of a chimera. But whenever the normal skin is replaced by the growth of the inner tissue, the previously hidden phenotype of the inner layer is expressed. This might take the form of an unexpected fleck or sector, or the whole surface colour of the organ might be altered. An instructive example, made easier by the regular, partial expression of LII, is the King Edward potato (*Solanum tuberosum*), which has a whitish skin, or periderm, with a strong splash of pink at the rose end and pinkish-white patches around the eyes (WWW) (Howard, 1959). A mutation to red in LI (RWW) produces a red periderm typical of Red King Edward with pinkish-white eye patches derived from the underlying LII. A mutation to red in LII (WRW) does not alter the periderm except that the eye patches derived from LII are now red. Finally, when both layers are genotypically red (RRW) the red periderm is without odd coloured patches. Confirmation of LII in each case is shown by the colour of the germination sprout in which the genotype expressed is now LII and not LI.

The commercially important apple, Delicious, produces frequent bud sports having an increased or decreased pigmentation, with or without a change in pattern from striped to solid colour. Suspecting that some trees might be periclinal chimeras, scions were irradiated with γ-rays to a limit of 80 Gy (10Gy = 1 krad) at the tip (Pratt *et al.*, 1972). Some of the sports were remarkably unstable, suggesting that there was indeed an alternative colour hidden in another layer. The changes were interpreted on the assumption that LI was normal, that the less stable sport had a mutant LII (NMN), or LIII (NNM), and that the more stable were mutant in both LII and LIII (NMM). Another series of sports, derivatives of the apple, Northern Spy, with increased anthocyanin pigments in either the epidermis alone, or in both epidermal and sub-epidermal layers, were thought to vary in the character of LII, and to have the periclinal chimera structures MNN or MMN (Pratt *et al.*, 1975).

A large collection of flower chimeras of carnation sports was investigated by Farestveit (1969). Among these were six cultivars in which all three germ layers had a different genotype. For example, after irradiation and propagation treatment to encourage layer changes, the cultivar Sunset Sim, with an orange–red flower, frequently produced yellow flowers, and when these yellow-flowered shoots were isolated and grown, red sectors occurred. The progressive revealing of colours (orange–red, yellow, red) clearly corresponded to the stepwise displacement of the irradiation-damaged outer tissues by regeneration from the core (OYR to YRR to RRR). The occurrence of such trichimeras means that although the flowers themselves are probably wholly derived from two layers, the vegetative shoots are three-layered and may therefore retain in LIII a genotype that is normally absent from the flowers, yet may be brought into the flower at any time the floral apex is damaged and LI or LII has to be replaced.

**16.1.3 Solid mutants**

Once the chimeral structure of a plant has been recognised, there are various techniques which may be tried in order to obtain a solid shoot from the mutant layer. Adventitious buds formed on true roots arise by endogenous growth from the central tissues which, pushing through the outer cortex, grow into plants exhibiting the character proper to the core. Therefore, whenever plants grown from root cuttings differ from those grown from stem cuttings, we may infer that the parental stocks are periclinal chimeras. Over a period of ten years Bateson (1916, 1921, 1926) was able to obtain adventitious buds on root cuttings taken from Bridesmaid, a cultivated hybrid between two species of *Bouvardia*, and from the fancy pelargoniums (*Pelargonium* × *domesticum*) Escot, Mrs Gordon and Pearl, and the zonal pelargonium, Salmon-fringed. In each case the shoots revealed the characteristics of the chimeral core, corresponding to the genotype of LIII. In a similar manner, Zimmerman and Hitchcock (1951) were able to demonstrate the periclinal nature of several rose (*Rosa multiflora*) sports. Cassells and Minas (1983) showed that when propagated by petiole explant culture the zonal pelargoniums Salmon-fringed, Double New Life and Kleiner Liebling reverted to the root-cutting types too.

Serving the same purpose, the method of eye-excision has been widely used for the determination, or isolation, of LIII from potato tubers. As a rule, a number of tubers belonging to the clone under test are cut longitudinally into halves. One half is planted directly as the control, while the other half is planted after removing all the eyes. On tubers with the eyes removed, growth is dependent on the formation of entirely new, adventitious buds on the cut surfaces, which usually develop from tissue of LIII origin (Asseyeva, 1927; Crane, 1936; Howard, 1969). The method has been successful for many potato cultivars chimeral for anthocyanin pigmentation, russet skin, and for leaf mutants, bolters and wildings (Tilney-Bassett, 1986).

Dermen (1948) obtained adventitious buds from the outer phloem region of the apple, Sweet and Sour, which gave rise to homogenous forms characteristic of the sour LIII tissue. He, likewise, obtained an entirely tetraploid shoot from the cytochimeral apple, Giant McIntosh (244) (Dermen, 1951).

In one survey, over 350 plants were believed to be able to develop adventitious buds, capable of developing into plantlets, on different parts of detached leaves (Broertjes et al., 1968). In the genera *Achimenes*, *Kalanchoe*, *Nicotiana*, *Saintpaulia* and *Streptocarpus* it was shown by mutation induction, and by cytological and histological examination, that the apices of adventitious buds formed near the cut end of leaf petioles invariably originated from a single cell, so mutants induced in this region were solid, and the problem of chimeral shoots was avoided (Broertjes and Van Harten, 1978). Moreover, in these plants the adventitious buds were derived from an epidermal cell from LI, not from LIII. In many other genera, too, detached leaves and bulb scales have been successfully used for the induction of solid mutants. Broertjes and Keen (1980) developed a stochastic model to describe the process of apex formation in adventitious buds, from which they could estimate the expected relative percentages of chimeras. What they found in practice was that the observed frequency was lower than expected, which led them to postulate that the new growing-point behaved as if ultimately derived from a single cell, often an epidermal cell. Other surrounding cells were probably involved in forming the connecting tissue between the outgrowing adventitious shoot and the explant, but were not directly involved in the formation of the apex.

In the case of the variegated leaves of *Peperomia glabella*, *P. obtusifolia* and *P. scandens*, the majority of adventitious shoots grew from the LIII layer of isolated leaves of the thick-skinned (GGW) and sandwich (GWG) chimeras, but there was a minority of shoots derived from LII, plus a few variegated shoots which must have included components from two or three layers (Bergann and Bergann, 1982). Three variegated cultivars of African violet (*Saintpaulia ionantha*) were Bold Dance, a GWG sandwich chimera, Marge Winters, a GGW thick-skinned chimera, and Calico Kitten, a WWG thick-skinned chimera. When leaf explants were grown on culture medium, the adventitious shoots that developed were just like their respective parental cultivars (Norris et al., 1983; Smith and Norris, 1983). The authors suggested that the successful propagation of the chimeras might be related to the lack of a callus intermediate stage which, when produced under the influence of growth regulators, resulted in the dissociation of the chimeral structures. The regular, true-to-type, development of chimeras from adventitious buds contrasted with the rare occurrence of such development in *Peperomia* species, and the finding that the adventitious shoots of many species originate from a single cell. The contrast has led Marcotrigiano and Stewart (1984) and Broertjes and Van Harten (1985) to object to the multicellular origin of the African violet buds on several grounds, one of which was that the three cultivars might not be chimeras at all! They suggested instead, that the cultivars might be pseudo-chimeras determined by stable figurative pattern genes (Kirk and Tilney-Bassett, 1978). It would surprise me if this was the case; if the three cultivars are true periclinal chimeras then it does appear that the single-cell origin of the growing-point of adventitious shoots is not a universal phenomenon. In that case, it would be valuable to assess the formation of adventitious shoots on explants from many more chimeras to really find out to what extent the origin of the shoot is determined by the species and by the culture medium. It would be very useful if one could regularly switch between the propagation of solid shoots or chimeral shoots simply by adjusting the cultural conditions.

Van Harten *et al.* (1981) irradiated explants of the potato, Désirée, and grew them on culture media. There was an increase in mutation frequency with increasing doses of X-rays from 15 to 27.5 Gy and, as the induction of chimeras remained low, a high proportion of the mutants were solid, a result comparable to the best obtained from experiments using adventitious sprouts (Van Harten *et al.*, 1972; Van Harten and Bouter, 1973). As the method was relatively simple, not too laborious, and kept the frequency of chimeras to a low level, it is very promising and should encourage the use of mutation breeding for the modification and improvement of potatoes.

The usual way of isolating a solid shoot from the sub-epidermal layer, LII, is to obtain seedlings. This is not possible, of course, if the plant is sterile. It is then necessary to displace the mutant layer into another position. Experiments with cytochimeral apples showed that X-ray irradiation of shoot apices was a useful means of altering the layer structure so that, for example, a $2x/4x$ periclinal chimera with a diploid LII might be altered to one with a tetraploid LII owing to the change 224 to 244 (Pratt, 1960). Pratt exposed dormant apple scions to 30 Gy X-rays or to 3 hours of thermal neutrons and grafted to untreated stocks. After damage to the central apical cells, shoot growth resumed either by the growth of adjacent cells which reformed the meristem, or by cells lateral to the damaged area forming a substitute meristem, or by the development of an axillary meristem at one or more nodes below the apical meristem. The chromosome constitution of the recovered shoots depended upon the ploidy of the cells taking part and their subsequent arrangement into layers. Generally speaking, increase in $2x$ outer tissue (224 to 222) was more frequent than increase in $4x$ inner tissues (224 to 244), but the direction of change appeared to be partly related to the length of the shoot and to the apple cultivar used. After radiation treatment, cytochimeras were more variable than those among untreated lines and included new types that had not been found by branch selection (Pratt et al., 1967).

Asseyeva (1931) used X-ray irradiation to transform the layer structure of potato chimeras. Irradiation of Red King Edward, a bud sport of King Edward in which LI is genetically red instead of white splashed pink, transformed it into a periclinal chimera (RWW to RRW) in which both LI and LII were genetically red (Howard, 1959, 1964). Other X-ray irradiated chimeras included Redskin Wilding (Howard, 1967) and Pink Arran Victory (Howard, 1969) among many.

De Loose (1979) irradiated the *Rhododendron, R. simsii* Mevr. R. de Loose (WR) with 20–30 Gy X-rays, or with recurrent treatment, and obtained a rearrangement of flower structure up to five times more frequently after a single treatment, and up to 40 times more frequently after recurrent treatment, than after no irradiation. About 93% of the structural changes were towards white (WR to WW) and 7% towards red (WR to RR). This suggests that LII suffered more severely from irradiation damage than LI, which contrasted with the effects of recurrent X-ray irradiation upon the cultivar de Waele's Favorite (WR), in which flower colour changes in both directions were equally represented.

Although irradiation treatment, as we have seen, is useful in altering layer structure and therefore increasing the chance of obtaining non-chimeral shoots, it has the disadvantage of itself inducing mutation. Dommergues and Gillot (1973) tackled this problem by inducing adventitious buds from apical or sub-apical *in vitro* cultures of the carnation, White Sim (WR), and by hormonal applications to the exposed surfaces of cut stems. They found that after *in vivo* propagation about 75% of the flowers were solid white (WR to WW) and 25% remained chimeral, whereas after *in vitro* propagation all were solid white, indicating that regeneration had most often been from LI alone. Johnson (1980) experimented with meristem culture, and with cultures of physically macerated shoot-tip explants, which he compared with material irradiated with a dose of 68 Gy X-rays. Meristem culture of S. Arthur Sim gave rise to 7% unchanged chimeral flowers, 27% flowers derived from LI and 60% flowers derived from LII. Macerating the shoot tips of Braun's Yellow Sim, Dusty and Pink Ice gave a rather variable separation of chimeral, LI, or mixed flowers, but there were no solid LII shoots.

Evidently, as with irradiation treatment, tissue-culture methods aid the separation of solid shoots from chimeral ones but, as with irradiation, the responses of plants are variable with either LI or LII being preferentially multiplied. There is, as yet, no single technique guaranteed to work for all plants; rather, there is a range of techniques available, and the individual breeder or grower must still experiment to determine which is the best method of isolating solid shoots from the plant material in hand. The existence of chimeras is still a problem.

## REFERENCES

Asseyeva, T. (1927). Bud variations in the potato and their chimerical nature. *J. Genet.* **19**, 1–26.
Asseyeva, T. (1931). Bud mutations in the potato. *Trudy prikl. Bot. Genet. Selek.* **27**, 135–217 (in Russian).
Bateson, W. (1916). Root cuttings, chimeras and 'sports'. *J. Genet.* **6**, 75–80.
Bateson, W. (1921). Root cuttings and chimeras II. *J. Genet.* **8**, 93–99.
Bateson, W. (1926). Segregation. *J. Genet.* **16**, 201–236.
Bergann, F., and Bergann, L. (1982). Zur Entwicklungsgeschichte des Angiospermblattes. 1. Über Periklinalchimären bei *Peperomia* und ihre experimentelle Entmischung und Umlagerung. *Biol. Zbl.* **101**, 485–502.
Bergann, F., and Bergann, L. (1984). Gelungene experimentelle Synthese zweier neuer Pfropfchimären—die Rotdornmispeln von Potsdam: *Crataegomespilus potsdamiensis* cv. "Diekto", cv. "Monekto". *Biol. Zbl.* **103**, 283–293.
Broertjes, C., and Keen, A. (1980). Adventitious shoots: do they develop from one cell? *Euphytica* **29**, 73–87.
Broertjes, C., and Van Harten, A. M. (1978). "Application of Mutation Breeding Methods in the Improvement of Vegetatively Propagated Crops". Elsevier, Amsterdam.
Broertjes, C., and Van Harten, A. M. (1985). Single cell origin of adventitious buds. *Euphytica* **34**, 93–95.
Broertjes, C., Haccius, B., and Weidlich, S. (1968). Adventitious bud formation on isolated leaves and its significance for mutation breeding. *Euphytica* **17**, 321–344.
Bush, S. R., Earle, E. D., and Langhans, R. W. (1976), Plantlets from petal segments, petal epidermis, and shoot tips of the periclinal chimera, *Chrysanthemum morifolium* 'Indianapolis'. *Am. J. Bot.* **63**, 729–737.
Cassells, A. C., and Minas, G. (1983). Beneficially-infected and chimeral pelargonium: Implications for micropropagation by meristem and explant culture. *Acta Hort.* **131**, 287–297.
Chittenden, R. J. (1927). Vegetative segregation. *Bibliogr. Genet.* **3**, 355–442.
Chittenden, R. J. (1928). Ever sporting races of *Myosotis*. *J. Genet.* **20**, 123–129.
Clayberg, C. D. (1975). Insect resistance in a graft-induced periclinal chimera of tomato. *HortScience* **10**, 13–15.
Crane, M. B. (1936). Note on a periclinal chimera in the potato. *J. Genet.* **32**, 73–77.
De Loose, R. (1979). Radiation induced chimeric rearrangement in flower

structure of *Rhododendron simsii* Planch. (*Azalea indica* L.). Use of recurrent irradiation. *Euphytica* **28**, 105–113.
Dermen, H. (1948). Chimeral apple sports and their propagation through adventitious buds. *J. Hered.* **39**, 235–242.
Dermen, H. (1951). Tetraploid and diploid adventitious shoots from a giant sport of McIntosh apple. *J. Hered.* **42**, 144–149.
Dommergues, P., and Gillot, J. (1973). Obtention de clones génétiquement homogènes dans toutes leur couche ontogénétiques a partir d'une chimère d'oeillet américain. *Ann. Amélior. Plantes* **23**, 83–95.
Farestveit, B. (1969). Flower colour chimeras in glasshouse carnations. *Dianthus caryophyllus* L. *Arssksrift K. Veterinaer-og Landbohkojskole, Kobenhaven* 19–33.
Gottschalk, W., and Wolff, G. (1983). "Induced Mutations in Plant Breeding". Springer-Verlag, Berlin.
Günther, E. (1961). Durch Chimärenbildung verursachte Aufhebung der Selbstinkompatibilatät von *Lycopersicon peruvianium* (L) Mill. *Ber. Dtsch. Bot. Ges.* **74**, 333–336.
Heichel, G. H., and Anagnostakis, S. L. (1978). Stomatal response to light of *Solanum pennellii, Lycopersicon esculentum*, and a graft-induced chimera. *Plant Physiol.* **62**, 387–390.
Howard, H. W. (1959). Experiments with a potato periclinal chimera. *Genetica* **30**, 278–291.
Howard, H. W. (1964). The use of X-rays in investigating potato chimeras. *Radiat. Bot.* **4**, 361–371.
Howard, H. W. (1967). The chimerical nature of a potato wilding. *Pl. Path.* **16**, 89–92.
Howard, H. W. (1969). A full analysis of a potato chimera. *Genetica* **40**, 233–241.
Imai, Y. (1935). Variegated flowers and their derivatives by bud variation. *J. Genet.* **30**, 1–13.
Imai, Y. (1937). Bud variation in a flaked strain of *Rhododendron obtusum*. *J. Coll. Agric., Tokyo* **14**, 93–98.
Johnson, R. T. (1980). Gamma irradiation and *in vitro* induced separation of chimeral genotypes in carnation. *HortScience* **15**, 605–606.
Junker, G., and Mayer, W. (1974). Die Bedeutung der Epidermis für Licht und temperatur-induzierte Phasenverschiebungen circadianer Laubblättbewegungen. *Planta* **121**, 27–37.
Kay, Q. O. N., Daoud, H. S., and Stirton, C. H. (1981). Pigment distribution, light reflection and cell structure in petals. *Bot. J. Linn. Soc.* **83**, 57–84.
Kirk, J. T. O., and Tilney-Bassett, R. A. E. (1978). "The Plastids: their Chemistry, Structure, Growth and Inheritance", 2nd edn. Elsevier/North-Holland, Amsterdam.
Klebahn, H. (1918). Impfversuche mit Propfbastarden. *Flora* **111**, 418–430.
Krenke, N. P. (1933). "Wundkompensation, Transplantation und Chimären bei Pflanzen". Springer-Verlag, Berlin.
Marcotrigiano, M., and Stewart, R. N. (1984). All variegated plants are not chimeras. *Science* **233**, 505.
Mayer, W., Moser, I., and Bunning, E. (1973). Die Epidermis als Ort der Lichtperzeption duer circadiane Laubblattbewegungen und photoperiodische Induktionen. *Z. Pflanzenphysiol* **70**, 66–73.
Moh, C. C. (1961). Does a coffee plant develop from one initial cell in the shoot

apex of an embryo? *Radiat. Bot.* **1**, 97–99.
Norris, R., Smith, R.H., and Vaughn, K. C. (1983). Plant chimeras used to establish *de novo* origin of shoots. *Science* 220, 75–76.
Péreau-Leroy, P. (1969). Effect de l'irradiation gamma sur une chimère complexe d'oeillet sim. *In* "Induced Mutations in Plants", pp. 337–344. Proc. Symp. FAO/IAEA. Vienna.
Péreau-Leroy, P. (1974). Genetic interaction between the tissues of carnation petals as periclinal chimeras. *Radiat. Bot.* **14**, 109–116.
Pratt, C. (1960). Changes in structure of a periclinal chromosomal chimera of apple following X-irradiation. *Nature* **186**, 255–256.
Pratt, C. (1963). Radiation damage and recovery in diploid and cytochimeral varieties of apples. *Radiat. Bot.* **3**, 193–206.
Pratt, C., Ourecky, D. K., and Einset. J. (1967). Variation in apple cytochimeras. *Am. J. Bot.* **54**, 1295–1301.
Pratt, C., Way, R. D., and Einset, J. (1975). Chimeral structure of red sports of 'Northern Spy' apple. *J. Am. Soc. hort. Sci.* **100**, 419–422.
Pratt, C., Way, R. D., and Ourecky, D. K. (1972). Irradiation of colour sports of "Delicious" and "Rome" apples. *J. Am. Soc. hort. Sci.* **97**, 268–272.
Smith, R. H., and Norris, R. E. (1983). *In vitro* propagation of African violet chimeras. *Hortscience* **18**, 436–437.
Sree Ramulu, K., Derreux, M., and de Martinis, P. (1976). Origin and genetic analysis of plants regenerated *in vitro* from periclinal chimeras of *Lycopersicon peruvianum. Z. Pflanzenzücht.* **77**, 112–120.
Steffensen, D. M. (1968). A reconstruction of cell development in the shoot apex of maize. *Am. J. Bot.* **55**, 354–369.
Tilney-Bassett, R. A. E. (1986). "Plant Chimeras". Edward Arnold, London.
Van Harten, A. M., and Bouter, H. (1973). Dihaploid potatoes in mutation breeding: some preliminary results. *Euphytica* **22**, 1–7.
Van Harten, A. M., Bouter, H., and Broertjes, C. (1981). *In vitro* adventitious bud techniques for vegetative propagation and mutation breeding of potato (*Solanum tuberosum* L.): II. Significance for mutation breeding. *Euphytica* **30**, 1–8.
Van Harten, A. M., Bouter, H., and Van Ommeren, A. (1972). Preventing chimerism in potato (*Solanum tuberosum* L.). *Euphytica* **21**, 11–21.
Verkerk, K. (1971). Chimerism of the tomato plant after seed irradiation with fast neutrons. *Neth. J. Agric. Sci.* **19**, 197–203.
Zimmermann, P. W., and Hitchcock, A. E. (1951). Rose 'sports' from adventitious buds. *Contrib. Boyce Thompson Inst.* **16**, 221–224.

# 17 World Strategies for Collecting, Preserving and Using Genetic Resources

J. G. HAWKES

*University of Birmingham, Birmingham*

Other chapters in this volume have stressed the need for a wide range of genetic diversity as the source material for breeding. Such a broad genetic base, as it is often called, offers a better chance of success in breeding for resistance to pests and pathogens, as well as adaptation to environmental stresses, higher yields and better protein and essential amino acid levels. With the soaring costs of fungicides, insecticides and other chemicals, in-built plant resistance can save great expense and enable crops to be grown more economically both in the developed countries and by the poor farmers of the Third World.

## 17.1 GENETIC DIVERSITY

The genetic diversity of our ancient crop plants and related wild species is of particular value in a search for various types of resistance, bearing in mind the difficulties encountered by breeders in finding lasting resistance to physiological races, or pathotypes of fungi, bacteria, viruses and nematodes.

Although very many successful cultivars have been produced in the past from crosses and selections based primarily on old cultivars and material derived from breeders' working collections, the situation has changed dramatically during the past few decades. Virtually all the useful genes in these materials, at any rate for the most important crops in temperate regions, have been evaluated and incorporated in the newer cultivars. We know that much exotic genetic diversity exists in certain parts of the world, particularly in those areas that the Russian geneticist,

N. I. Vavilov, designated "Centres of Origin and Diversity" of crop plants. Vavilov, and many others after him, began to collect the genetic diversity in these ancient centres during the 1920s and 1930s. These Centres of Diversity were mostly to be found in the mountains of the tropics and warm temperate regions, but subsequent studies have shown that diversity for certain crops also exists in the lowland tropics and elsewhere, even outside the gene centres. Some of this "exotic" material had already been collected by breeders and others in the 1940s and 1950s to help broaden the genetic base, though the use made of it at that time was not very great.

### 17.1.1 Genetic erosion

Better farming methods and the introduction of "green revolution" and other high-yielding cultivars, particularly in the developing world, have caused considerable erosion of crop genetic diversity during the past few decades, to the extent that the old land races and primitive forms of crop plants are now rapidly disappearing. The extension of farming and cattle grazing into grassland, savanna and forest is also destroying wild species of potential or actual use to breeders. Even if these species are not actually destroyed completely, their range of genetic diversity is greatly diminished through the destruction of their habitats, thus confining them to very small areas in comparison with their former distribution.

Because of breeders' needs, either now or in the future, it has become clear that genetic diversity, or genetic resources as they are generally called, must be conserved in a living state and in such a way that they can be made easily available to breeders. Furthermore, materials that are of no apparent value for present needs should not be discarded. It is quite possible that they might be of value in 20, 50 or even 100 years' time, even though they may seem to be valueless today. Breeders' needs are bound to change from decade to decade.

### 17.1.2 Genetic resources development

The need for genetic resources conservation began to be felt in the 1950s by FAO (Food and Agriculture Organization of the United Nations). Technical conferences were convened at FAO headquarters in Rome in 1961, 1967, 1973 and 1981; their aim was to define the problems of genetic conservation and discuss ways of solving them. The second and third conferences, convened jointly by FAO and the International Biological Programme, established the subject as a discipline in its own right and developed clear guidelines for future action (Frankel and Bennett, 1970;

Frankel and Hawkes, 1975). At the fourth conference, convened by the FAO and the IBPGR (International Board for Plant Genetic Resources), further steps were taken, particularly concerning international collaboration (Holden and Williams, 1984). These conferences, the FAO Advisory Committees, and the work of IBPGR and other organisations have now brought us to the point of identifying the organisational framework and the considerable research initiatives which are still needed, but which were scarcely envisaged in the early days.

### 17.1.3 Genetic resources activities

Genetic resources activities may be separated into seven categories, as follows: (a) exploration, (b) conservation, (c) evaluation, (d) data management, (e) utilisation, (f) training, and (g) international co-ordination. In the rest of this chapter I shall deal with each in turn, placing emphasis on those aspects that are particularly involved with vegetatively propagated plants. Most of the early thought and discussion revolved round cereal crops, and only during the past decade have the vegetatively propagated crops been receiving the attention they merit.

## 17.2 EXPLORATION

This section considers the survey and collection of wild and cultivated materials in the field. In the early days, it was thought sufficient to take a sample of seeds, add a few locality notes to the label, together with a cultivar name if this was available, and put them all in a bag. However, studies of the population genetics of wild species (Allard, 1970; Bennett, 1970) showed that more sophisticated sampling methods were needed if a reasonable amount of the genetic diversity of a species was to be captured. Marshall and Brown (1975), Bradshaw (1975) and Jain (1975) developed sampling methods for wild species and stated that these methods also held good for the old land races and primitive forms of cultigens, since all these were held to possess some kind of population structure.

### 17.2.1 Seed sampling

It is generally agreed that sampling on a population basis should be random or non-selective, since selective sampling will pick out only those genes with a clear morphological expression in the phenotype and will be likely to leave out those controlling disease resistance and other physiological characteristics. By sampling 50 seeds from 50 plants collected at random and bulking these together as a single sample, Marshall and Brown have calculated that at least one sample of each allele present in the population at a frequency of 5% or over will be collected at a 95% level of certainty. Thus, only rare alleles may be missed, but even many of these seem to be picked up, as judged from sampling exercises undertaken by some of my students at Birmingham and in Indonesia during practical training courses.

The slogan of "maximum diversity for minimum sample size and number" would appear to be a reasonable one. Yet it must not be forgotten that interpopulation diversity should also be collected by sampling a group of populations in any one geographical locality. Furthermore, diversity which is often of a clinal nature linked to geographical, climatic and biotic factors must also be collected by means of selecting sampling sites throughout the distribution area of a species. This latter objective is often accomplished by making a coarse grid survey (Jain, 1975) during a preliminary trip, followed by a fine grid survey in areas of interest on subsequent visits. If genetic erosion is very intense, there may be no time for several collecting trips before the material disappears for ever, and in such cases a "now-or-never" strategy should be adopted.

### 17.2.2 Sampling vegetatively propagated plants

The problem associated with sampling these crops has been discussed by Hawkes (1975, 1980). Most vegetatively propagated crops possess the capacity to reproduce sexually, or at least they did so before the advent of plant breeding and selection; and the very primitive forms and old land races probably still do so. Related wild species certainly retain both methods of reproduction as useful alternative strategies according to whether uniformity (asexual) or diversity (sexual) is needed for species survival (Mather, 1966).

In all vegetatively propagated crops, farmers appear to have carried out such strong artificial selection that the original population structure has virtually disappeared. For instance, in the ancient Centre of Diversity of potatoes in the central Andes of South America, each market area or district seems to contain not more than 50–100 distinct morphotypes. The number of morphotypes might have been as high as 200 or more in some areas before genetic erosion took place, though it is difficult to be certain of this. We assume that each of these morphotypes represents a distinct genotype, but it is possible that some morphotypes may conceal several different genotypes. Adjacent market areas or districts seem to possess much the same range of morphotypes, though this range would gradually change as one went further away, and other morphotypes would gradually appear more and more frequently.

This knowledge helps one to build up a sampling strategy for potatoes, cassava, sweet potatoes, yams and all other crops that are predominantly vegetatively propagated. It is recommended for these that selective sampling of every distinct morphotype in an area be carried out. In the adjacent market area another complete set of morphotypes is collected, and so on. These are all grown together in the gene bank or introduction station and the duplicates eliminated. At that stage also, it is possible to identify slight differences in what were, at first sight, thought to have been identical clones.

Where sexual reproduction takes place in a vegetatively propagated crop, and if the crop is largely outcrossing, the seeds should be sampled on a population basis, even though the plants have undergone much artificial selection and do not constitute a true population. Most vegetatively propagated crops are very heterozygous and release much diversity when they reproduce from seed; in addition, the seeds represent a sample of the local gene-pool when cross-pollination has taken place.

When collecting related wild species the methods are slightly different. With these, the population structure is still present, and thus seed sampling will follow the example given above for seed crops. When seeds are not available it is sufficient to take a bulk sample of whatever vegetative propagules are present. These should then be grown as a single collection in the introduction station and seeds then obtained by bulk pollination; again, these seeds will constitute a single sample.

Vegetatively propagated fruit trees also require some comment. My experience in Indonesia has shown that some species in certain areas are vegetatively propagated by cuttings, etc. This agrees with the views of Zohary and Spiegel-Roy (1975) for fruit trees in the Near and Middle East. In these instances the collector must sample selectively by morphotype collection, as shown above. On the other hand, not all fruit trees are propagated vegetatively, at least in Indonesia. In fact, many are grown from seed, and should be sampled accordingly. The farmer may possess only one or two trees in his orchard garden or backyard; how then can the collector sample a population? The method here is to consider the whole village as containing the population, which it almost certainly does, and take random samples from this area. The sampling of large-seeded materials, such as coconuts, on a population basis brings with it logistic problems which will not be considered here. Suffice it to say, however, that even small population samples are better than none at all.

Finally, we come to data recording. The International Board for Plant

Genetic Resources has advocated the acceptance of minimum data recording sheets for field collections. I reproduce a generalised one here (Fig. 17.1), which I developed on the basis of many years' experience. Very complex sheets, prepared for individual crops, can be useful but may perhaps have the disadvantage that the time spent in filling up the data sheets might be better spent in making more collections.

## 17.3 CONSERVATION

Storage of genetic resources material is normally undertaken in a genetic resources centre, gene bank, seed bank, or germplasm station, to cite various names for this kind of institute. Strategies have generally been worked out for seeds, since they are the easiest to store. However, other techniques need to be established for vegetatively propagated plants and even some seed crops, as will be seen later. The material is usually kept in two groups, a *base collection* for long-term storage under optimum conditons, and an *active collection* for shorter-term storage, distribution, evaluation tests and the like. This latter is not to be confused with breeders' working collections which, as the name implies, are the stocks on which breeders are currently working.

### 17.3.1 Orthodox seeds

Generalised seed-storage strategies have been worked out for those species which store best under low temperatures and at low moisture content. So far as we know, all temperate and warm temperate crop species and related wild species belong to this category, which for obvious reasons has been termed "orthodox". These orthodox seeds store best at temperatures of 0 to $-20°C$ or even lower. Seed moisture contents of about 4%, combined with low temperatures, give us reason to expect that they will have a very long storage life (Roberts, 1975). Periodic germination tests are needed but, if no decrease in germination seems to be taking place, the tests can be scheduled at intervals of 5 years or perhaps more. Although some 400 seeds are required for a proper test, the method of sequential testing (Ellis *et al.*, 1980; Roberts and Ellis, 1984) requires not more than 40, unless the viability has fallen.

When viability falls to 85–90% of that when the seeds were first placed in store, the material must be regenerated by growing out a new seed generation, so as to prevent genetic changes taking place in storage when viability falls to lower levels (Roberts, 1975).

Although little genetic change should take place in self-pollinating crops during regeneration, there is considerable risk attending regeneration of cross-pollinated species. Care then must be taken to prevent pollination from other samples and to ensure that the gene pool is preserved intact (Porceddu and Jenkins, 1982). Such maintenance of the genetic integrity of the sample is difficult; even in self-pollinated species, interplant competition may result in the loss or suppression of genotypes that are less well adapted to the climate or soil of the gene bank. Ideally, material ought to be regenerated in the area whence it originally came, but this is a council of perfection, seldom realised in practice. However, the reduction of interplant competition is very helpful. Reasonably large numbers of plants should also be grown, so as to prevent inbreeding depression or genetic drift due to the use of too small numbers of plants. Numbers suggested are from 50 to 100, but often smaller numbers seem to be used. More research is urgently needed in this difficult matter of regeneration. Meanwhile, longer regeneration intervals will help to prevent loss of genetic integrity until more precise regeneration methods have been established.

### 17.3.2 Recalcitrant seeds

Many tropical and some temperate fruit and nut trees belong to this group, as do certain tropical plantation crops such as tea, coffee, cocoa, oil palm, coconut palm and others. The seeds die very quickly when they dry to less than 20% moisture content, and they are not able to withstand low temperatures. Although research on controlled drying or preservation in a fully imbibed state is in progress, the results have not yet reached the point when they can be recommended for standard practice. Material of these "recalcitrant species" can, of course, be stored in meristems (see G. G. Henshaw, this volume), though here again more research is still needed (see also Roberts et al., 1984; Hanson, 1984).

At present, the standard way of storing recalcitrant species is in orchard or forest plantations. This is very wasteful of space, since every tree is a genotype occupying perhaps as much as 10 × 10 m. Compare this with an orthodox seed collection of 2500 seeds in a small receptacle in a seed store. Each is a distinct genotype, but to conserve that same number of genotypes in an orchard collection we should need an area of 25 h. Clearly, numbers of samples in orchard plantations must be very restricted in comparison with those stored as seed.

Another possible method is to store material as *seedlings* rather than as seeds or mature plants (Hawkes, 1982). Although research has not yet begun it would seem that the storage of seedlings or young plants has

| | |
|---|---|
| Expedition/Organization: | |
| Country: | |
| Team/Collector(s): | Collector's Number: |
| Date of Collection: | Photo Number(s): |
| Species Name: | |
| Vernacular/Cultivar Name: | |
| Locality: | |

Latitude: _____ ° _____ '    Longitude: _____ ° _____ '    Altitude: _____ m

Material:   Seeds    Inflorescences    Roots/tubers    Live Plants    Herbarium

Sample:    Population    Pure Line    Individual        Random    Non-Random

Status:    Cultivated    Weed    Wild

Source:    Field    Farm Store    Market    Shop    Garden    Wild vegetation

Original Source of Sample: _____

Frequency:    Abundant    Frequent    Occasional    Rare

Habitat: _____

Descriptive Notes: _____

| Collector | Collector | Collector | Collector |
|---|---|---|---|
| No: | No: | No: | No: |

(a)

**Figure 17.1.** Suggested standard recording sheet for field collections: (a) front of page; (b) back of page.

*Uses:* ........................................................

........................................................................................................................................................

*Cultural Practices:*         Irrigated        Dry

*Season:*         *Approximate sowing dates:*............................................................

                *Approximate harvesting dates:*..........................................................

*Soil Observations:*   *Texture:* ........................................................................

                *Stoniness:* ......................................................................

                *Depth:* .............................................................................

                *Drainage:* ......................................................................

                *Colour:* ..........................................................................

*Soil pH:* .............

*Land Form:*         *Aspect:* ..........................................................................

                *Slope:* ............................................................................

*Topography:*       Swamp    Flood Plain    Level    Undulating

                Hilly    Hilly Dissected    Steeply Dissected

                Mountainous    Other (specify)

*Plant Community:* ........................................................................................

........................................................................................................................................................

*Other Crops grown near or in rotation:* ..................................................

........

*Pests/Pathogens:* ........................................................................................

*Name and Address of Farmer:* ...........................................................

........................................................................................................................................................

*Taxonomic Identification:* ...............................................................................

*by* ................................................................ *Date:* ..........................................

*Name of Institution:* ............................................................ *Accession No.*..............

**(b)**

possibilities, since it would be making use of the evolutionary strategy of tropical rainforest trees, the seeds of which are adapted to rapid germination and establishment, growing at very low light levels on the forest floor and waiting for a gap in the canopy. Once this occurs the seedlings grow very rapidly, to reach the upper levels where light is permanently available. Thus, storage of recalcitrant species in seedling banks may provide the key to their conservation.

### 17.3.3 Vegetatively propagated crops

The great value of preserving exact genotypes in a disease-free condition is widely acknowledged (see Henshaw, this volume; Withers, 1984). The time taken to regenerate long-lived woody perennials is a drawback, but it is an even greater problem if their seeds are to be stored, when a complete sexual cycle of perhaps 50–80 years would be needed when the seeds lost their viability (Hawkes, 1982). On the other hand, subculturing of meristems only takes a few minutes. Of course, this problem does not arise with short-lived perennials such as potatoes, yams, cassava, sweet potatoes and many other important tropical starch crops.

### 17.3.4 *In situ* conservation

The methods previously discussed are all based on *ex situ* conservation. However, one of the obvious ways of conserving wild species is by *in situ* biosphere reserves or conservation areas. These possess the additional advantage that whole ecosystems are preserved, not just the single species in which we happen to be interested at present, but all those that may very possibly have some value for us in the future.

Disadvantages to the nature or biosphere reserve do exist, however. First, we can only conserve parts of the total geographical and ecological diversity of the species, since the reserves can never be of sufficient size to encompass its total distribution area unless it has a very restricted range. One must admit, however, that the ranges of many tropical forest species are now so restricted because of widespread forest destruction that even a single biosphere reserve would be all that was needed to save their last remnants.

Various suggestions have been made for conserving cultivated materials in a "natural" state by setting aside regions where the traditional agricultural methods and cultivars are preserved. There are many difficulties involved here with the control of such activities, and in any case the reserves would only contain a very small part of the total genetic diversity of any one crop.

Nevertheless, gene parks, as they are sometimes called, could be of great educational value as examples of a nation's cultural heritage. They could also be used as experimental areas for research into the genetic balance between crops, weeds and related wild species, and also to monitor the relations between the crops on the one hand, and the races of their pests and pathogens on the other. When nearly all crop diversity is stored in the gene bank there will be no evolutionary equilibrium between it and its parasites. Thus, gene parks could provide much useful information and insights into host–parasite relationships for the future.

## 17.4 EVALUATION

Screening or evaluation of gene bank materials is an essential part of genetic resources work. It is useless to conserve museum collections (for that is what they would be) of accessions whose properties are unknown and are thus of no value to breeders. A great deal of money is spent on collecting and conserving these materials so as to use them to improve existing cultivars, and a knowledge of their valuable qualities is thus essential.

At this point we come to a divergence of opinion. If an accession is found not to possess the required characters, should it be discarded? Some breeders will say "yes", because this is what they are accustomed to do with material found to be valueless in their working collections. However, when genetic resources materials are being screened ("active" rather than "base" collections, it will be remembered) and found to be useless for a particular breeding programme, they should still be conserved. Future breeding needs may be very different from present ones, and qualities needed then may be contained in accessions that we at present consider useless and are tempted to discard. To take an example from potatoes, resistance to various races of root-knot and cyst nematodes have been discovered in wild species that 30 years ago were thought to be useless. Virus M resistance has recently been found in *Solanum gourlayi*, and bacterial wilt resistance in *S. sparsipilum*. Neither of these species was formerly considered to be of any value to breeders, since their yield and quality are very poor. Now, however, their resistance to pests and pathogens makes them very valuable.

### 17.4.1 Categories of screening

There are various stages of evaluation that have been recognised (Erskine and Williams, 1980). First, we have "initial characterisation", which is taken to include the scoring of highly heritable characters, generally of a morphological or agronomic nature. Care must be taken not to expend too much effort and time on this initial work because although scoring for morphological features may be valuable for eliminating duplicates (see below), it may be of less value to breeders than the second and third evaluation stages.

"Primary evaluation" follows as the second stage, in which features of agronomic value are scored. This type of evaluation is clearly of much greater use to breeders when dealing with the introduction of breeding lines or obsolete or modern cultivars from other countries.

The third stage is "secondary" or "in depth" evaluation. This is considered to come much later in the evaluation and to include screening for resistance to pests and pathogens, temperature extremes and special soil conditions such as drought and salinity. However, it must be emphasised here that many breeders, especially those using wild species in attempts to find resistances that have not been found in the cultigen itself, need to start immediately with secondary evaluation, since characterisation and primary evaluation do not really interest them. Thus, to take an example from wild potatoes again, breeders go straight into screening for pest or pathogen resistance, often choosing materials known to have been collected in regions where resistance has previously been found.

### 17.4.2 Elimination of duplicates

Let us now return to morphological, or morphometric characters as they are often called. The need for selective collections of vegetatively propagated crops, in which each distinguishable morphotype is collected within one region, has already been stressed. Similarly, in all other regions, a complete collection of morphotypes is made. Thus, there will inevitably be much duplication, and the duplicates will need to be identified and eliminated.

To take yet another example from potatoes, the practice of CIP (International Potato Centre) has been to grow the whole collection of some 15 000 lines in the field, but with quantities of that magnitude it is clearly impossible to eliminate all duplicates by eye. It was therefore decided (Huamán et al., 1977) to score all morphometric characters on a previously agreed plan of characters (descriptors) and character states (descriptor states) and put the data into a computer, which was programmed to sort out identical groups. The following year the individuals of each group were planted together, and similar but not identical groups were planted nearby. A second check by eye was then carried out, and this was followed in certain cases by a third check by means of slab electrophoresis of tuber proteins and isozymes. Thus for potatoes, an initial collection of some 15 000 accessions has been reduced to about one-third of that number. It should be mentioned, however, that not all duplicates were discarded; for instance, where 15–30 duplicates were identified, some 3–5 were retained, and true seed was obtained from every duplicate as an insurance against loss.

To sum up this section, evaluation should in general be related to breeders' needs, having in mind both present and future ones. Taxonomic identification is, of course, one part of evaluation, enabling breeders to formulate ideas on the species and geographical regions of greatest use to them. Thus, they will be in a position to request material of a certain type and from a certain region. If biosystematic studies have been carried out, they will also be able to request promising wild species which are known to be capable of crossing and exchanging genes with the cultivated forms.

## 17.5 DATA MANAGEMENT

So much material is now being deposited in gene banks, and so much information is forthcoming as a result of evaluation, that we are faced with something of an information explosion. Yet this information should somehow be made available to breeders to enable them to choose carefully the initial materials for their breeding programmes. The obvious answer must be to use electronic data processing systems for the storage and retrieval of information. This should include not only the results of various types of evaluation but also collection and gene banking details (so-called "passport data"), as was mentioned in an earlier section. Recent experience has shown that standard formats for setting out the data for each crop are essential, and the IBPGR has now established a number of crop committees charged with this task. The data are sorted out into a series of descriptors (characters) and descriptor states (character states) that are readily coded into machine-readable form. Standardisation at this level is essential, even though the data management systems and types of computers can be quite varied. Some 10 years ago we believed that these programming systems and computers ought also to be standardised. However, this was quickly seen to be impossible because of the breakneck speed of development of new computers and new and even better programming systems. So long as the data are set out in some kind of standard format, it was felt that the rest could take care of itself, and this is precisely what is now happening.

## 17.6 UTILISATION

The end-point of all genetic resources work is the utilisation of the living materials in breeding and selection programmes. Yet, in many instances, there is still a large gap between the assembling and screening of material on the one hand, and its use by breeders on the other. Gene-bank managers often feel that their responsibilities cease when they have sent printed lists or computer printouts to breeders telling them what species are present in their gene banks and what are the useful characters that these collections possess.

### 17.6.1 Pre-breeding

Many breeders are satisfied with such lists but many others are more than a little wary of what is often called "exotic germplasm". To combat this attitude and at the same time to provide useful genetic characters in a more readily available form yet another activity is needed, namely, pre-breeding. The difficulty of introducing characters from wild species into breeding lines is often very great. On the other hand, it is not so difficult to breed with highly bred lines or cultivar introductions whose yield and quality characters are already satisfactory. Breeders are quite understandably cautious about using materials from wild species or primitive cultivated forms with poor yields and poorer quality which when crossed with current cultivars will transfer not only the one valuable gene but a host of undesirable ones also. In any case, it takes some 10–15 years to breed a new cultivar and pass it through all the trials until it arrives as a marketable commodity. If one starts with wild species, it may take twice as long.

Hence, it is of considerable importance that either the gene-bank manager or others involved with the crop should take responsibility for incorporating useful genes from wild species or primitive forms into useful breeding lines with a "cultivated" rather than a "wild" genetic background. Such pre-breeding or "germplasm enhancement" should become a more and more important function of the gene-bank manager, his colleagues, or scientists in other institutes. The various ways in which this should be done are too long to be considered here in detail, but could form the basis of a subsequent discussion on germplasm utilisation.

## 17.7 TRAINING

It so happens that the greatest diversity of crop plants and related wild species is found in the developing countries of the tropics and subtropics.

Thus, an urgent need has arisen for training young scientists from those countries in the various aspects of genetic resources work. An M.Sc. course in genetic conservation has been offered at the University of Birmingham since 1969 and, during the past seven years, three-month modules have been offered for trainees who are unable to spend a whole year away from their institute or gene bank. Both of these courses are supported by the IBPGR.

Special courses of about 2–4 weeks' duration, generally dealing with the practical aspects of exploration and conservation of germplasm, have been offered in the gene centres, as for instance in The Philippines, Indonesia, India, Nigeria, Argentina, Peru and Colombia. Special training in data management has been offered in the USA, West Germany and Italy, while many gene banks offer practical experience in the day-to-day activities of gene-bank management. All of these training programmes are helping to give scientists and technical staff in developing and developed countries a good practical training, as well as a sound theoretical background in gene bank work.

## 17.8 INTERNATIONAL COLLABORATION AND FUNDING

Mention has been made previously of FAO's initiatives in germplasm work during the 1950s and 1960s. In 1972, a genetic resources plan was developed by an *ad hoc* committee at Beltsville, USA, sponsored by the Technical Advisory Committee (TAC) of the Consultative Group on International Agricultural Research (CGIAR). This latter body also established such well-known agricultural centres and institutes as IITA (International Institute of Tropical Agriculture) in Nigeria, CIP (International Potato Centre) in Peru, and IRRI (International Rice Research Institute) in the Philippines. The plan put forward by the Beltsville Committee was adopted by TAC and CGIAR, and as a result the International Board for Plant Genetic Resources (IBPGR) was established in 1974. This body is funded by donor countries through the World Bank, and has provided guidance and funding for a wide range of genetic resources activities. Thus, it has funded surveys; collecting expeditions; gene-bank facilities such as refrigerated seed stores and computers; evaluation; training courses; research; technical conferences, either on its own or jointly with FAO, Eucarpia, Sabrao and others and meetings of crop committees and other specialist committees. Also of importance are its establishment of regional and crop priorities for collecting, and its encouragement of the development of regional programmes, run co-operatively between groups of countries. Successful programmes of this type exist for South-east Asia, South Asia, and Europe. Plans for East Asia, Africa, Meso-America and the Andean zone of South America are well under way.

## 17.9 CONCLUSIONS

As a result of the plans and activities described above, considerable progress has been made in conserving crop germplasm for the present and the future. Yet it can clearly be seen that progress is still rather uneven. Much thought has been given to the conservation of seed crops of the orthodox type, and especially of cereals—admittedly the world's most important crops. Yet the almost equally important vegetatively propagated crops have not attracted the attention they deserve. This is partly because, apart from potatoes, they are cultivated in the tropics where less research has been carried out in comparison with crops of temperate countries. This gap in our knowledge should speedily be rectified. We need also more research and the bringing together of intellectual resources on the so-called "recalcitrant species" (see Withers and Williams, 1982), many of which are vegetatively propagated. We also need research on timber, nut and fruit crops, of which, again, many are vegetatively propagated. We need to bridge the gaps by promoting research in germplasm collection and conservation and, by so doing, help to strengthen research on the tropical crops and closely related wild species.

## REFERENCES

Allard, R. W. (1970). Population structure and sampling methods. *In* "Genetic Resources in Plants—their Exploration and Conservation" (O. H. Frankel and E. Bennett, eds), pp. 97–113. Blackwell Scientific, Oxford.
Bennett, E. (1970). Tactics of plant exploration. *In* "Genetic Resources in Plants—their Exploration and Conservation" (O. H. Frankel and E. Bennett, eds), pp. 157–179. Blackwell Scientific, Oxford.
Bradshaw, A. D. (1975). Population structure and the effects of isolation and selection. *In* "Crop Genetic Resources for Today and Tomorrow" (O. H. Frankel and J. G. Hawkes, eds), pp. 37–51. Cambridge University Press, Cambridge.
Ellis, R. H., Roberts, E. H., and Whitehead, J. (1980). A new, more economic approach to monitoring the viability of accessions during storage in seed banks. *Plant Genet. Resour. Newslett.* **41**, 3–18.
Erskine, W., and Williams, J. T. (1980). The principles, problems and responsibilities of the preliminary evaluation of genetic resources samples of seed-propagated crops. *Plant Genet. Resour. Newslett.* **41**, 19–33.

Frankel, O. H., and Bennett, E. (1970). "Genetic Resources in Plants—their Exploration and Conservation". Blackwell Scientific, Oxford, 554 pp.

Frankel, O. H., and Hawkes, J. G., eds (1975). "Crop Genetic Resources for Today and Tomorrow." Cambridge University Press, 492 pp.

Hanson, J. (1984). The storage of tropical tree fruits. *In* "Crop Genetic Resources: Conservation and Evaluation" (J. H. W. Holden and J. T. Williams, eds), pp. 53–62. Allen and Unwin, London.

Hawkes, J. G. (1975). Vegetatively propagated crops. *In* "Crop Genetic Resources for Today and Tomorrow" (O. H. Frankel and J. G. Hawkes, eds), pp. 117–121. Cambridge University Press.

Hawkes, J. G. (1980). "Crop Genetic Resources Field Collector's Manual". IBPGR/Eucarpia, 37 pp.

Hawkes, J. G. (1982). Genetic conservation of "recalcitrant species"—an overview. *In* "Crop Genetic Resources: the Conservation of Difficult Material" (L. A. Withers and J. T. Williams, eds), pp. 83–92. IUBS, Série B42, Paris.

Holden, J. H. W. and Williams, J. T., eds (1984). "Crop Genetic Resources: Conservation and Evaluation". Allen and Unwin, London, 296pp.

Huamán, Z., Williams, J. T., Salhuana, W., and Vincent, L. (1977). "Descriptors for the Cultivated Potato". IBPGR, Rome.

Jain, S.K. (1975). Population structure and the effects of breeding system. *In* "Crop Genetic Resources for Today and Tomorrow" (O. H. Frankel and J. G. Hawkes, eds), pp. 15–36. Cambridge University Press.

Marshall, D. R., and Brown, A. H. D. (1975). Optimum sampling strategies in genetic conservation. *In* "Crop Genetic Resources for Today and Tomorrow" (O. H. Frankel and J. G. Hawkes, eds), pp. 53–78. Cambridge University Press.

Mather, K. (1966). Breeding systems and response to selection. *In* "Reproductive Biology and Taxonomy of Vascular Plants" (J. G. Hawkes, ed.), pp. 13–19. Pergamon/BSBI, Oxford.

Porceddu, E., and Jenkins, G. (1982). "Seed Regeneration in Cross-pollinated Species". Balkema, Rotterdam, 293 pp.

Roberts, E. H. (1975). Problems of long-term storage of seed and pollen for genetic resources conservation. *In* "Crop Genetic Resources for Today and Tomorrow" (O. H. Frankel and J. G. Hawkes, eds), pp. 269–295. Cambridge University Press.

Roberts, E. H., and Ellis, R. H. (1984). The implications of the deterioration of orthodox seeds during storage for genetic resources conservation. *In* "Crop Genetic Resources: Conservation and Evaluation" (J. H. W. Holden and J. T. Williams, eds), pp. 18–37. Allen and Unwin, London.

Roberts, E. H., King, M. W., and Ellis, R. H. (1984). Recalcitrant seeds: their recognition and storage. *In* "Crop Genetic Resources: Conservation and Evaluation" (J. H. W. Holden and J. T. Williams, eds), pp. 38–52. Allen and Unwin, London.

Withers, L. A. (1984). Germplasm conservation *in vitro*: present state of research and its application. *In* "Crop Genetic Resources: Conservation and Evaluation" (J. H. W. Holden and J. T. Williams, eds), pp. 138–157. Allen and Unwin, London.

Withers, L. A., and Williams, J. T. (1982). "Crop Genetic Resources. The Conservation of Difficult Material". IUBS, Série B42, Paris, 134pp.

Zohary, D., and Spiegel-Roy, P. (1975). Beginnings of fruit growing in the old world. *Science* **187**, 319–327.

# 18 New Techniques for Germplasm Storage

G. G. HENSHAW

School of Biological Sciences, University of Bath, Bath

Clones of vegetatively propagated plants are frequently maintained on a long-term basis for use as germplasm in breeding programmes and as healthy, genetically stable nuclear foundation stocks from which plants are ultimately derived for commercial multiplication. The same rules apply for both purposes: genetic stability is of prime importance and the exclusion of pests and pathogens is critical because of the lack of a developmental stage, equivalent to the seed, which can act as a barrier to the transmission of disease. Ideally, the material should also be in a form in which it can be made available quickly for rapid multiplication and in which it is convenient and acceptable for international distribution.

Traditionally, vegetative germplasm is maintained as normal plants, propagated by conventional methods, either under protected conditions in insect-proof glass or screen houses or as plantations in the field, possibly with intermittent periods of storage if suitable propagules are available. To varying degrees, depending on the species and the particular circumstances, these methods are not entirely satisfactory because they can be expensive and there is the risk of losses through disease and other natural disasters. The continuity of the genetic line is generally maintained by the propagation of organs produced from non-adventitious shoot buds, which essentially form a germ line in which genetic stability is favoured by diplontic selection. However, no propagation methods guarantee trueness-to-type, and it is necessary to take precautions to ensure that off-types can be identified and eliminated by roguing. Even so, roguing procedures are based on observations of the phenotype and more cryptic changes can escape notice with the result that the original genotype could be lost unless adequate sampling procedures are employed at the time of propagation.

IMPROVING VEGETATIVELY PROPAGATED CROPS
ISBN 0-12-041410-4

Copyright © 1987 Academic Press Limited
All rights of reproduction in any form reserved

Without question, the most satisfactory methods for the long-term conservation of genotypes would not involve continuous growth and, ideally, some means of storing material in a state of suspended animation should be available. At present, cryopreservation involving long-term storage at the temperature of liquid nitrogen ($-196°C$), seems to be the only feasible means of achieving this state, although such techniques have not yet been developed to the extent that they can be used routinely with a wide range of vegetative material. Basically, it is the high water content of many plant cells which presents difficulties with regard to cryopreservation. Only those cells with a relatively low water content can survive the freezing process without close and elaborate attention being paid to their individual requirements. In the normal plant, such cells appear to be readily available only in pollen, seeds and dormant meristems, but only the last would be suitable for use with clonal material. Generally speaking, however, it is necessary to separate these freeze-tolerant cells from other more susceptible tissues, but the conventional methods of propagation could not be applied to such isolated meristems. On the other hand, meristems or, rather, shoot-tips with a limited number of leaf primordia, are used routinely in many of the *in vitro* procedures which are being used increasingly for the rapid multiplication of clonal material. Like the conventional methods of propagation, these procedures can also be adapted for the storage of germplasm.

## 18.1 GERMPLASM STORAGE *IN VITRO*

*In vitro* propagation methods fall into two basic categories: those involving the production of non-adventitious shoot meristems and those based on the adventitious production of organs or embryos, either from cells present in the original explant or from cells which have proliferated *in vitro*. In principle, the former methods involving meristem or shoot-tip cultures are very similar to conventional propagation procedures in which new shoots are produced non-adventitiously so that some continuity with the original apical meristems is maintained. These methods are now used widely for commercial micropropagation and their record for genetic stability seems to be good, although the evidence does not seem to have been fully collated.

However, the genetic stability of systems in which adventitious organs or embryos are produced from explants or from disorganised callus or suspension cultures may be poor. This could be due to a lack of genetic uniformity among the cells of the explant, since it is known that many plant species are polysomatic (D'Amato, 1978) and there is increasing suspicion that more cryptic genetic changes might occur in somatic cells, possibly as part of the developmental process (Larkin and Scowcroft, 1981). In addition, there is good evidence that the *in vitro* conditions themselves can be mutagenic and there have been many reports of chromosomal abnormalities in callus and suspension cultures (D'Amato, 1978). The *in vitro* conditions could also act in a selective manner if a heterogeneous population of cells is produced, with the result that there is likely to be a gradual shift in the genetic composition of such cultures if they are maintained under conditions which allow cell proliferation or cell death. Despite these potential problems concerning the trueness-to-type of plants regenerated by an adventitious process, it has been suggested that with certain species this type of culture is stable (Sheridan, 1968; Davies and Heslop, 1973) and, indeed, with some species, such as the oil palm, it probably represents the only practical proposition if *in vitro* techniques are to be used for propagation and germplasm storage.

In view of these differences between the two basic types of *in vitro* propagation procedures, it must be concluded that as a general rule it is desirable that germplasm storage procedures should be based on non-adventitious shoot-tip culture systems, although there will be exceptions to this rule where such systems do not exist or when it is known that acceptable levels of genetic and morphogenetic stability exist in callus or suspension cultures. Cryopreservation techniques might overcome some of the problems with the latter type of culture and, indeed, the technical problems are generally less daunting with isolated cells than with meristems. However, if the cultures are known to be unstable, there could be some risk of genetic change during the periods of growth before and after cryopreservation. A further argument in favour of using shoot-tip cultures for germplasm storage would be that, for many species, the same basic procedure could also be used for the elimination of systemic pathogens, especially viruses, and for rapid multiplication when the conserved material is to be utilised (Henshaw, 1979).

Assuming that genetically stable culture systems are available for germplasm storage purposes, there remain questions of their suitability with regard to plant health problems and economics. From the point of view of plant health, shoot-tip cultures would seem to have several advantages compared with normal plants grown in the field or even under protected conditions. They provide the basis, generally in combination with heat therapy, of what is often the most effective method of eliminating systemic pathogens from vegetative plant material (Walkey, 1978, 1985), although the method can never be assumed to be totally effective and it must be used in association with appropriate indexing procedures. Many other contaminants would be also detected and eliminated by the standard aseptic procedures used for the initiation and maintenance of cultures, but certain organisms, especially bacteria, might remain undetected unless suitable selective media are employed routinely (Langhans *et al.*, 1977). The risk lies in assuming too much about the basic *in vitro* techniques, which certainly do not eliminate the need for routine plant health precautions. Even so, once healthy cultures have been established, the risk of re-contamination is relatively low, compared with a situation in which plants are grown under conventional conditions.

The question concerning economics is complex, the answer probably varying according to the species and circumstances. If there are few plant health problems arising from the use of conventional methods and the plants can be grown inexpensively in the field as, perhaps, with some of the long-lived perennial species, there might be little advantage in establishing elaborate *in vitro* facilities for the specific purpose of germplasm storage. If, on the other hand, such facilities are likely to be established for other purposes, such as propagation or breeding, or if there are formidable plant health problems which might be alleviated by the use of *in vitro* techniques, then the economic argument is likely to be different. Either way, any procedure leading to a reduction in the need for the routine management of stored cultures would have considerable economic advantages. This is most easily achieved by the use of techniques which reduce the growth rates of cultures, preferably to zero, which would be the case if cryopreservation techniques could be employed. Such procedures might have added advantages with regard to the genetic and morphogenetic stability of cultures, since the risk of mutation is also likely to be reduced.

Unfortunately, although it would seem that fully hydrated somatic cells in certain plants can survive for very long periods of time in a non-dividing state (more than a century has been suggested (MacDougal, 1926) for pith cells of the giant cactus *Cereus giganteus*), the mechanisms that operate in such circumstances are not understood and little success has been achieved in prolonging the life of non-dividing cultures beyond a limited number of years, except by cryopreservation. It might eventually be possible to simulate natural dormancy mechanisms in suitable embryos and meristems produced *in vitro*, so that they can be stored in conditions similar to those used for long-term storage of seeds, but at present the most practical storage procedures for cultures generally permit some growth, unless they are sufficiently hardy to be stored at temperatures close to 0°C. Such growth is not necessarily a disadvantage because it is usually possible to keep a visual check on the viability of the cultures, and material that is taken from store is fairly quickly replaced. However, in very large germplasm collections, possibly containing several thousand accessions, even the relatively small effort involved in maintaining cultures under restricted growth conditions would be time-consuming and costly and it is highly desirable that alternative methods, that allow cultures to be stored for many years without attention, should also be available. For the reasons stated earlier, cryopreservation techniques seem to offer the only possible solution to this problem which has a chance of being widely applicable.

## 18.2 RESTRICTED-GROWTH STORAGE PROCEDURES

Since the genetic and morphogenetic stability of many cultures that produce shoots or embryos adventitiously cannot be guaranteed on a long-term basis, it is advisable that restricted-growth storage procedures for the purpose of germplasm conservation should, where possible, be employed with meristem or shoot-tip cultures. Investigations with such cultures from a range of crop species indicate that subculture intervals of at least 1 year, which would be fairly convenient, seem to be quite readily achieved, and in some cases the intervals can be extended to several years.

The various restricted-growth options can be grouped into three main categories involving manipulation of the environmental conditions (temperature, light, osmolarity, etc.), the nutrient conditions or the supply of growth regulators. In general, the most successful and also the most practical procedures have involved changes in environmental conditions. Reduced temperatures, for example, can probably be used effectively with a wide range of species. The optimal temperatures for long-term storage vary considerably from species to species but it can be assumed that they range above freezing temperatures because ice formation, except at ultra-low temperature, is not compatible with long-term storage. Cultures from hardy species such as strawberry have been stored at normal refrigerator/cold room temperatures (4°C) for as long as 6 years, with very small additions of medium every 6 months to replenish nutrients and water (Mullin and Schlegel, 1976), and less hardy material from a range of potato species has been stored for more than 1 year at temperatures between 6° and 10°C (Westcott, 1981a). Tropical species such as sweet potato can suffer chilling damage at such low temperatures, although a shift away from the optimal growing temperatures, to 18°C, usefully extends the subculture interval (Alan, 1979). On the other hand, cocoyam (*Xanthosoma sagittifolium*), another tropical species, has been stored successfully for more than 1 year at 10°C (Acheampong, 1982).

Reduced temperatures can be expensive to maintain on a large scale and alternative procedures permitting storage at higher temperatures are valuable. Growth rates of cultures are reduced substantially when the osmolarity of the medium is raised by the addition of compounds such as mannitol or sorbitol in the concentration range 2–5% (w/v). Since this has been shown for species as diverse as potato (Westcott, 1981b), yam (*Dioscorea rotundata*) (Mathias, 1980) and cocoyam (Acheampong, 1982), the technique might have wide applicability.

Adjustments to the concentrations of nutrients in the medium are probably less useful than the procedures already described. Growth-limiting concentrations of nutrients can be employed but deficiency symptoms and senescence are rarely long-delayed. Sucrose concentrations can sometimes be reduced if the cultures are at least partially autotrophic and this can have the effect of preventing excessive growth which might exhaust other nutrients (cocoyam, Acheampong, 1982). Also, sucrose is probably the only nutrient whose concentration can be raised usefully without causing toxicity problems (carrot embryos, Jones, 1974; potato, Westcott, 1981a), although this treatment alone does not necessarily reduce growth rates.

Growth rates of cultures can certainly be reduced by the addition of various inhibitory growth regulators (Westcott, 1981b) to the medium, but their use for germplasm storage is questionable in view of the possibility of genetic damage. This might be less of a problem with the naturally occurring growth regulators such as abscisic acid, which has been shown to inhibit growth reversibly in shoot-tip cultures of potato (Westcott, 1981b), sweet potato (Alan, 1979) and cocoyam (Acheampong, 1982), and gibberellic acid, which has similar effects with cultures of yam (Mathias, 1980) and cocoyam (Acheampong, 1982).

## 18.3 CRYOPRESERVATION

The higher water contents of most plant cells, except some found in pollen and seeds, seem to preclude the use of cryopreservation techniques for their long-term storage unless suitable combinations and concentrations of cryoprotective substances and cooling rates can be identified. Some of the principles governing the use of these procedures have been established but, unfortunately, the different cell types, that are assembled in organs, seem to have such diverse properties that appropriate cryopreservation protocols can be devised only in an empirical manner. At present, only the relatively small cells found in callus and suspension cultures during the exponential phase of cell division and in meristems, and the dormant cells in certain hardy tree species, have responded to these techniques.

Because of the specific requirements of individual cell types and the difficulty in establishing uniform conditions throughout a large mass of tissue, the greatest success has been achieved with the dispersed population of relatively uniform cells that can occur in suspension cultures (see Withers, 1980a). It is very likely that satisfactory cryopreservation protocols can be established for such cultures from a wide range of species and undoubtedly these techniques will be extremely valuable in the future for the conservation of the less conventional type of germplasm produced by *in vitro* breeding and genetic engineering programmes. This type of culture is less satisfactory for the storage of conventional vegetative germplasm, for reasons that have already been stated, and it is essential that more universally applicable procedures should be devised for use with meristems from which plants can be regenerated. So far, only meristems from a relatively small number of species have responded and within those species the responses of different genotypes can be quite varied, probably to the extent that the expense involved in establishing facilities for the large-scale use of the procedures would not yet be justified.

The difficulties concerning the establishment of successful cryopreservation procedures arise from the complexity of the freezing process. A detailed description of the sequence of events would be inappropriate here; accounts have been given by Mazur (1969) and Meryman and Williams (1980) and there have been extensive reviews of cryopreservation procedures by Withers (1980a,b,c) and various authors (see Kartha, 1985).

Essentially, it has been found that the relationship between cell survival and the cooling rate is bimodal, with one optimum occurring at relatively slow rates, approximately in the range $0.1$–$1.0°C$ $min^{-1}$, and the other at rapid rates, in excess of $1000°C$ $min^{-1}$, although there is a limit to the size of specimen with which the latter rates can be achieved. The slow cooling procedure apparently involves the osmotic dehydration of the cells in response to extracellular ice formation, so that intracellular ice crystal development is reduced or completely avoided; rapid cooling does not permit dehydration but it favours the formation of amorphous ice, rather than large crystals which are likely to cause damage if they develop within the cells. Thawing rates can also be critical, and if the cell contains large amounts of amorphous ice, as after rapid cooling, thawing must be rapid to avoid ice recrystallisation at certain critical temperatures. Slow thawing, on the other hand, might help to avoid damage caused by the rapid rehydration of dehydrated cells.

The mode of action of cryoprotectants, which are almost invariably required for the successful cryopreservation of plant cells, is complex, with different mechanisms almost certainly operating according to particular circumstances. Theoretically, any solute which can penetrate the cell in sufficiently high concentrations without being toxic, and even some non-penetrating substances, can have cryoprotectant properties, but in practice relatively few compounds are used, with dimethylsulphoxide and glycerol seeming to have the widest applicability with plant cells.

Since the ice present in cryopreserved cells is in a metastable state, it is essential that temperatures below −120°C should be employed for long-term storage, otherwise recrystallisation will lead to the development of large injurious ice crystals. Such temperatures are most readily achieved by the use of liquid nitrogen refrigerators which, being no more than insulated containers, are relatively cheap and reliable as long as liquid nitrogen supplies are guaranteed. At the storage temperature of −196°C most physical and chemical processes cease to operate, and there should be little risk of deterioration apart from what are believed to be the relatively minor effects of background radiation (Whittingham *et al.*, 1977).

The meristematic regions of plants do not, of course, contain uniform populations of cells, and the extent of vacuolation can vary quite considerably, to the extent that it might be difficult to satisfy the requirements of every cell type. Further, in addition to the need to maintain the viability of the individual cells, there is also the need to maintain organisation within the overall structure. Nevertheless, meristems from a number of species have been successfully cryopreserved (see Kartha, 1985), and further investigations of this potentially important technique are certainly warranted.

At present, there seems to be some consensus that the most suitable cryopreservation procedures for suspension cultures involve some form of slow cooling which would permit at least a degree of "protective" dehydration of the cells. On the other hand, the initial successes with meristems were achieved by rapid cooling methods (carnation, Seibert, 1976; potato, Grout and Henshaw, 1978) but, more recently, two-stage methods, involving an initial slow cooling sequence, have been successfully employed (strawberry, Sakai *et al.*, 1978; Kartha *et al.*, 1980; pea, Kartha *et al.*, 1979; potato, Towill, 1981; Henshaw *et al.*, 1985a,b).

The rapid cooling method has the considerable advantage of extreme technical simplicity since the specimen is plunged directly into liquid nitrogen on, for example, the tip of a hypodermic needle (Grout and Henshaw, 1978), but there is little scope for fine adjustment, and difficulty has been encountered in applying the technique to different species or even to different genotypes of the same species. Slow cooling, on the other hand, requires more sophisticated apparatus but there is the possibility of rates being finely adjusted and, indeed, this might be necessary, as shown by Kartha *et al.* (1980) for strawberry meristems where the optimal procedure involved slow cooling at 0.84°C min$^{-1}$ to −40°C before transfer to liquid nitrogen; alternative rates of 0.56°C min$^{-1}$ and 0.95°C min$^{-1}$ were considerably less effective. It is quite clear from the limited successes so far achieved with meristems that cryopreservation procedures have not yet been developed sufficiently for their routine use in a large gene bank with responsibility for many accessions from a range of species. The feasibility of storing shoot meristems in liquid nitrogen has, however, been demonstrated and it remains to be seen whether techniques which are less sensitive to the genotype and the physiological state of the meristem can eventually be devised.

## 18.4 CONCLUSIONS

*In vitro* germplasm storage techniques based on the application of restricted-growth procedures to non-adventitiously multiplying shoot-tip cultures, or possibly callus cultures, could now be used with a wide range of vegetatively propagated species of economic importance, although suitable cultures may not yet be available, particularly from some of the tree species. There is also the prospect that cryopreservation techniques will eventually be a practical proposition for long-term storage of vegetative germplasm. However, cryopreservation techniques are unlikely to replace restricted-growth procedures since, although it should be possible, theoretically, to store material indefinitely at the temperature of liquid nitrogen with the minimum of attention, this is really an act of faith. As the viability of such stored material could not be checked visually, it would be necessary to arrange regular sampling for viability tests. Material removed from store would not be replenished by growth, so there would probably be a need for regular growth cycles to replenish stocks.

Caution must be expressed about the sampling procedures accompanying the use of *in vitro* germplasm storage techniques, since it would be possible, without proper precautions, to take explants which were not representative of the clonal material, either initially or during the routine maintenance of cultures. The precautions must also take into account the fact that roguing would not be possible with material actually in store, since the *in vitro* phenotype has little or no resemblance to that which is produced *in vivo*. It is now becoming evident that considerable importance should be attached to the development of routine methods for the monitoring of genetic stability in stored germplasm, so that the need for expensive regeneration and testing procedures might be reduced. Such methods based on the analysis of isozymes or DNA have been discussed recently by Simpson and Withers (1986) and Anon. (1986).

## REFERENCES

Acheampong, E. (1982). Multiplication and Conservation of Cocoyam (*Xanthosoma sagittifolium*) Germplasm by the Application of Tissue Culture Methods. Ph.D. thesis, University of Birmingham.
Alan, J. J. (1979). Tissue Culture Storage of Sweet Potato Germplasm. Ph.D. thesis, University of Birmingham.
Anon. (1986). "Design, Planning and Operation of *In Vitro* Genebanks". Advisory Committee on *In Vitro* Storage, International Board for Plant Genetic Resources, Rome, 17 pp.
D'Amato, F. (1978). Chromosome number variation in cultured cells and regenerated plants. In "Frontiers of Plant Tissue Culture, 1978" (T. A. Thorpe, ed.), pp. 287–296. International Association for Plant Tissue Culture, Calgary.
Davies, D. R., and Heslop, P. (1973). *In vitro* propagation of Freesia. *Rep. John Innes Inst. for 1972*, p. 64.
Grout, B. W. W., and Henshaw, G. G. (1978). Freeze-preservation of potato shoot-tip cultures. *Ann. Bot.* **42**, 1227–1229.
Henshaw, G. G. (1979). Plant tissue culture: its potential for dissemination of pathogen-free germplasm and multiplication of planting material. In "Plant Health" (D. L. Ebbels and J. E. King, eds), pp. 139–147. Blackwell Scientific, Oxford.
Henshaw, G. G., Keefe, P. D., and O'Hara, J. F. (1985a). Cryopreservation of potato meristems. In "*In Vitro* Techniques: Propagation and Long Term Storage" (A. Schäfer-Menuhr, ed.), pp. 155–160. Martinus Nijhoff/Dr W. Junk, Dordrecht (for the EEC).
Henshaw, G. G., O'Hara, J. F., and Stamp, J. A. (1985b). Cryopreservation of potato meristems. In "Cryopreservation of Plant Cells and Organs" (K. K. Kartha, ed.), pp. 159–170. CRC Press, Florida.
Jones, L. H. (1974). Long term survival of embryos of carrot (*Daucus carota* L.). *Pl. Sci. Lett.* **2**, 221–224.
Kartha, K. K., ed. (1985). "Cryopreservation of Plant Cells and Organs." CRC Press, Florida, 276 pp.
Kartha, K. K., Leung, N. L., and Gamborg, O. L. (1979). Freeze preservation of pea meristems in liquid nitrogen and subsequent plant regeneration. *Pl. Sci. Lett.* **15**, 7–15.
Kartha, K. K., Leung, N. L., and Pahl, K. (1980). Cryopreservation of strawberry meristems and mass propagation of plantlets. *J. Am. Soc. hort. Sci.* **105**, 481–484.
Langhans, R. W., Horst, R. K., and Earle, E. D. (1977). Disease-free plants via tissue culture propagation. *HortScience*, **12**, 25–26.
Larkin, P. J., and Scowcroft, W. R. (1981). Somaclonal variation—a novel source of variability from cell cultures for plant improvement. *Theoret. appl. Genet.* **60**, 197–214.
MacDougal, D. T. (1926). Growth and permeability of century-old cells. *Am. Nat.* **60**, 393–415.
Mathias, S. F. (1980). *Dioscorea rotundata* POIR, Storage and Propagation through Tissue Culture. M.Sc. thesis, University of Birmingham.
Mazur, P. (1969). Freezing injury in plants. *Ann. Rev. Pl. Physiol.* **20**, 419–448.
Meryman, H. T., and Williams, R. J. (1980). Mechanisms of freezing injury and natural tolerance and the principles of artificial cryoprotection. In "Crop

Genetic Resources: the Conservation of Difficult Material" (L. A. Withers and J. T. Williams, eds), pp. 5–37. Série B42, Paris.

Mullin, R. H., and Schlegel, D. E. (1976). Cold storage maintenance of strawberry meristem plantlets. *HortScience*, **11**, 100–101.

Sakai, A., Yamakawa, M., Sakato, D., Harada, T., and Yakuwa, T. (1978). Development of a whole plant from an excised strawberry runner apex frozen to −196°C. *Low Temp. Sci.* Ser. B **36**, 31–38.

Seibert, M. (1976). Shoot initiation from carnation shoot apices frozen to −196°C. *Science* **191**, 1178–1179.

Sheridan, W. F. (1968). Tissue culture of the monocot *Lilium*. *Planta* **82**, 189–192.

Simpson, M. J. A., and Withers, L. A. (1986). "Characterization of Plant Genetic Resources using Isozyme Electrophoresis: a Guide to the Literature". IBPGR, Rome, 102 pp.

Towill, L. E. (1981). *Solanum tuberosum*: a model for studying the cryobiology of shoot-tips in the tuber-bearing *Solanum* species. *Pl. Sci. Lett.* **20**, 315–324.

Walkey, D. G. A. (1978). *In vitro* methods for virus elimination. *In* "Frontiers of Plant Tissue Culture, 1978" (T. A. Thorpe, ed.), pp. 245–254. International Association for Plant Tissue Culture, Calgary.

Walkey, D. G. A. (1985). "Applied Plant Virology". 329 pp. London, Heinemann.

Westcott, R. J. (1981a). Tissue culture storage of potato germplasm: I. Minimal growth storage. *Potato Res.* **24**, 331–342.

Westcott, R. J. (1981b). Tissue culture storage of potato germplasm: II. Use of growth retardants. *Potato Res.* **24**, 343–352.

Whittingham, D. E., Lyon, M. E., and Glenister, P. H. (1977). Long term storage of mouse embryos at −196°C: the effect of background radiation. *Genet. Res.* **29**, 171–181.

Withers, L. A. (1980a). Low temperature storage of plant tissue cultures. *In* "Advances in Biochemical Engineering", vol. 18 (A. Fiechter, ed.), pp. 102–150. Springer-Verlag, Berlin.

Withers, L. A. (1980b). Preservation of germplasm. *In* "Perspectives in Plant Cell and Tissue Culture". *Int. Rev. Cytol.*, Suppl. IIB (I. K. Vasil, D. E. Murphy, eds), pp. 101–136. Academic Press, New York.

Withers, L. A. (1980c). Tissue Culture Storage for Genetic Conservation. Tech. Rep., IBPGR, Rome, 91 pp.

# Part V Mutation Breeding

# 19 The Genetic Basis of Variation

W. GOTTSCHALK

*Institute of Genetics, University of Bonn, FRG*

Many papers have been published dealing with the possibilities of increasing genetic variability by means of mutation, selection and recombination. Despite this work, some important problems of experimental mutation research have not yet been solved, and in other fields very little reliable information is available. This is true for both basic research and mutation breeding. The main purpose of this chapter is to discuss those problems of sexually propagated crops, in which agronomically useful mutant genes are positively or negatively influenced by other genes or by environmental factors. Studies on the combined action of mutant genes, displaying specific interactions, have been made using our comprehensive collection of *Pisum* mutants and the reactions of useful mutants to different climates and to controlled phytotron conditions are discussed. A particularly important problem of applied mutagenesis is pleiotropic gene action, because this is the main reason why many mutants with characters of agronomic interest are not utilised in plant breeding.

## 19.1 THE SELECTION VALUE OF MUTANT GENES AND ITS IMPROVEMENT THROUGH RECOMBINATION

The selection value of a mutant gene is composed of many individual characters, most of which have not yet been analysed in detail but which directly or indirectly influence the competitiveness of the mutant. This is also true for certain physiological traits such as the duration of seed germeability, speed of germination and growth, time of flowering, duration of the life cycle, among others. Also, the reaction of the mutant gene in different genotypic backgrounds or changed environmental conditions has to be considered. The most important selection criterion, however, is the seed fertility of the mutant compared with its mother variety and/or of other mutants. Our work during the past 25 years on induced *Pisum* mutants is relevant here.

After X-ray and neutron irradiation of pea seeds, more than 800 mutants have been selected. About one-third of them are fertile enough to be propagated directly by seed. The other mutants show lethal changes or are sterile or so weakly fertile, that they can be propagated only by segregations from plants heterozygous for the respective genes. The seed production of 253 *Pisum* mutants is shown in Fig. 19.1. Most of these genotypes have been tested over several generations, many with several replications. The mean number of seeds per plant is related to the value for the mother variety grown in the same year at the same location.

The relative seed production of most mutants differs considerably from year to year in relation to the controls. This may be because many mutants differ from their mother variety in their adaptability to different environmental conditions. The seed production of the high-yielding fasciated mutants, for instance, is influenced negatively by rainfall during flowering and early ripening but the mother variety is not. An early flowering mutant of our collection shows a certain degree of heat tolerance which is not important in mid-Europe but it is a valuable trait in India. The yield of waxless mutants or of desynaptic genotypes depends likewise on climatic conditions. Our findings demonstrate that reliable conclusions on the productivity of mutants can be made only with data from many generations.

Only about 15% of the 253 fertile *Pisum* mutants show seed production similar to or better than that of the mother variety (cf. Fig. 19.1): this is 4.5% of the total number of mutants selected. Not all of the mutants of this small group are of interest for pea breeding. Most of them are too tall or mature too late; or the favourable yielding properties are negatively influenced by pleiotropic action or reduced penetrance of the genes involved. Only about 1% of our mutant *Pisum* material may be of agronomic interest. A similiar proportion of barley mutants was obtained by Gustafsson (1954). The true proportion may be somewhat higher if the micromutants are included, but evidence for them is difficult to obtain. Data is available for barley (Gaul *et al.*, 1969). Thus, mutation breeding in sexually propagated crops can only be done in a prospective way if a large number of different mutants is available, because only a very small proportion of them can be actually utilised for breeding purposes. The situation is more favourable in many vegetatively propagated ornamentals.

Until 1985, a total of 688 commercial varieties of many crops, developed by means of induced mutations, had been recorded at the International Atomic Energy Agency in Vienna (Micke *et al.*, 1985); 39% of them were cereals. The following examples of the numbers of mutant cultivars may demonstrate the effectiveness of the methods used:

| | |
|---|---|
| *Oryza sativa* | 103 |
| *Hordeum vulgare* | 70 |
| *Triticum aestivum* | 56 |
| *Glycine max* | 17 |
| *Chrysanthemum* sp. | 99 |
| *Dahlia* sp. | 34 |
| *Begonia* sp. | 21 |
| *Streptocarpus* sp. | 18 |

The constantly increasing number of mutant varieties during the past decades becomes discernible in Fig. 19.2. The mutagens used are known for 544 mutant varieties; 91.4% have been produced by means of irradiation, predominantly by X- and γ-rays, whereas only 7.5% arose after treatment with mutagenic chemicals (for details, see Table 19.1).

One of the next steps in mutation breeding will be to combine useful genes derived from different mutants, to produce recombinants of specific genotypic constitution. Many findings show that the accumulation of mutant genes in the same plant often leads to further reduction of the seed production. This is not, however, generally valid. On the contrary. many of our *Pisum* recombinants, homozygous for two to more than ten mutant genes, have an unexpectedly high seed fertility.

We have crossed a large number of mutants and now have more than 700 recombinants of different genotypic constitution. Seed production of 649 of them is compared with the 253 *Pisum* mutants (Fig. 19.3). Most of the mutants tested produce between 5 and 80% as much seed as the mother variety, and only a small proportion produce more than 80% (Fig. 19.3a). However, the recombinants show higher seed productivity, more than 60% of these genotypes producing seed in quantities ranging between 65 and about 140% of the control values (Fig. 19.3b). These findings demonstrate that the selection value of many mutant genes can be considerably improved by combining them with other specific mutant genes.

## 19.2 GENOTYPES HOMOZYGOUS FOR MANY MUTANT GENES

Quite frequently, more than one gene mutates during the mutagenic treatment in the initial cells of the embryonic growing points, giving rise to the $M_1$ plants. In this case, genotypes homozygous for several mutant

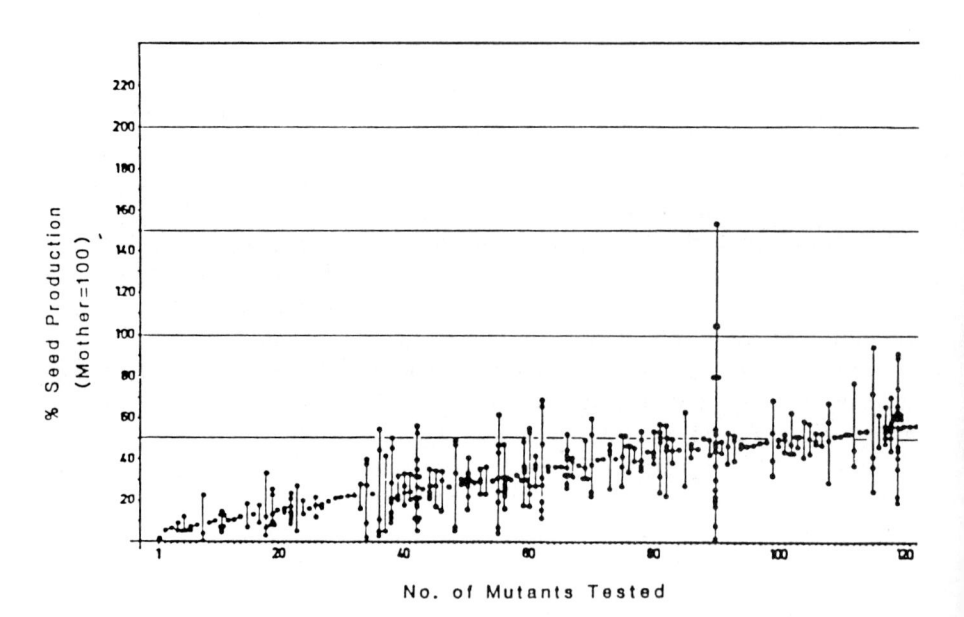

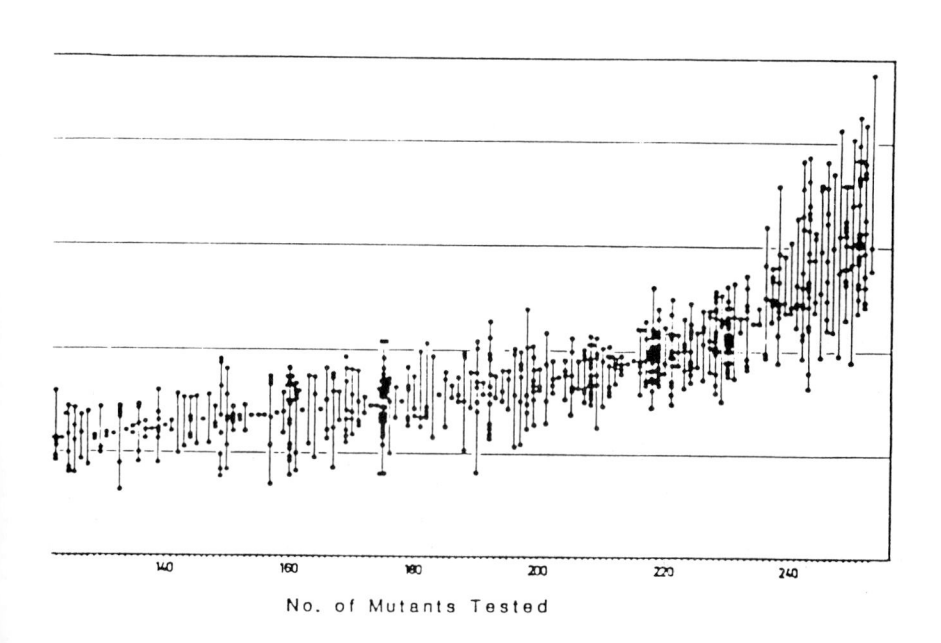

**Figure 19.1.** Seed production of 253 radiation-induced *Pisum* mutants. Each point gives the mean value for the trait "number of seeds per plant" for one generation as a percentage of the mother variety (= 100%). Means connected by vertical lines belong to the same mutant.

**Table 19.1.** Mutagens used for obtaining 544 mutant varieties (evaluation of a list given by Micke et al. 1985)

| Mutagen | Number of varieties | Percentage of total |
|---|---|---|
| γ-rays | 277 | 50.9 |
| X-rays | 158 | 29.0 |
| Neutrons | 35 | 6.4 |
| Rays (no details given) | 11 | 2.0 |
| β-rays | 2 | |
| $^{32}$P | 2 | |
| Combination of different rays | 7 | |
| Combination of γ-rays + UV | 1 | |
| Combination of γ-rays + microwaves | 3 | |
| Combination of γ-rays + laser | 1 | |
| Rays in total | 497 | 91.4 |
| Mutagenic chemicals | 37 | |
| Colchicine | 4 | |
| Chemicals in total | 41 | 7.5 |
| Combination of rays + chemicals | 3 | |
| Combination of rays + colchicine | 3 | |
| Combined trials | 6 | 1.1 |
| Grand total | 544 | 100.0 |

genes are formed. If they are crossed with other mutants of this category, highly heterozygous $F_1$ plants arise giving complicated segregations in their progenies. In this way, collections of recombinants with a high degree of genetic variability can be produced quickly.

A clear example is provided by a group of six X-ray- and neutron-induced fasciated mutants in our *Pisum* collection. They are homozygous for more than 20 genes mutated during irradiation and most of them are identified in the six genotypes mentioned (Gottschalk, 1977, 1981c). This is also true for a spontaneous fasciated mutant from which Scheibe (1965) developed the fasciated fodder pea cultivar, Ornamenta. The simultaneous mutation of such a large group of identical genes in seven different embryos cannot have arisen randomly. We must assume that a single basic change (equal in the six embryos involved) was responsible for the

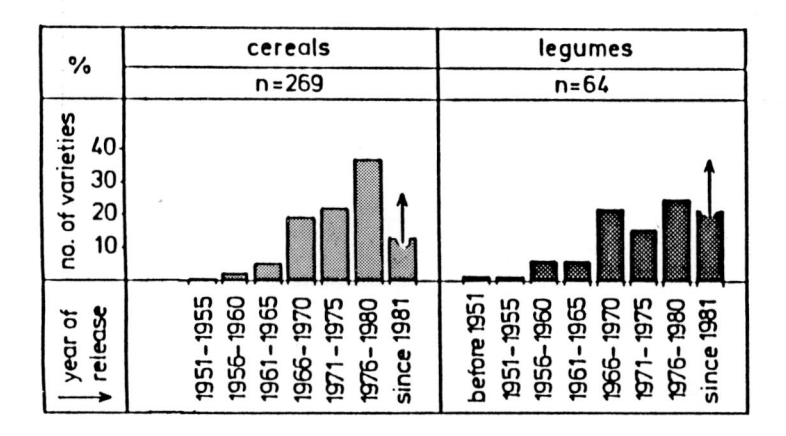

**Figure 19.2.** The number of released mutant varieties of different crops and ornamentals in relation to the year of release. So far, a total of 688 cultivars has been recorded at the International Atomic Energy Agency (Micke et al., 1985) but in seven of them the year of release is not mentioned.

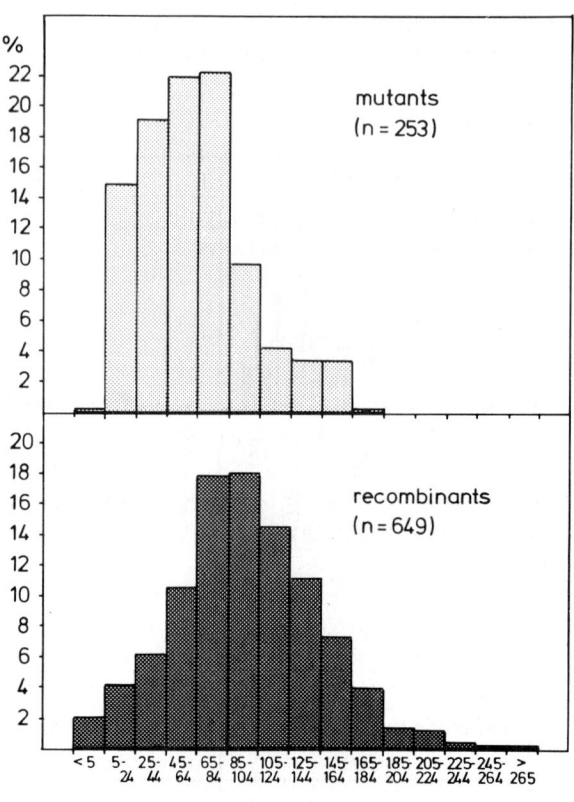

**Figure 19.3.** Distribution of the mean values for the number of seeds per plant of: (a) 253 *Pisum* mutants and (b) 649 recombinants. Most of the mutants are homozygous for only one mutant gene whereas the recombinants contain 2–10 mutant genes.

mutations in the same 20–25 loci. This phenomenon has not yet been clarified but it may be that the primary event was mutation in a kind of a "mutator gene" which caused secondary mutations in a large group of specific genes more-or-less randomly distributed over the seven chromosomes of the genome.

These fasciated mutants are of practical value as most of their mutant genes control characters of agronomic interest, such as internode length, flowering and ripening time, stem bifurcation and different kinds of stem fasciation, resulting in high seed production. Moreover, they are homozygous for gene *fis* which controls the photoperiodic behaviour; these six mutants need long-day conditions for flowering whereas the mother variety is day-neutral. After crossing the fasciated mutants with non-fasciated genotypes, more than 600 recombinants of different genotypic constitution were selected and developed into pure lines. This very broad genetic variability has been obtained within a few years. In these recombinants, the useful traits of the parental mutants are combined to produce a number of different kinds of improvement of the parental genotypes. This material is useful for studying the co-operation of mutant genes and positive or negative interactions that exist between them.

### 19.3 THE PROBLEM OF DESIRED MUTATIONS

Many mutation experiments are not performed to increase the genetic variability of a species in general but to meet a specific breeding aim. In these cases, the selection of great numbers of different mutants, which might be of interest for basic research, is not the object of the programme. The aim is to select a very limited number of distinct mutants of direct economic value. What are the chances for inducing a specific mutation?

It is not yet possible to induce specific "desired" mutations in higher organisms; on the contrary, the mutational event is still a matter of chance. But, as many genes of a crop are involved in the control of shoot structure, it is not difficult to alter the shoot structure by means of mutant genes. It is, for instance, certain that mutants with reduced plant height or altered branching will result. Most of them will have no agronomic interest because of their poor yielding properties, but useful genotypes do appear in low frequencies. This is also true for early flowering and ripening genotypes derived from late maturing mother varieties.

The problem is much more difficult when seeking a specific mutation, e.g. pest resistance. Some examples follow:

(a) More than 2.5 million $M_2$ barley plants were tested in the German Democratic Republic for the selection of 95 mildew-resistant mutants (Hentrich, 1977).
(b) Seven mildew-resistant barley mutants were selected from 1.2 million $M_2$ plants in Japan (Yamaguchi and Yamashita, 1979).
(c) Screening 951 000 $M_2$ barley plants in Denmark resulted in selecting five $M_2$ plants which gave rise to powdery mildew-resistant $M_2$ seedlings (Jörgensen, 1975).
(d) Three blast-resistant rice mutants were isolated in Japan from an $M_2$ generation of 51 530 plants (Yamasaki and Kawai, 1968).
(e) From more than 6 million $M_2$ plants of *Mentha piperita*, seven wilt-resistant strains were developed in the USA (Murray, 1969, 1971).

(f) A single mutant with improved winter-hardiness was derived in the USA from the irradiation of 500 000 stems with dormant buds of Coastcross-1 Bermuda grass hybrids (Burton et al., 1980).

According to Yonezawa (1975), at least 200 000–250 000 $M_2$ plants are necessary for the reliable selection of a mutation for a distinct quantitative character. Model experiments in the USA, using sodium azide to produce mutations in barley, gave frequencies for specific locus mutations of 2.7 mutations per 10 000 $M_2$ seedlings for "waxy endosperm" and 1.0 mutations for 10 000 $M_2$ seedlings for "vine (gigas)" (Kleinhofs et al., 1978).

Similar low frequencies have been obtained with cereals in efforts to alter the seed protein composition. Twenty barley mutants with increased lysine content were selected in Denmark from 14 776 $M_2$ plants (Doll et al., 1974); some bread-wheat mutants with improved protein quality were selected from 25 000 $M_2$ plants by Parodi and Nebreda (1979), whereas no such mutants were found in 15 000 samples studied by Johnson et al. (1973).

These results are not discouraging; on the contrary, they demonstrate that it is possible to get a desired mutation, but a considerable expenditure in time, space and money is necessary.

### 19.4 PLEIOTROPIC GENE ACTION

One of the most important disadvantages of using mutants in plant breeding is the pleiotropic action of mutant genes. In higher plants, it is practically impossible to find a gene having no pleiotropic effect. The useful character of a mutant is usually combined with a negative trait, both being part of the pleiotropic pattern of the respective gene. This is the main reason why many mutants with prospective characters, which have existed for a long time in our collection, cannot be used agronomically.

It has been shown in barley that the relative strength of the different components can be influenced by transferring the gene into different genotypic backgrounds (Gaul et al., 1968; Gaul and Grunewaldt, 1971; Grunewaldt, 1974; Gaul and Lind, 1976; Lind and Gaul, 1976). However, this procedure cannot be specifically directed, i.e. it is not known in which background a given gene displays that part of its pleiotropic spectrum which is of agronomic interest.

In exceptional cases, climatic factors can influence a pleiotropic pattern positively or negatively. The effect of gene *ion* of the *Pisum* genome is to increase the number of seeds per pod but to decrease the number of pods per plant. This response appears regularly in each generation when the mutant is grown in Germany. In northern India (Haryana), however, the positive part of the pattern is realised but not the negative one, and the plants are richly branched producing a high number of pods per plant. This effect is not observed in the mother variety; it is a specific reaction of the mutant to the climate in northern India (Gottschalk and Kaul, 1975). In the hot and dry parts of western India (Rajasthan) and in Ghana, not only the number of pods per plant but also the number of seeds per pod is reduced; thus, both traits of the pleiotropic pattern are negative. The same situation is found if the gene *ion* is combined with other specific mutant genes. The mutant, homozygous for *ion*, is of little interest in Germany, Ghana and Rajasthan, but it is of great interest in Haryana. The total reaction of the gene to environment and genotypic background is presented schematically in Fig. 19.4.

The main problem, however, is whether a so-called pleiotropic pattern results really from the action of a single gene. Theoretically, it could be due to the action of several genes which are very closely linked and which mutate more or less simultaneously during mutagenic treatment. In an organism heterozygous for the respective gene pairs, the mutant genes cannot be separated from each other because the crossing-over point will not lie between them because of their close linkage. Consequently, the whole group of mutant genes will remain together, causing a block of deviating characters if this group of genes is homozygous. The whole block behaves as a single trait. In this way, the pleiotropic action of a single gene is simulated, but in reality, a whole group of mutant genes is responsible for this kind of "pleiotropism".

For methodological reasons, it is not possible to test pleiotropic genes systematically in order to clarify their genetic situation in the mutant. But it was possible to confirm this hypothesis by means of time-consuming crossing experiments between similar mutants of our *Pisum* collection. We have been able to split up "pleiotropic patterns" of this kind and identify specific traits with independently acting but very closely linked genes (Gottschalk, 1965, 1967, 1976).

As an example, mutations in the so-called *dim*-segment of the *Pisum* genome are presented in Fig. 19.5. Altogether, 12 mutants are available in our collection which are homozygous for single genes or for gene-groups of this segment, some of them being identical. Each of them shows monohybrid segregation for the complex of abnormal characters, as expected for pleiotropic gene action. If we cross the different mutants of the group with each other, it becomes obvious that these patterns are due to the action of single, non-pleiotropic genes. The number of these cases is increasing, not only in pea but also in barley. Thus, this situation,

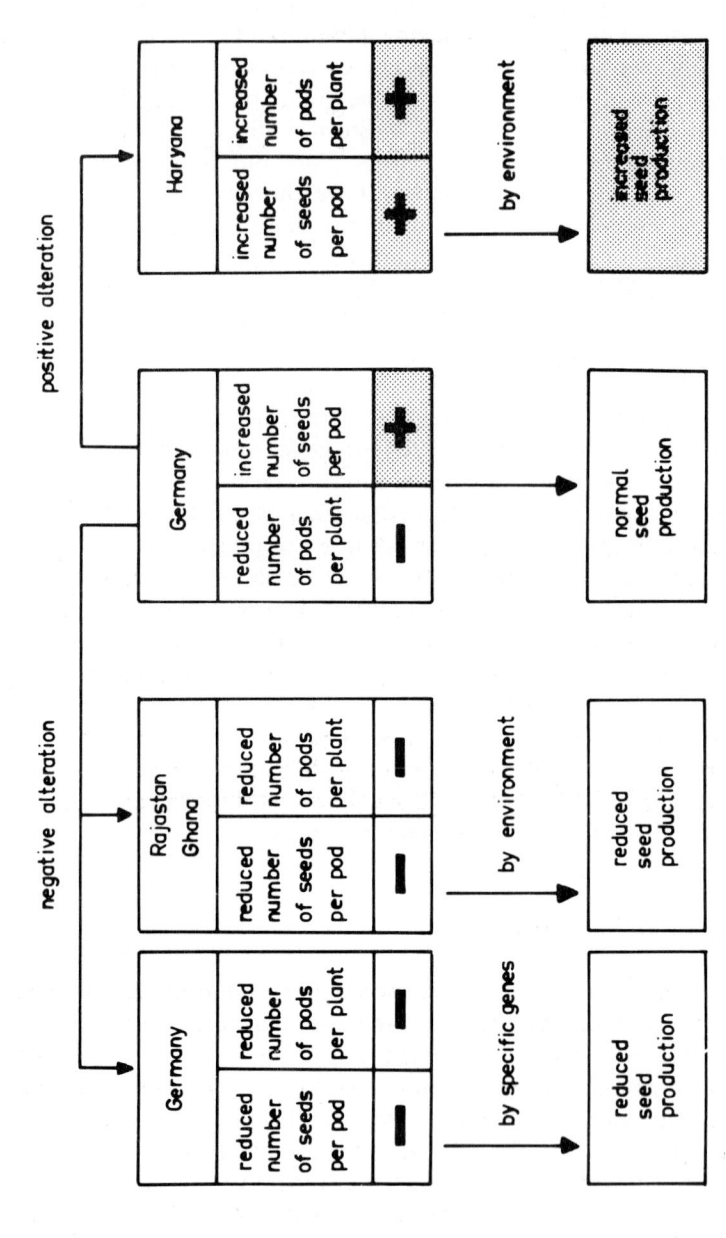

**Figure 19.4.** The reaction of gene *ion* of the *Pisum* genome to environment and genotypic background.

# The Genetic Basis of Variation

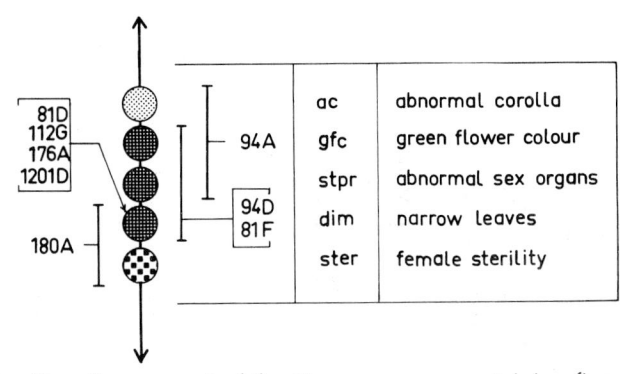

**Figure 19.5.** The *dim*-segment of the *Pisum* genome containing five very closely linked genes. Simultaneous mutation of several genes of the segment leads to different but partially agreeing mutants showing monohybrid segregations for the whole complex of diverging characters.


observed several times in *Pisum*, is not a rare phenomenon and it must be considered with the problems of pleiotropic gene action.

The situation discussed above has serious consequences for the use of mutants in plant breeding (Fig. 19.6). If an agronomically useful trait is part of a pleiotropic pattern, it cannot be used for breeding purposes in those cases in which a negative trait belongs to the same pattern (Fig. 19.6a). If, however, two traits are controlled by two different but closely linked genes, it is possible, in principle, to separate them from each other and use the positive one (Fig. 19.6b and c). In this situation, it is necessary to produce a large $F_2$ generation because of the small chance of obtaining a crossing over between the gene pairs involved.

## 19.5 PROBLEMS OF "GENECOLOGY"

"Genecology" is a new interest in experimental mutation research and has relevance to mutation breeding. An agronomically useful gene can be utilised if it is expressed fully. This is frequently not the case because many mutants differ considerably from their mother varieties with regard to their expression under different climatic conditions. This fact has been known for a long time, but experimental data are limited. The subject is best studied by growing mutants under a range of specific controlled conditions in a phytotron, as has been done successfully by Gustafsson's group with barley mutants (Dormling *et al.*, 1966; Dormling and Gustafs-

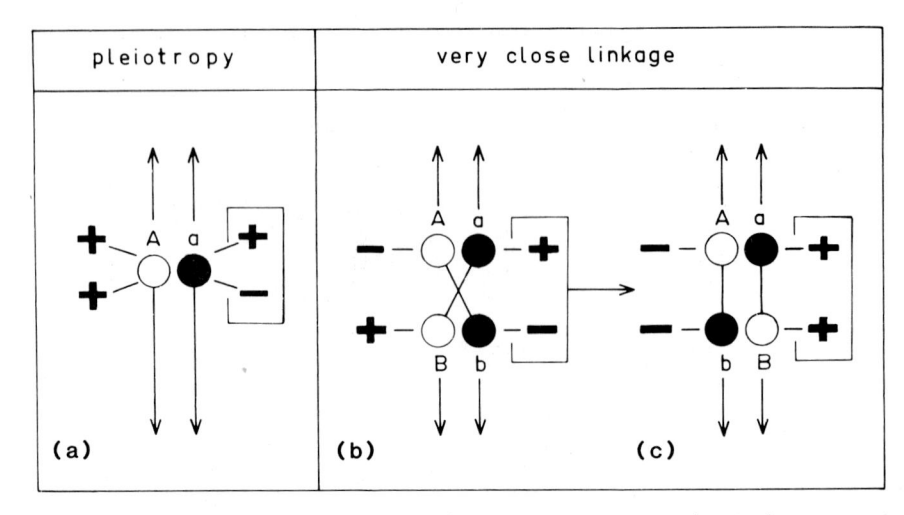

**Figure 19.6.** Differences between true pleiotropic gene action of a single gene and simultaneous mutation of two neighbouring genes with regard to the utilisation of a useful trait.

son, 1969; Gustafsson, 1969; Gustafsson and Dormling, 1972; Gustafsson *et al.*, 1973a,b,c, 1974, 1975; Gustafsson and Lundqvist, 1976).

We are undertaking similar experiments with *Pisum* mutants and recombinants in our phytotron. As an example, I will describe the reaction of a useful mutant gene to the environment and genotypic background using the gene for earliness *efr*. This gene is homozygously present in recombinant R 46C and in more than 100 different recombinant types selected after crossing R 46C with other mutants. The flowering behaviour of these genotypes was examined in short-day conditions (12 h dark, 12 h light). The reaction of gene *efr* in R 46C and in 13 different recombinant types homozygous for *efr* and for other mutant genes is shown in Fig. 19.7 and compared with the mother variety (Dippes Gelbe Viktoria).

Under long-day field conditions in West Germany, recombinant R 46C begins flowering 10–14 days earlier than the mother variety, due to the formation of the first flowers at very low nodes. Under short-day phytotron conditions, about the same differences between the two genotypes is observed. Most of the other early-flowering recombinants tested show similar flowering behaviour to R 46C, demonstrating that the additional mutant genes present in their genomes do not influence gene *efr*. They are not shown in Fig. 19.7. However, some recombinants are even earlier than R 46C, due to a positive effect of one of the other mutant genes on gene *efr*. In a different group of recombinants, the opposite effect is

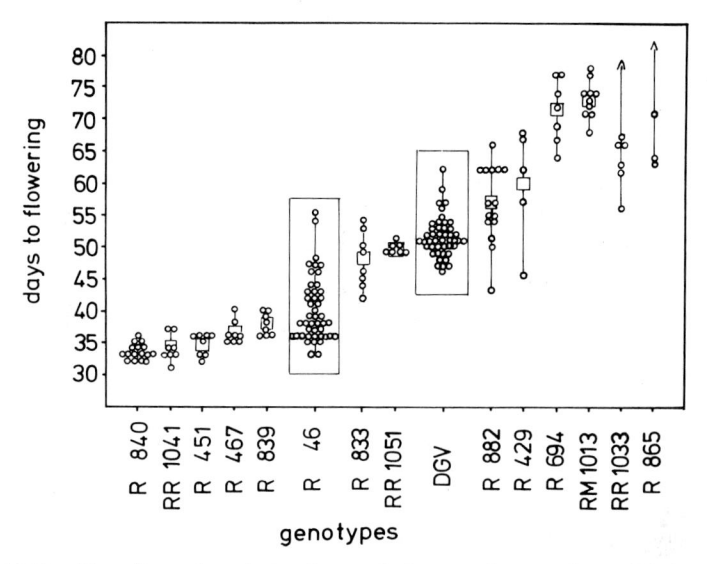

**Figure 19.7.** The flowering behaviour of the mother variety, Dippes Gelbe Viktoria (DGV) of *Pisum sativum*, recombinant R 46C—the donor of gene *efr* for earliness—and of 13 different recombinants homozygous for *efr* and for other mutant genes under short-day phytotron conditions. Each point gives the value for one plant.

observed; they are not only considerably later than R 46C but also later than the mother variety not having gene *efr* for earliness. In these cases, a negative effect of other mutant genes on *efr* is apparent. The respective genotypes are genetically early because of the presence of *efr*, but this gene is not expressed at the normal stage of ontogenetic development because its action is curtailed by the influence of other specific genes.

Of particular interest is a group of 16 different recombinants, homozygous for *efr* and other genes, which are not able to produce functional flowers under short-day conditions. Transition from the vegetative to the reproductive stage occurs at the normal stage of development and minute flower buds are formed. However, those buds are not able to develop further. None of the plants of these 16 recombinant types produce any fully developed flowers in short-days, but when conditions are changed to long-day, normally developed flowers are formed within a few days. These recombinants have two genes for stem fasciation and stem bifurcation, deriving from a group of genetically similar fasciated mutants of complicated genotypic constitution. One of these two genes is obviously responsible for the suppression of the full action of gene *efr*.

This is an interesting example of interactions between different genes as well as between genes and environmental factors. A specific mutant gene suppresses the action of another gene, but only under short-day conditions. In long days, the suppressor is not effective.

A completely different situation is found in a group of fasciated mutants which do not flower in short days. Flower initiation does not occur in these genotypes; they remain vegetative. Under long-day conditions, they show normal flowering behaviour. The specific gene *fis* is responsible for controlling the photoperiodic behaviour of the plants. Plants homozygous for this gene require long days for flowering. These mutants would be of agronomic interest for developing countries because of the high seed yield as a consequence of stem fasciation but they cannot be used because of their short-day reaction. After having crossed them with other genotypes in our collection, high-yielding fasciated recombinants without gene *fis* were selected. They flower under short-day phytotron conditions and are just being tested in India in natural short days.

Further details on the reaction of our *Pisum* mutants and recombinants to controlled long- and short-day conditions have been published elsewhere (Gottschalk, 1981a, 1982, 1985). Finally, it should be mentioned, that some genes of our collection are not able to manifest their action under certain climatic conditions. We have two polymeric *bif* genes causing dichotomous stem bifurcation and, in turn, improved seed production. In India, Brazil, Egypt and in some other countries, the action of these genes is not discernible, for reasons which are not yet known (Gottschalk, 1981b).

## REFERENCES

Burton, G. W., Constantin, M. J., Dobson, J. W., Hanna, W. W., and Powell, J. B. (1980). An induced mutant of Coastcross 1 Bermudagrass with improved winter hardiness. *Environ. Exptl. Bot.* **20**, 115–117.

Doll, H., Koie, B., and Eggum, B. O. (1974). Induced high lysine mutants in barley. *Radiat. Bot.* **14**, 73–80.

Dormling, I., and Gustafsson, Å. (1969). Phytotron cultivation of early barley mutants. *Theoret. appl. Genet.* **39**, 51–61.

Dormling, I., Gustafsson, Å., Jung, H. R., and von Wettstein, D. (1966). Phytotron cultivation of Svalöf's Bonus barley and its mutant Svalöf's Mari. *Hereditas* **56**, 221–237.

Gaul, H., and Grunewaldt, J. (1971). Independent variation of culm length and spike-internode length of a barley *erectoides* mutant in a changed genetic background. *Barley Genet.* **II**, 106–118.

Gaul, H., and Lind, V. (1976). Variation of the pleiotropy effect in a changed genetic background demonstrated with barley mutants. In "Induced Mutations in Cross-Breeding", pp. 55–69. IAEA, Vienna.

Gaul, H., Grunewaldt, J., and Hesemann, C. U. (1968). Variation of character expression of barley mutants in a changed genetic background. *In* "Mutations in Plant Breeding II", pp. 77–95, IAEA, Vienna.

Gaul, H., Ulonska, E., Zum Winkel, C., and Braker, G. (1969). Micromutations influencing yield in barley: studies over nine generations. *In* "Induced Mutations in Plants", pp. 375–398, IAEA, Vienna.

Gottschalk, W. (1965). A chromosome region in *Pisum* with an exceptionally high susceptibility to X-rays. *Radiat. Bot. Suppl.* **5**, 385–391.

Gottschalk, W. (1967). Neue Aspekte zum Problem der pleiotropen Genwirkung. *Ber. Dtsch. Bot. Ges.* **80**, 545–553.

Gottschalk, W. (1976). Genetically conditioned male sterility. *In* "Induced Mutations in Cross-Breeding", pp. 71–78. IAEA, Vienna.

Gottschalk, W. (1977). Fasciated peas—unusual mutants for breeding and research. *J. Nucl. Agric. Biol.* **6**, 27–33.

Gottschalk, W. (1981a). Induced mutations in gene-ecological studies. *In* "Induced Mutations—a Tool in Plant Research, pp. 411–436, IAEA, Vienna.

Gottschalk, W. (1981b). The suppression of gene actions through environmental factors. *Egypt. J. Genet. Cytol.* **10**, 159–174.

Gottschalk, W. (1981c). Genetic constitution of seven fasciated pea mutants. A mutator gene in *Pisum? Pulse Crops Newslett.* **1** (1), 54–55.

Gottschalk, W. (1982). The flowering behaviour of *Pisum* genotypes under phytotron and field conditions. *Biol. Zbl.* **101**, 249–260.

Gottschalk, W. (1985). Phytotron experiments in *Pisum*: 1. Influence of temperature on the flowering behaviour of different genotypes. *Theoret. appl. Genet.* **70**, 207–212.

Gottschalk, W., and Kaul, M. L. H. (1975). Gene-ecological investigations in *Pisum* mutants. I. The influence of climatic factors upon quantitative and qualitative characters. *Z. Pflanzenzücht.* **75**, 182–191.

Grunewaldt, J. (1974). Untersuchungen an dem *erectoides*-Komplex von Gerstenmutanten, I, II. *Z. Pflanzenzücht.* **71**, 193–207, 330–340.

Gustafsson, Å. (1954). Mutations, viability, and population structure. *Acta agric. scand.* **4**, 601–632.

Gustafsson, Å. (1969). A study of induced mutations in plants. *In* "Induced Mutations in Plants", pp. 9–31. IAEA, Vienna.

Gustafsson, Å., and Dormling, I. (1972). Dominance and overdominance in phytotron analysis on monohybrid barley. *Hereditas* **70**, 185–216.

Gustafsson, Å., and Lundqvist, U. (1976). Controlled environment and short-day tolerance in barley mutants. *In* "Induced Mutations in Cross-Breeding", pp. 45–53. IAEA, Vienna.

Gustafsson, Å., Dormling, I., and Ekman, G. (1973a). Phytotron ecology of mutant genes. I. Heterosis in mutant crossings of barley. *Hereditas* **74**, 119–126.

Gustafsson, Å., Dormling, I., and Ekman, G. (1973b). Phytotron ecology of mutant genes. II. Dynamics of heterosis in an intralocus mutant hybrid of barley. *Hereditas* **74**, 247–258.

Gustafsson, Å., Dormling, I., and Ekman, G. (1973c). Phytotron ecology of mutant genes. III. Growth reactions of two quantitative traits in barley. *Hereditas* **75**, 75–82.

Gustafsson, Å., Dormling, I., and Ekman, G. (1974). Phytotron ecology of mutant genes. V. Intra- and inter-locus overdominance involving early mutants of Bonus barley. *Hereditas* **77**, 237–254.

Gustafsson, Å., Dormling, I., and Ekman, G. (1975). Phytotron ecology of mutant genes. VI. Clima reactions of the *eceriferum* mutations *cer-i*$^{16}$ and *cer-c*$^{36}$. *Hereditas* **80**, 279–290.

Hentrich, W. (1977). Tests for the selection of mildew-resistant mutants in spring barley. *In* "Induced Mutations against Plant Diseases", pp. 333–341. IAEA, Vienna.

Johnson, V. A., Mattern, P. J., Schmidt, J. W., and Stroike, J. E. (1973). Genetic advances in wheat protein quantity and composition. *Proc. 4th Int. Wheat Genet. Symp., Univ. Columbia, Miss.*, p. 547.

Jörgensen, J. H. (1975). Identification of powdery mildew resistant barley mutants and their allelic relationship. *Barley Genet.* **III**, 446–455.

Kleinhofs, A., Warner, R. L., Muehlbauer, F. J., and Nilan, R. A. (1978). Induction and selection of specific gene mutations in *Hordeum* and *Pisum*. *Mutat. Res.* **51**, 29–35.

Lind, V., and Gaul, H. (1976). Studies of pleiotropic genes and their character complexes in *erectoides* mutants. *Barley Genet.* **III**, 171–180.

Micke, A., Maluszynski, M., and Donini, B. (1985). Plant cultivars derived from mutation induction or the use of induced mutants in cross breeding. *Mutat. Breed. Rev.* **3**, 1–92.

Murray, M. J. (1969). Successful use of irradiation breeding to obtain *Verticillium*-resistant strains of peppermint, *Mentha piperita* L. *In* "Induced Mutations in Plants", pp. 345–371. IAEA, Vienna.

Murray, M. J. (1971). Additional observations on mutation breeding to obtain *Verticillium*-resistant strains of peppermint. *In* "Mutation Breeding for Disease Resistance", pp. 171–195. IAEA, Vienna.

Parodi, P. C., and Nebreda, I. M. (1979). Protein and yield response of six wheat (*Triticum* ssp.) genotypes to gamma radiation. *In* "Seed Protein Improvement in Cereals and Grain Legumes *II*", pp. 201–209. IAEA, Vienna.

Scheibe, A. (1965). Die neue Mähdrusch-Futtererbse "Ornamenta". *Saatgutwirtschaft* **17**, 116–117.

Yamaguchi, I., and Yamashita, A. (1979). Resistant mutants of two-rowed barley to powdery mildew (*Erysiphe graminis* f. sp. *hordei*). *Techn. News, Inst. Radiat. Breed., Ohmiya, Jpn.* **20**.

Yamasaki, Y., and Kawai, T. (1968). Artificial induction of blast-resistant mutations in rice. *In* "Rice Breeding with Induced Mutations", pp. 65–73. IAEA, Vienna.

Yonezawa, K. (1975). Method and efficiency of mutation breeding for quantitative characters. *Gamma Field Symp.* **14**, 39–58.

# 20 Application of Mutation Breeding Methods

C. BROERTJES[1] and A. M. VAN HARTEN[2]

[1]*Research Institute ITAL, Wageningen, The Netherlands*
[2]*Department of Plant Breeding (IvP), Agricultural University, Wageningen, The Netherlands*

The induction of mutations is one of the means the plant breeder has for creating genetic variability in order to obtain new and better genotypes by subsequent selection procedures. This method has proved to be of considerable success in the production of an increasing number of commercial mutants of vegetatively propagated crops, and in particular of ornamentals (Broertjes and Van Harten, 1978; Van Harten, 1982).

The main advantage of mutation breeding in vegetatively propagated plants is the ability to improve one or a few characteristics of an otherwise outstanding genotype, theoretically without altering the remaining part of it. Leading cultivars can thus be further perfected relatively quickly and at low cost.

An important limitation of mutation breeding is that most mutations are of a recessive nature. This "one-way-situation" must be taken into account when trying to decide whether mutation breeding or another breeding method should be applied. The main practical bottleneck, chimera formation after mutagenic treatment of the multicellular apex, is a consequence of the fact that the occurrence of a specific mutation is a one-cell event. The complications of chimera formation can be avoided by the use of different methods, such as adventitious bud techniques, either *in vivo* or *in vitro*. Both (periclinal) chimeras and solid, non-chimeric mutants have their advantages and disadvantages.

The present importance of mutation breeding in vegetatively propagated crops is illustrated in Table 20.1, which supports the view that mutation breeding has been particularly successful in ornamental plants. This is

[1] Present address: Eykmanstraat 11, 6706 JT Wageningen, The Netherlands.

**Table 20.1** Commercial mutants of various vegetatively propagated crops.

| Plant group | Number of commercial mutants ||||||Total number of mutants | Crops of which induced mutants have been commercialised (number in parenthesis) |
| --- | --- | --- | --- | --- | --- | --- | --- |
| | Before 1950 | 1950–1959 | 1960–1969 | 1970–1979 | 1980–1986 | | |
| Root and tuber crops | 1 | — | — | — | — | 1 | Potato (1) |
| Ornamentals | | | | | | | |
| Tuber and bulb crops | 1 | 1 | 13 | 36 | 5 | 56 | *Dahlia* (45), *Gladiolus* (1), *Lilium* (2), *Polyanthus* (2), *Tulipa* (6) |
| Pot plants | — | — | 7 | 54 | 12 | 73 | *Achimenes* (8), *Azalea* (10), *Begonia* (21), *Calathea* (1), *Ficus* (1), *Guzmania* (1), *Hoya* (4), *Kalanchoë* (3), *Streptocarpus* (24) |
| Cut flowers | — | — | 18 | 121 | 40[a] | 179[a] | *Alstroemeria* (24), *Carnation* (12), *Chrysanthemum* (130),[a] *Euphorbia fulgens* (1) *Rose* (12) |
| Other ornamentals | — | — | — | 20 | 4 | 24 | *Abelia* (1), *Azalea* (1), *Bougainvillea* (5), *Forsythia* (2), *Malus* (1), *Populus* (1), *Portulaca* (11), *Weigelia* (2) |
| Fruit crops | — | — | 4 | 14 | 4 | 22 | Apple (4), apricot (1), blackcurrant (1), cherry (7), olive (1), peach (4), fig (1), grapefruit (1), pomegranate (2) |
| Other crops | — | — | — | 14 | 3 | 17 | Peppermint (3), sugarcane (11), grasses (1), matgrass (2) |
| Totals | 1 | 1 | 42 | 260 | 68[a] | 372[a] | |

*Source*: Van Harten and Broertjes, 1986; Table 1.

[a] In *Chrysanthemum*, induction of mutations is everyday practice, and the exact number of commercial mutants can no longer be determined. It must exceed several hundred, or even many more, by now.

reflected by the high proportion of commercial mutants as well as by the number of references about such crops in the literature. Most of the methods used in ornamentals can also be used in other crops like root and tuber crops, fruit crops, grasses, etc. However, results in the latter categories require more time and energy. Details concerning material, methods and results for some selected crops are presented in Table 20.2.

A different method to describe mutants as compared to Table 20.2 is on the basis of characteristics which have been changed, such as foliage and flower colour, plant structure, resistances, chemical composition, etc. This approach was followed by Konzak (1984) for seed-propagated crops. Foliage and flower mutations (in ornamentals) are the largest and most important types of mutations, but successes have been obtained in other groups of plants and for other characteristics.

In recent years, the reader has been confronted with a fast increasing number of publications about the potential value for plant breeding of a number of techniques that together are indicated, more or less correctly, as "genetic engineering" or "plant biotechnology". As there are many parallels between the now established science of mutation breeding and the new techniques which, for the major part are still in their infancy, we will deal briefly with this topic in one of the last paragraphs.

## 20.1 WHEN SHOULD MUTATION BREEDING BE USED?

Conventional cross-breeding undoubtedly is, and in most cases will remain, the most important way to produce improved cultivars, not only in seed-propagated plants but also in crops which belong to the group of vegetatively propagated plants, as has been shown for many years with potato and other root and tuber crops in temperate and tropical regions, with fruit crops and many ornamentals. Among the factors reducing the efficiency of cross-breeding in this particular group of crops is the often high degree of heterozygosity which, like polyploidy, causes a complicated system of inheritance and decreases the chance of selecting improved genotypes. A long juvenile phase, crossing-barriers, apomixis or sterility are other factors which may make cross-breeding unattractive or sometimes even impossible.

It is therefore not surprising that there is considerable interest in the improvement of *existing* genotypes which, though outstanding, could be improved further to meet changing demands: for instance, in relation to mechanization, scaling-up, taste, energy costs, etc. The possibility of improving only one characteristic in an otherwise outstanding cultivar remains *the main advantage* of mutation breeding.

Table 20.2. Examples of successful mutation breeding programmes in some vegetatively propagated plants.

| Crop | Material used for mutagenic treatment | Mutagenic treatment | Nature of the mutant plant | Characteristics improved | Number of commercial mutants reported until April 1985 |
|---|---|---|---|---|---|
| *Achimenes* spp. (ornamental pot plant) | Freshly detached leaves | 20–30 Gy[a] X-rays | Solid mutants | Flower (colour, size, shape, etc.) Plant habit Earliness | 8 |
|  | Freshly detached leaves | Colchicine solutions: 0.05–0.1%, 7 h, 20 °C | Solid tetraploids | Flower (colour, size, firmness) Plant habit (compactness) | 4 |
| *Alstroemeria* spp. (ornamental cut flower) | Stolons of young plants | 3.5–4.5 Gy X-rays | Solid mutants(?) | Flower colour Plant habit (reduced height) | 24 |
| Apple | Dormant scions | 50–100 Gy γ-rays | Periclinal chimeras | Fruit characteristics (colour, texture) Compact growth Better cropping Self-fertility | 4 |
| *Cynodon* spp. (Bermuda grass) | Dormant rhizomes | 70 Gy γ-rays | Periclincal chimeras or solid mutants | Density and quality of turf Resistance, e.g. to nematodes Frost tolerance | 1 |

| | | | | | |
|---|---|---|---|---|---|
| Chrysanthemun morifolium | (a) Rooted cuttings | 17.5 Gy X- or γ-rays | Periclinal chimeras ("sports") | Flower (colour, form) Growth habit Low-temperature tolerance Male-sterility | Hundreds |
| | (b) Pedicel explants | 9 Gy X-rays | Solid mutants | Same | Few |
| | (c) Cell suspension | 12–18 Gy X-rays or chemical mutagens | Solid mutants | Low-temperature tolerance | 0 |
| Mentha piperita | Dormant stolons | 60 Gy γ-rays $22 \times 10^{12}$ $N_{th}$ cm$^{-2}$ sec$^{-1}$ | Solid mutants | *Verticillium* wilt resistance | 2 |
| Saccharum spp. (sugarcane) | (a) Growing plants | Chronic γ-radiation | Periclinal chimeras | Yield (cane, sugar) Tillering | 1 |
| | (b) Stem cuttings | 100 Gy γ-rays | Periclinal chimeras | Yield Disease resistance | |
| | (c) Callus *in vitro* | Untreated or chemical mutagens | Solid mutants | Same | |

Sources: *Mutation Breeding Newsletter* (IAEA), vols 1–21; Broertjes and Van Harten (1978).

[a] 10 Gy (= 10 gray) = 1 krad.

It can yield commercial results, often within a relatively short period of time. Examples are found in many crops (Broertjes and Van Harten, 1978; Van Harten, 1982), such as the all-year-round glasshouse crop chrysanthemum, for which a whole series of flower colour mutants have been produced in The Netherlands from the cvs Horim and Miros (Broertjes *et al.*, 1980). Similar results were obtained with the English chrysanthemum cv. Snapper, radiation-induced mutants of which penetrated rapidly into the important Dutch market, primarily because of their tolerance to a relatively low temperature (A. G. Sparkes, 1982, pers. comm.). The same method is being applied in other ornamental crops such as *Alstroemeria* where outstanding genotypes are irradiated routinely to produce desired variations in, for example, flower colour, sometimes even before the original cultivar reaches the market.

Theoretically, a single gene can be mutated without altering the remaining genotype, but in practice more than one characteristic is usually involved. This is sometimes explained on the basis of pleiotropy, but probably is caused more often by (small) chromosome aberrations in which several genes are involved. At a relative high dose of radiation there may be more than one independent mutagenic event per cell. Whichever explanation is applicable, the fact remains that a desirable mutated characteristic often is accompanied by other mutations which may result in an unacceptable genotype. It then becomes a matter of statistics (numbers!) to trace plants with the desired mutation and without accompanying undesirable mutations for important characteristics. The numerous mutants in commercial use show that this can be achieved.

## 20.2 THE GENETIC CONSTITUTION OF THE PARENT VARIETY

One of the considerations to keep in mind when a choice between the use of conventional and mutation breeding methods must be made is that most mutational events are of a recessive nature. It is therefore of utmost importance to collect information about the number and kind of genes controlling the specific character whose change is sought in order to choose the right cultivar with which to start a mutation breeding programme. In general, it is unrealistic to irradiate genotypes that are homozygous recessive or dominant for the characteristic in question, since the frequency of dominant mutations (aa → Aa) is very low and estimated to be somewhere between 1 and 5% (Brock, 1979). In addition, we generally cannot select for heterozygotes (AA → Aa or A–) as they cannot be distinguished from the dominant starting material. An exception could be made where large numbers can be handled, a very selective screening method is available and an adventitious bud technique can be used. Examples are the work with peppermint (*Mentha piperita*), where a dominant gene for resistance to *Verticillium* wilt was probably induced after irradiation of large numbers of dormant stolons (Murray, 1969). Likewise, four cold-resistant plants of Coastcross-1 Bermuda grass (*Cynodon dactylon*) were found after irradiation of 500 000 green stems (each containing several dormant buds) and subsequent exposure to winter conditions after planting (Burton *et al.*, 1980).

It is remarkable that even when certain characters most probably show polygenic inheritance, a considerable number of mutations can be observed in practice.

Many vegetatively propagated crops have a high degree of heterozygosity and consequently yield a high frequency of recessives, (Aa → aa or –a) after irradiation. In polyploids, the situation is more complicated but generally polyploidy is favourable for mutation breeding (and disadvantageous for conventional breeding). Such crop plants are octoploid dahlias, hexaploid chrysanthemums and tetraploid potatoes. They all show a considerable frequency of spontaneous and induced mutations both for qualitative (monogenic) and quantitative (polygenic) characteristics. To choose the right cultivar for irradiation is often a matter of experience and sometimes of knowing the genetics of the attribute to be improved. For example, we know from statistical analysis as well as from experience that most of the flower colours existing in chrysanthemum can be obtained by irradiating a pink, pinkish or nearly white-flowering genotype.

The irradiation of brown or yellow genotypes is non-productive. Similarly, the irradiation of white-flowered cultivars of rose, dahlia, carnation and many other crops will never yield flower colour mutations.

Mutational events with a *dominant* expression have been put forward to explain the 20–40-fold increase in mutation frequency in autotetraploid *Achimenes* genotypes, compared with their corresponding diploid cultivars (Broertjes, 1976).

When nothing is known about the genetics of the characteristic(s) to be improved, a study of the spectrum of spontaneous mutations might help the breeder in choosing the genotype for irradiation. So far, there are no indications that the spectrum of spontaneous natural mutations (i.e. "sports") differs from that of induced ones.

## 20.3 CHIMERA FORMATION

Chimera formation following a mutagenic treatment of plant parts such as cuttings, bulbs, tubers, rhizomes, etc. (which all have multicellular apices), is one constraint in mutation breeding of vegetatively propagated crops. Chimera formation results automatically because a mutation is a unicellular event. Apices of phanerogams contain a number (mostly three) of rather autonomous groups of cell-layers. When the mutated cell is situated in the right part of the apex and its multiplication is not limited by surrounding cells (a kind of competition, often indicated as diplontic selection), it forms a group of cells within one layer (a mericlinal chimera) and ultimately, through axillary bud formation within the mutated area, a complete periclinal chimera (a "sport").

Some of the *disadvantages* of chimeras are that:

(a) mericlinal chimeras (often, though incorrectly, indicated as "sectorial chimeras") are easily overlooked and, consequently, selection during the early stages is generally impossible, particularly for non-visible characteristics;
(b) even if mericlinal chimeras can be recognised, it is often difficult to secure the complete periclinal chimera (by axillary bud formation or in other ways);
(c) periclinal chimeras are sometimes unstable and may revert back to the original cultivar (uncovering or "back-sporting").

There are, however, *advantages* as well, such as:

(a) the production of "sports" is sometimes fast and easy, and therefore economically valuable, such as in chrysanthemum;
(b) "sports", such as flower-colour mutants, generally conform closely with the original cultivar and react similarly to climatic factors such as temperature and daylength, which has several advantages for the grower.

The disadvantageous aspects tend to predominate in crops that are slow growing and hard to propagate, like woody (fruit) crops. It is not surprising, therefore, that for those and many other crops, breeders have tried to develop methods to avoid chimera formation. In principle, this could be rather simple if it were possible to develop complete plants from a single cell or from a few vegetative daughter cells of a single cell. If such a cell is normal, a normal plant will result (at least in theory: in practice always a low percentage of aberrant types may be expected). If that cell had mutated, a large majority of complete non-chimeral mutants would be formed.

There are several *advantages* of solid mutants:

(a) early and easy selection is possible also for most non-visible characteristics (provided a selective screening method is available);

(b) mutants are stable;
(c) mutants can be propagated vegetatively by any method; and
(d) mutants can be used in conventional breeding programmes.

A *disadvantage* of solid mutants could be that all (accompanying) mutations due to pleiotropic effects and close linkage of genes are expressed as well, so that conformity with the original cultivar is not as complete as is desirable. This, however, is predominantly a matter of statistics and generally it is possible to obtain the desired mutation without associated unfavourable mutations for important characteristics.

Mutants which are complete and genetically homogeneous, can be obtained through the production *in vivo* or *in vitro* of adventitious plantlets from irradiated plant parts or explants, or from cell cultures *in vitro*. The meristem on which the apex of an adventitious shoot develops is produced by undifferentiated cells. The apex apparently is formed by one or a few vegetative daughter cells derived from a mutated or unmutated (epidermal or callus) cell, as can be tested experimentally. Supporting evidence comes from a computer model (Broertjes and Keen, 1980).

It has been demonstrated for various crops that adventitious shoots *in vitro*, as those *in vivo*, ultimately originate from a single (mutated) cell, since by using this method almost exclusively, normal plants and completely mutated plants are produced. This has been demonstrated in *Begonia* (Roest *et al.*, 1981), chrysanthemum (Broertjes *et al.*, 1976) and potato (Van Harten *et al.*, 1981).

Not all researchers agree with the single-cell concept of adventitious buds (see, for example, Norris *et al.*, 1983 about African violet), but the supporting evidence for a multicellular origin, so far, is not very convincing (Marcotrigiano and Stewart, 1984; Broertjes and Van Harten, 1985).

## 20.4 *IN VIVO* ADVENTITIOUS BUD TECHNIQUES

The first crop for which it was demonstrated that adventitious shoots ultimately originate from a single cell was African violet, *Saintpaulia ionantha* H. Wendl. Sparrow *et al.* (1960) obtained virtually only normal plants or complete mutants, following the irradiation of detached leaves and the subsequent rooting and production of plantlets. Later, many other plants were used to study this phenomenon, such as *Achimenes*, *Begonia*, *Lilium*, *Muscari*, *Ornithogalum*, *Streptocarpus*, etc. (Broertjes and Van Harten, 1978). It was found that the mutants were solid, non-chimeral with only few exceptions, and that by the use of such an *in vivo* adventitious bud technique a range of mutants could be produced rapidly.

Early selection and fast multiplication by the same technique provide a relatively rapid means of producing commercially interesting genotypes. Similarly, this method can be used to produce non-cytochimeral auto-tetraploids of diploid plants (Broertjes, 1974). However, the technique is of limited application because most plants cannot be propagated this way, despite the fact that hundreds of plants can be reproduced from adventitious shoots or plantlets (Broertjes et al., 1968).

## 20.5 IN VITRO ADVENTITIOUS BUD TECHNIQUES

Fortunately, methods of multiplication *in vitro* have become available and a large and still increasing range of plants can now be propagated under sterile conditions. For mutation breeding, those methods by which adventitious shoots originating from as few cells as possible are produced on irradiated explants, either directly from epidermal cells or indirectly on callus, are of importance. We even may expect this method to increase rapidly in importance, particularly when associated with selection *in vitro* at the cellular level with the subsequent production of adventitious plantlets (see also section 20.7).

## 20.6 OTHER PROPAGATION METHODS USEFUL IN MUTATION BREEDING

Where adventitious bud techniques are not available or cannot be applied, it is necessary to look for other methods of propagation in order to limit the disadvantageous effects of chimera formation.

For *Alstroemeria*, actively growing rhizomes of young plants are irradiated, generally with 3–4 Gy of X-rays (1 Gy = 100 rad). After planting and regrowth (flower colour) mutants can be observed almost always in the form of completely mutated shoots. Chimeras are rarely found. Whether the plants are solid mutants or are complete periclinal chimeras is not known, but "back-sporting" has never been observed. This indicates that the "sports" most probably are solid, non-chimeral mutants. This phenomenon possibly has to do with the sympodial growth pattern of *Alstroemeria*. Several commercial cultivars were produced in co-operative programmes (Broertjes and Verboom, 1974) and this method has been adopted as a routine procedure in recent breeding programmes with this increasingly important cut flower in The Netherlands.

Dormant scions are used in Bermuda grass (*Cynodon dactylon*) of which solid mutants as well as periclinal chimeras are obtained after treatment with X-rays (Burton et al., 1980).

Stolons can also be used, as is demonstrated by the successful mutation breeding programme in peppermint (Murray, 1969; Broertjes and Van Harten, 1978).

Dahlia provides an example in which the irradiation of ontogenetically very young buds gives optimum results. Tubers are irradiated immediately after harvest in the autumn, when no visible buds are present. After winter storage, the developing shoots are detached in the spring, and rooted and planted as usual. Depending on the cultivar used, many mutations can be observed generally during flowering, and up to 50% of the mutants obtained appeared to be solid. We do not know if these mutants are periclinal chimeras or really homogeneous mutants but, since "back-sporting" has not been reported they could be either. The many commercial dahlia cultivars that have been produced demonstrate the usefulness of this method for practical breeders (Broertjes and Ballego, 1967).

When no such techniques are available, the irradiation of "normal" plant material, containing fully developed buds is the only alternative. The economically important cut flower crop chrysanthemum is an example that has illustrated how this method may rapidly lead to a high frequency of potentially interesting mutants. This is linked to decapitating the plant to induce the rapid formation of periclinal chimeras ('sports'), the quick rooting and growth of cuttings and the year-round manipulation with long- and short-day conditions for vegetative growth and flower bud initiation. Commercial production of flower colour mutants is a routine matter, involving successive irradiations of radiation-induced mutants (Broertjes *et al.*, 1980). Low-temperature tolerant mutants can also be produced in a similar manner, as has been demonstrated by the commercial release of three T-mutants of chrysanthemum (Broertjes *et al.*, 1983).

Irradiation of normal plant material is time consuming in crops which cannot be manipulated in the same way as chrysanthemum, e.g. tulip (bulbs), fruit trees (scions), sugarcane (stem cuttings), grapevine (scions), etc. An example of the laborious work on one such crop, namely apple, is discussed by Lacey and Campbell (see Chapter 21).

## 20.7 THE APPLICATION OF NEW TECHNIQUES

In addition to the production *in vitro* of non-chimeral mutants, monocell cultures or cell suspension cultures have been reported as potentially valuable new techniques in relation to mutation breeding. The possibility exists of selecting from among large numbers of cells for rare mutation events (like dominant mutations) on a single cell basis, provided (a) an adequate screening method is available for the direct or indirect identification and preferential selection of the cell(s) in question; and (b) methods are available to regenerate cell-lines, shoots and plantlets from these cells.

There are two prerequisites. The first is that the selected characteristic expresses itself not only under *in vitro* conditions, but also in the plant through, for example, disease resistance or saline tolerance, or is linked with other important characteristics of the adult plant. The second is that the method itself does not result in an unacceptably high genetic variability, since the object is to induce a specific mutation in an otherwise unchanged genotype.

At present much attention is paid to the use of somaclonal variation (Larkin and Scowcroft, 1981) and induced mutations *in vitro*. Despite some optimistic preliminary reports about, for example, induced as well as spontaneously (somaclonally) arising resistance for late blight (*Phytophthora infestans*) in potato (Behnke, 1979 and Shepard et al., 1980, respectively), there is as yet little convincing evidence of these mutant plants being of commercial value. It may be worthwhile to mention here that the use of modern techniques of genetic engineering is not much further advanced. The best result obtained so far is, that a vector-induced gene for tolerance to the herbicide glyphosate in tobacco could be expressed at the plant stage (Comai et al., 1985).

To our knowledge, no examples exist of important mutants obtained by indirect selection. A supra-optimal concentration of benzyladenine in the culture medium could possibly be used to select for compactness in apples, as suggested by Lane and Looney (1982). The development of such indirect, two-step, selection systems seems to be of great interest since many important characteristics cannot be selected on a one-cell basis. These include yield, quality, resistance to many diseases, growth pattern, colour, form, size, fruit characteristics, seed-set and low-temperature tolerance.

Mutation breeding in vegetatively propagated crops has been successful, especially in easily propagated and fast-growing ornamental crops. The number of mutants in commerce is increasing steadily and will continue to do so now that the method is routine with many crops. During the selection of the starting material, such as seedlings resulting from selfing or interspecific crosses, etc., the breeder takes into account what can be done later on with induced mutations to improve the best cultivars. Sometimes there have been negative criticisms of some results, such as the introduction of numerous minor flower colour changes in ornamentals. However, critics overlook the commercial importance of such flower colour mutants for the ornamental industry in a country like The Netherlands where only genotypes which have identical climatic and other requirements are often grown in a single large glasshouse.

The need for improvement through mutation breeding of other characteristics, such as disease resistance, and a low requirement for energy, temperature or light in glasshouse crops is becoming more important. For example, the mutants of chrysanthemum which succeed at lower glasshouse temperatures (Broertjes *et al.*, 1983; Broertjes and Lock, 1985).

In this chapter we have not discussed the relative advantages of various mutagenic treatments. In general, radiation by X-rays or γ–rays has been most used for obvious reasons, such as general availability and considerable experience of suitable dosimetry and reproducibility. In addition, the method is easy to handle and is easily applied to large quantities of plant material. Chemical mutagens, on the other hand, have been less favoured so far, because of the problems of penetration and uncertain dosimetry and reproducibility. However, with the recent developments in techniques *in vitro*, the application of chemical mutagens should be reconsidered.

**REFERENCES**

Behnke, M. (1979). Selection of potato callus for resistance to culture filtrates of *Phytophthora infestans* and regeneration of resistant plants. *Theoret. appl. Genet.* **55**, 69–71.

Brock, R. D. (1979). Mutation plant breeding for seed protein improvement. *In* "Seed Protein Improvement in Cereals and Grain Legumes". Proc. Symp. IAEA/FAO/GSF, Neuherberg, BRD, 1978, pp. 43–45. IAEA, Vienna.

Broertjes, C. (1974). The production of polyploids using the adventitious bud technique. *In* "Proceedings of the Eucarpia FAO/IAEA Conference on Mutations and Polyploidy, Bari, 1972", pp. 29–35. IAEA, Vienna.

Broertjes, C. (1976). Mutation breeding of autotetraploid *Achimenes* cultivars. *Euphytica* **25**, 297–304.

Broertjes, C., and Ballego, J. M. (1967). Mutation breeding of *Dahlia variabilis*. *Euphytica* **16**, 171–176.

Broertjes, C., and Keen, A. (1980). Adventitious shoots: do they develop from one cell? *Euphytica* **29**, 73–87.

Broertjes, C., and Lock, C. A. M. (1985). Radiation-induced low-temperature tolerant solid mutants of *Chrysanthemum morifolium* Ram. *Euphytica* **34**, 97–103.

Broertjes, C., and Van Harten, A. M. (1978). "Application of Mutation Breeding Methods in the Improvement of Vegetatively Propagated Crops". Elsevier, Amsterdam, 316 pp.

Broertjes, C., and Van Harten, A. M. (1985). Single cell origin of adventitious buds. *Euphytica* **34**, 93–94.

Broertjes, C., and Verboom, H. (1974). Mutation breeding of *Alstroemeria*. *Euphytica* **23**, 39–44.

Broertjes, C., Haccius, B., and Weidlich, S. (1968). Adventitious bud formation on isolated leaves and its significance for mutation breeding. *Euphytica* **17**, 321–344.

Broertjes, C., Koene, P., and Van Veen, J. W. H. (1980). A mutant of a mutant of a mutant of a .... Irradiation of progressive radiation-induced mutants in a mutation breeding programme with *Chrysanthemum morifolium* Ram. *Euphytica* **29**, 525–530.

Broertjes, C., Koene, P., and Pronk, Th. (1983). Radiation-induced low-temperature cultivars of *Chrysanthemum morifolium* Ram. *Euphytica* **32**, 97–101.

Broertjes, C., Roest, S., and Bokelmann, G. S. (1976). Mutation breeding of *Chrysanthemum morifolium* Ram. using *in vivo* and *in vitro* adventitious bud techniques. *Euphytica* **25**, 11–19.

Burton, G. W., Constantin, M. J., Dobson, J. W. Jr, Hanna, W. W., and Powell, J. S. (1980). An induced mutant of Coastcross-1 bermuda-grass with improved winterhardiness. *Environ. Exptl. Bot.* **20**, 115–117.

Comai, L., Faciotti, D., Hiatt, W. R., Thompson, G., Rose, R. E., and Stalker, D. M. (1985). Expression in plants of a mutant *aroA* gene from *Salmonella typhimurium* confers tolerance to glyphosate. *Nature* **317**, 741–744.

Konzak, C. F. (1984). Role of induced mutations. *In* "Crop Breeding, a Contemporary Basis" (P. B. Vose and S. G. Blixt, eds), pp. 216–292. Pergamon Press, Oxford.

Lane, W. D., and Looney, N. E. (1982). A selective tissue culture medium for growth of compact (dwarf) mutants of apple. *Theoret. appl. Genet.* **61**, 219–223.

Larkin, P. J., and Scowcroft, W. R. (1981). Somaclonal variation—a novel source of variability from cell cultures for plant improvement. *Theoret. appl. Genet.* **60**, 197–214.

Marcotrigiano, M., and Stewart, R. N. (1984). All variegated plants are not chimeras. *Science* **223**, 505.

Murray, M. J. (1969). Successful use of irradiation breeding to obtain *Verticillium*-resistant strains of peppermint, *Mentha piperita* L. *In* "Induced Mutations in Plants", Proceedings of IAEA/FAO Symposium, Pullman, USA, 1969, pp. 345–370. IAEA, Vienna.

Norris, R., Smith, R. H., and Vaughn, K. C. (1983). Plant chimeras used to establish *de novo* origin of shoots. *Science* **220**, 75–76.

Roest, S., Van Berkel, M. A. E., Bokelmann, G. S., and Broertjes, C. (1981). The use of an *in vitro* adventitious bud technique for mutation breeding of *Begonia* × *hiëmalis*. *Euphytica* **30**, 381–388.

Shepard, J. F., Bidney, D., and Shahin, E. (1980). Potato protoplasts in crop improvement. *Science* **208**, 17–24.

Sparrow, A. H., Sparrow, R. C., and Schairer, L. A. (1960). The use of X-rays to induce somatic mutations in *Saintpaulia*. *Afr. Violet Mag.* **13**, 32–37.

Van Harten, A. M. (1982). Mutation breeding in vegetatively propagated crops with emphasis on contributions from the Netherlands. *In* "Eucarpia Conference on Induced Variability in Plant Breeding", Wageningen, 1981, pp. 22–30. PUDOC, Wageningen.

Van Harten, A. M., and Broertjes, C. (1986). Mutation breeding: a stepping-stone between Gregor Mendel and genetic manipulation (a treatise for vegetatively propagated crops). *In* "Genetic Manipulation in Plant Breeding" (W. Horn, C. J., Jensen, W. Odenbach and O. Schieder, eds), Proceedings of Eucarpia meeting, Berlin, 1985, pp. 3–15. Walter de Gruyter, Berlin.

Van Harten, A. M., Bouter, H., and Broertjes, C. (1981). *In vitro* adventitious bud techniques for vegetative propagation and mutation breeding of potato (*Solanum tuberosum* L.): II. Significance for mutation breeding. *Euphytica* **30**, 1–9.

# 21 Selection, Stability and Propagation of Mutant Apples

C. N. D. LACEY[1] and A. I. CAMPBELL[2]

[1]*AFRC Institute of Plant Science Research, Plant Breeding Institute, Cambridge*
[2]*Long Ashton Research Station, Bristol*

The genetics of apples is such that slight improvements of established cultivars cannot be produced by normal breeding. In most countries the named cultivars were originally chance seedlings, found in the last century, and these continue to dominate all new orchards. Thus in England, over 75% of all the new trees planted in the 1980s have been the cultivars Cox's Orange Pippin and Bramley's Seedling, both of which are well over 100 years old. This is partly due to the reluctance of growers and markets to introduce any of the numerous new cultivars in quantity as there is assumed to be a consumer preference for known and accepted cultivars with clearly defined and well-liked attributes.

Existing cultivars have sometimes been improved by selecting natural sports or mutations in which important characters such as skin colour or tree habit have changed. When this improvement has been stable and successful, the original cultivar has gradually been superseded by new named mutants. In the USA, Red Delicious has been replaced by several compact, spur or good-coloured types such as Oregon Spur, Starking and Red Chief. In the past, natural sports have not been so well exploited in the UK, but in 1983 about half of the Cox's Orange Pippin trees planted were the EMLA clone (East Malling–Long Ashton) of Queen Cox, a red-skinned selection, and the trend away from normal Cox is increasing (Turnbull, 1983).

Similar trials to assess and evaluate the characteristics of natural mutations in apple cultivars are in progress in most apple-growing countries. For example, recent reports from Italy (della Strada *et al.*, 1983), from the Federal Republic of Germany (Silbereisen, 1986) and from Denmark (Grausland, 1986) have emphasised the benefits that can

IMPROVING VEGETATIVELY PROPAGATED CROPS
ISBN 0–12–041410–4

*Copyright © 1987 Academic Press Limited*
*All rights of reproduction in any form reserved*

be obtained from good selections and the faults that can be found in bad ones.

In addition, recent work on pome fruit cultivars and rootstocks in The Netherlands by Wertheim and van Oosten (1986), and in Belgium by Gilles et al. (1986), have shown the importance of plant health when assessing clonal effects. The dangers of comparing natural mutants of cultivars with different virus infections is emphasised. Campbell (1983) has discussed various aspects of tree health and quality in relation to the EMLA fruit tree scheme in the UK and shown that interactions between viruses and genetic variants frequently cause assessment problems.

It is well known that some of the natural sports are less stable than others and this can be a serious fault in an otherwise excellent mutant (Faedi and Rosati, 1985). Various attempts have been made to compare and improve the stability of mutants, so far without much success (Lacey et al., 1982). Although it is hoped that micropropagation techniques may offer a solution to the stability problems in the future.

Instead of searching for natural mutants or waiting for them to occur, it seemed appropriate to induce greater numbers of mutations of widely planted cultivars and to select the best of these mutants for commercial use. The Long Ashton mutation breeding programme began in 1968 with the major objective of improving the leading British apple and pear cultivars and some aspects of this programme will be discussed later. Results from similar mutation breeding programmes with apples have been published recently by Blazek (1985) working in Czechoslovakia, by Li et al. (1986) in Taiwan, by Seth et al. (1982) in India and by Yoshida et al. (1981) in Japan.

The methods used to produce mutant forms of vegetatively propagated plants, such as apple, are well established (Broertjes and Van Harten, 1978). For some species, particularly ornamentals, where the demand for new cultivars is great and plants are short-lived, new mutants may make an immediate impact on the market. For other crop plants there has been more caution in using mutagenesis as a tool for plant improvement.

Apple trees, along with many other fruit crops, live for many years and represent a large investment of capital, land and labour. A grower needs good evidence of improvement before he will change from the old tried and tested cultivars, and while this may give mutants of existing cultivars the edge over fruit cultivars bred conventionally, new mutants still require long-term trials of all aspects of their behaviour, including storage, disease resistance and quality as well as yield.

While meristem culture and *in vitro* proliferation are possible with apple, production of plants from single cells has not yet been achieved, and adventitious buds, if they can be induced, may come from multicellular initials (Lacey et al., 1980). Mutation is a single cell event and hence mutagenesis of multi-cell initials would be expected to lead to the formation of vegetatively maintained chimeras that are not completely stable (Tilney-Bassett, 1986). A selected chimeral mutant will, on occasion, apparently revert to the original form. This may not be very important in relatively short-term crops such as ornamental flowers, but for fruit trees it is potentially more damaging and may limit the use of some mutants. An orchard of self-fertile clones that reverted would need a pollinator added, while reversion of a compact to the original vigorous type would cause orchard overcrowding and loss of fruit quality. Such problems have been noted with some naturally occurring spur forms of apple cultivars (Maas, 1970). Treatment of a bud initial with a mutagen such as gamma radiation may result not only in the production of mutant cells within the meristem but also in the death of all or part of that meristem (Pratt, 1963). Some mutant plants may therefore be derived from severely damaged buds with very few surviving cells actively dividing. As very few cells (perhaps nine in normal apple buds) are ultimately responsible for the formation of the meristem and hence the whole plant (Stewart and Dermen, 1970) their number may be reduced still further by the mutagenic treatments, giving rise to homohistant plants (see Fig. 21.2).

This chapter discusses the production and selection of mutant apples and describes the tests used so far to determine the nature of the mutated plants, together with a brief description of the experimental evidence on which this is based (see also Lacey and Campbell, 1979; Lacey, 1982a).

## 21.1 PRODUCTION AND SELECTION OF MUTANT CLONES

The mutant clones were selected representatives of the $MV_2$ generation (i.e. second vegetative generation from mutagen treatment) of Bramley's Seedling apple (Fig. 21.1). The mutagenic treatment was approximately 60 Gy (1 Gy = 100 rad) of gamma radiation at 10 Gy h$^{-1}$ from a Co$^{60}$ source given to dormant shoots in the winter. This dose did not kill any material, but the extension growth of the treated buds was very variable and averaged about half that of the untreated controls. When the $MV_2$ generation was produced by budding all available material from the $MV_1$ shoots, groups of compact trees could be seen in the nursery rows. Some of these were selected for further comparison, re-multiplied and planted as an $MV_3$ orchard (Lacey and Campbell, 1980).

If the radiation treatment had resulted, as expected, in small sectors of mutated material within the treated buds, then at this stage either single

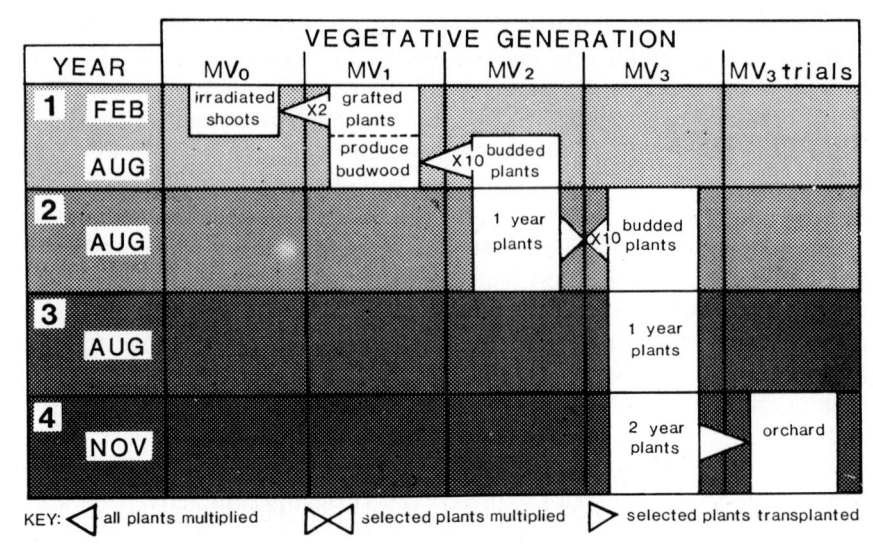

**Figure 21.1.** Scheme of apple mutation breeding programme at Long Ashton Research Station.

mutant trees or small groups of mutant trees would have been found within the vegetative $MV_2$ families formed by sequential budding from each $MV_1$ plant. Instead, the population contained many vegetative families which were either entirely or mainly compact. It was relatively more rare to find a family with mutant trees evenly spaced as they would be from the propagation of a mericlinal chimera (Lacey and Campbell, 1979). The apparently completely mutated families must have come from an $MV_1$ plant with all the buds on the shoot equally mutated, i.e. a periclinal chimera or homohistant plant rather than a mericlinal or sectoral chimera (Fig. 21.2). Therefore at least one histologic layer in these plants must have been derived from a single mutated cell, though observation could not distinguish between periclinal chimeras and homohistant plants.

The mutated clones with improved pomological characters were retained, re-multiplied to check they were genuine mutants and grown as an orchard ($MV_3$) for at least 7 years to compare their performance with the original clone. In the case of Bramley's Seedling, 12 of approximately 100 compact $MV_3$ clones were selected from the mutants found among the 6000 $MV_2$ plants (about 400 vegetative families and $MV_1$ shoots). Selection was for decreased growth combined with yield maintained at or near the original amount and unchanged fruit quality and size. A further selection criterion must be stability of the new clones, and hence the

Propagation of Mutant Apples

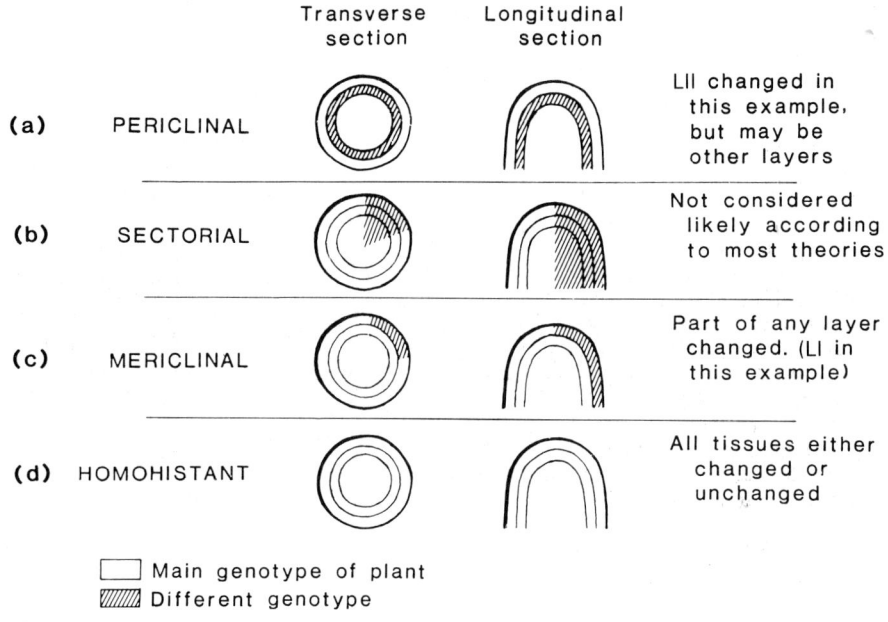

|     |           | Transverse section | Longitudinal section |                                                            |
|-----|-----------|--------------------|----------------------|------------------------------------------------------------|
| (a) | PERICLINAL|                    |                      | LII changed in this example, but may be other layers       |
| (b) | SECTORIAL |                    |                      | Not considered likely according to most theories           |
| (c) | MERICLINAL|                    |                      | Part of any layer changed. (LI in this example)            |
| (d) | HOMOHISTANT|                   |                      | All tissues either changed or unchanged                    |

☐ Main genotype of plant
▨ Different genotype

**Figure 21.2.** Diagram of chimeral forms.

chimeral status of the 12 selected clones was examined (Lacey and Campbell, 1980; Lacey, 1982b).

## 21.2 PRELIMINARY TESTS FOR PERICLINAL CHIMERISM

Radiation can be used as a tool to disrupt the meristem of a natural periclinal chimera of apple to reveal the nature of its composite layers (Decourtye, 1967; Pratt et al., 1972). Five hundred trees of each Bramley clone were propagated by bud-grafting without re-irradiation to assess their stability following conventional multiplication, and a further group of $MV_5$ trees was propagated following re-irradiation of the $MV_4$ graft wood to compare the stability of the mutants.

### 21.2.1 Conventional propagation

Dormant scion wood, selected from seven-year-old trees of each compact mutant clone and a control clone was grafted on to pot-grown rootstocks in an unheated glasshouse. The summer shoots ($MV_4$) from the grafts provided the buds for further propagation in the field ($MV_5$). Each bud

**Table 21.1** Heights, as percentages of the control, standard errors and coefficients of variation, of 12 compact mutant clones of Bramley's Seedling apple and control (100 trees per clone).

| Clone | Height (% of control) | S.E. of mean | Coefficient of variation (%) |
|---|---|---|---|
| Control | 100 | 3.4 | 3.4 |
| 20 | 89 | 2.6 | 2.9 |
| 36 | 81 | 3.4 | 4.2 |
| 44 | 79 | 2.7 | 3.4 |
| 58 | 88 | 3.3 | 3.7 |
| 68 | 81 | 2.7 | 3.4 |
| 91 | 84 | 4.1 | 4.9 |
| 94 | 96 | 5.0 | 5.2 |
| 10N[a] | 99 | 5.2 | 5.2 |
| 10M[a] | 74 | 4.8 | 6.5 |
| 27N | 100 | 3.6 | 3.6 |
| 27M | 80 | 5.3 | 6.6 |
| 42N | 101 | 4.0 | 4.0 |
| 42M | 90 | 3.7 | 4.1 |
| 46N | 102 | 3.9 | 3.8 |
| 46M | 87 | 3.7 | 4.1 |
| 69N | 96 | 5.9 | 6.1 |
| 69M | 81 | 3.2 | 4.0 |

[a] N, apparently normal trees; M, compact mutant.

could therefore be traced back to its original parent tree. Each clone was assigned to two adjacent nursery rows of graded MM106 (EMLA) rootstocks. The budsticks from different mother-trees within each clone were propagated at random along the rows.

The control clone produced extremely uniform trees (Table 21.1). The mutant clones could be separated into two groups; seven clones appeared as uniform as the control, while the other five produced two distinct types of tree. The first type, comprising 50–80% of the trees, appeared identical to the control trees, while the second type was like the originally selected compact mutant (Fig. 21.3). In Table 21.1, which shows the mean heights, standard errors and coefficients of variation for 100 trees of each clone of Bramley's Seedling, the two types of tree from the unstable clones are shown separately to demonstrate the similarity in height of the reverted forms to the control trees. The difference between the two types of tree within each unstable clone was significant in every case ($P = 0.01$).

Propagation of Mutant Apples

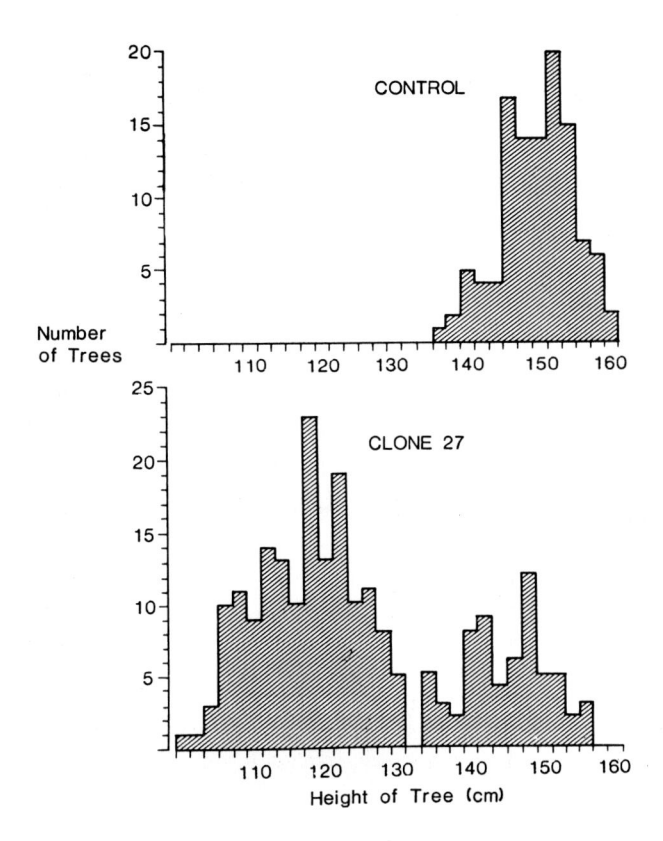

**Figure 21.3.** Distribution of one-year-old tree heights (cm) for control and an unstable mutant clone of Bramley's Seedling (clone 27).

### 21.2.2 Propagation after re-irradiation

Two dormant scions from each mutant clone were submitted to one of six irradiation doses (20–70 Gy), the buds grafted and plants grown in the same way except that the summer growth of the surviving plants was propagated with two buds on each rootstock to save space.

The trees grown from this re-irradiated budwood were very variable, probably due to the damaging effect of the treatment, and the results are therefore difficult to summarise numerically. The most obvious effect of irradiation was that the control clone and the stable compact mutant clones produced very few trees larger than the original form, while the

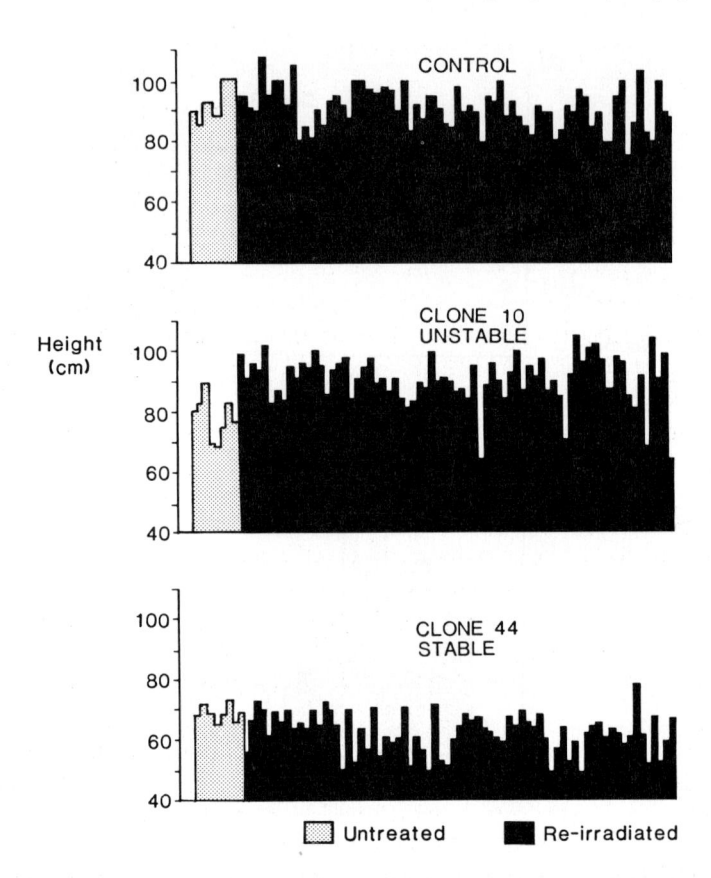

**Figure 21.4.** Size of individual one-year-old mutant trees of Bramley's Seedling with and without a second radiation treatment (each vertical bar represents one tree).

unstable compact mutants produced trees that were almost all taller than the original mutant and similar to the control trees (Fig. 21.4). No clones that had behaved in a stable way during normal propagation produced larger trees when re-irradiated and there was no apparent difference in the effect of the different doses of radiation.

Among the 12 induced compact mutant clones tested, seven appeared at this stage to be as stable as the original clone. These clones may have been derived from only one mutant cell surviving in a meristem that had been severely damaged by the original radiation treatment, or they could be exceptionally stable periclinal chimeras.

With the unstable clones, only two types of tree were produced. There are four possible rearrangements of mutant tissue even if only two semi-stable layers of cells are involved in a periclinal chimera, since it should be possible to derive both the homohistant compact and normal forms as well as the inverse of the original periclinal form. As only two forms were produced either some of these alternative arrangements do not occur or they appear similar to each other.

The most stable layers is reported to be the epidermis (LI) (Lapins and Hough, 1970); therefore a possible explanation of our results would be that the epidermis is normal in the mutant trees of the unstable clones. The rest of the tree would be all mutated tissue as the inner layers are more likely to be of the same genetic constitution following the relatively more frequent anticlinal cell divisions in LII and LIII (Dermen, 1965). Such a periclinal chimera with LI normal and LII/III mutant would appear mutated and often propagate in an apparently stable way. If the replacement of cells in the LII/III by cells from the LI occurred in the meristem, then an area would result that consisted of more vigorous, normal cells. This replacement could be expected to occur after irradiation and the results indicate that it may happen frequently without radiation treatment. The difference between an LII/III mutant and its homohistant derivative could be slight, and undetectable, as so little extra tissue is involved.

If, on the other hand, LI is mutant and LII/III normal in the unstable clones it seems unlikely that such a small amount of tissue could have the profound effect on growth which is seen in the compact mutants. Homohistant mutants resulting from the replacement of normal LII/III by mutant tissue would contain so much more mutated tissue that one would expect them to appear more dwarf. Such plants were not found.

If it is the case that the LII/III is mutated, then homohistant mutant trees should result from the growth of endogenous adventitious buds (Dermen, 1948). These have to be produced from root cuttings of Bramley's Seedling, as disbudded trees of this cultivar died before producing adventitious shoots, so an experiment to assess the growth of plants from root cuttings was started in 1979 (Lacey et al., 1980).

## 21.3 TESTS FOR HOMOHISTANT MUTANTS THROUGH ROOT CUTTINGS

It has been hypothesised that trees derived from endogenous buds, arising from tissue without the original LI, would show the growth habit of the inner tissues (Dermen, 1965). Roots on Bramley's Seedling apple appear to arise from callus at the base of the cutting, probably derived from the cambium. Shoots from these roots arise deep in the cortex close to the stele, and so are of definite endogenous origin (probably LIII or lower) (Lacey et al., 1980).

If the compact unstable mutants have genetically compact inner layers, then all the trees derived from the roots will be homohistant compact, while if LI or possibly LII was the only compact tissue, then, presumably, the LIII-derived root cuttings will produce normal trees. In the case of homohistant compact mutants all propagation systems will produce compact trees, barring further mutation.

To test this hypothesis, we produced trees from root cuttings. Mature one-year-old shoots, taken from mother-trees of EMLA Bramley's Seedling and from each of the mutant clones in the autumn, were rooted during the winter using the warm rooting bin technique (Child and Hughes, 1978). Those propagated successfully were transferred to a nursery plot where they were kept until roots over 1 cm in diameter could be harvested. These roots were washed, surface sterilised and 15 cm lengths laid horizontally on sand in a propagation bench, as described by Child (1975). Adventitious shoots were produced within a few weeks and tip-grafted on to small seedling rootstocks to overcome establishment problems. To examine the growth characteristics of these plants they needed to be grown from budwood on uniform clonal MM106 rootstock, where two comparisons were made; with samples of the original mutant types propagated from conventionally grown budwood and with the original EMLA Bramley clone as a standard.

Four of the clones, 46, 69, 91 and 94, did not survive all stages of the treatment to produce root-derived trees. However, we succeeded in raising trees from eight clones. Their mean heights, measured in August, are compared with corresponding shoot-derived trees and with the standard from conventional propagation in Table 21.2.

Taken together with the results of conventional multiplication (Table 21.1) it can be seen that two clones, 44 and 68, remained stable throughout. These therefore, do appear to be homohistant, though there is still a slight risk that LI, which cannot be tested directly, may be normal and hence there could be reversion later.

Clones 27 and 42, found to be unstable in normal propagation and after re-irradiation, gave unclear results. In this small experiment their control clones were not significantly different from the standard. In clone 42, all the root-derived trees were larger than their control (though not significantly) as were most of the root-derived trees of clone 27. It appears in these cases that LIII may be normal while LII and, possibly, LI are compact. Clone 10 was represented by only one root-derived clone, which was similar to the standard and significantly larger than the control

**Table 21.2.** Mean height (*n* plants) of maiden trees of Bramley compact mutants grown from root or shoot propagation.

| | Shoot-derived | | | Root-derived | | | |
|---|---|---|---|---|---|---|---|
| Clone | *n* | Mean (cm) | Difference from "standard" | *n* | Mean (cm) | Difference from "standard" | "control" |
| Standard (EMLA) | 13 | 125.5 | | 13 | 124.2 | n.s. | |
| | | | | 11 | 133.3 | n.s. | |
| 10 | 6 | 78.2 | *** | 12 | 131.9 | n.s. | *** |
| 20 | 6 | 98.3 | *** | 7 | 138.0 | ** | *** |
| | | | | 9 | 130.0 | n.s. | *** |
| 27 | 6 | 116.2 | n.s. | 8 | 117.9 | * | n.s. |
| | | | | 11 | 132.3 | n.s. | * |
| | | | | 7 | 126.6 | n.s. | n.s. |
| | | | | 7 | 123.9 | n.s. | n.s. |
| | | | | 9 | | n.s. | n.s. |
| | | | | 9 | | ** | *** |
| 36 | 6 | 93.5 | *** | 11 | 116.2 | n.s. | *** |
| | | | | 14 | 128.2 | n.s. | *** |
| 42 | 6 | 117.5 | n.s. | 6 | 121.8 | n.s. | n.s. |
| | | | | 12 | 120.0 | n.s. | n.s. |
| | | | | 10 | 123.7 | n.s. | n.s. |
| | | | | 6 | 121.2 | n.s. | n.s. |
| | | | | 6 | 117.5 | n.s. | n.s. |
| 44 | 6 | 113.3 | * | 7 | 106.6 | *** | n.s. |
| | | | | 13 | 107.3 | *** | n.s. |
| | | | | 7 | 106.6 | *** | n.s. |
| | | | | 11 | 103.4 | *** | n.s. |
| | | | | 13 | 110.5 | *** | n.s. |
| | | | | 9 | 113.3 | *** | n.s. |
| | | | | 10 | 104.3 | *** | n.s. |
| | | | | 9 | 108.7 | *** | n.s. |
| | | | | 6 | 114.3 | * | n.s. |
| | | | | 8 | 118.5 | * | n.s. |
| 58 | 6 | 106.7 | *** | 13 | 126.9 | n.s. | *** |
| | | | | 12 | 126.2 | n.s. | *** |
| | | | | 8 | 128.6 | n.s. | *** |
| | | | | 8 | 130.1 | n.s. | *** |
| | | | | | | n.s | |
| 68 | 6 | 90.3 | *** | 11 | 118.5 | * | *** |
| | | | | 7 | 114.7 | n.s. | *** |
| | | | | 8 | 118.7 | n.s. | *** |

*, **, *** Indicate significantly different from control (i.e. same clone grown by conventional bud-grafting) or standard (normal EMLA Bramley grown by convention bud-grafting) at $P$ = 0.5, 0.1 or 0.01. Significance of each comparison by *t*-test.

mutant. This clone was unstable on normal propagation and appears to be of a similar histological constitution to clones 27 and 42.

The other three clones, 20, 36 and 58, which appeared to be stable compact forms under conventional multiplication and re-irradiation, produced normal-sized trees from root cuttings. Therefore, some deep layers of the mutants, perhaps LIII or lower, must have been normal but in a very stable configuration, while LII and possibly LI were compact. Changes to normal may have occurred under re-irradiaton but have been masked by damage symptoms; a further vegetative multiplication may have been worthwhile.

## 21.4 CONCLUSIONS

It appears, in the case of these intensively studied mutants of Bramley's Seedling apple, that useful mutants can readily be produced by irradiation, given a realistic objective (Campbell and Sparks, 1986). Other mutants produced by the programme included self-fertile mutants of Cox's Orange Pippin (Campbell and Lacey, 1982; Lacey *et al.*, 1982; Sparks and Campbell, 1986). However, due to the system used of treating multicellular initials with a mutagen, the majority of such clones will be chimeras, some so unstable as to be useless commercially (e.g. Bramley clones 10 and 27). Others may have sufficient stability to be commercially acceptable but warrant special care and checking of one-year-old trees during propagation (e.g. Bramley clones 20, 36 and 58). A few clones (e.g. Bramley clones 44 and 68) may be found that appear completely stable and hence homohistant and derived from a single mutated cell. Our final trial with root-derived plants appears to indicate that at least three discrete layers are involved in the chimeras.

From a practical point of view, this implies that programmes of mutation production must be large enough to enable, in the case of vegetatively propagated crops, the stable forms to be selected from within a range of useful mutants. For seed-propagated crops, unless single cells are the mutagen targets (i.e. pollen or egg mother-cells rather than seed) a further seed propagation must be used to screen out chimeras.

This caution could apply equally to the more recently developed techniques such as protoclone or somaclone production, where it is possible that the mutagenic event may occur during the multicellular callus stage. We have seen obvious periclinal and mericlinal chlorophyll mutants in the products of such a programme.

However, the fact remains that given an adequate population size, effective selection techniques and checks for chimerism, mutation breeding can be successful. There is no other way in which such useful trees as compact forms of Bramley or self-fertile Cox could be produced.

**REFERENCES**

Blazek, J. (1985). Spur type growth habit in apples. *Acta Hort.* **159**, 69–76.
Broertjes, C., and Van Harten, A. M. (1978). "Application of Mutation Breeding Methods in the Improvement of Vegetatively-Propagated Crops". Elsevier, Amsterdam, 316 pp.
Campbell, A. I. (1983). Apple tree health and quality. *In* "Apples and Pears" (Elspeth Napier, ed.), *Rep. R. hort. Soc. Conf., Lond.*, pp. 19–30.
Campbell, A. I., and Lacey, C. N. D (1982). Induced mutants of Cox's Orange Pippin apple with increased self-compatibility: 1. Production and selection, *Euphytica* **31**, 469–475.
Campbell, A. I., and Sparks, T. R. (1986). Compact mutants: production, selection and performance in Bramley's Seedling apple. *Acta Hort.* **180**, 11–18.
Child, R. D. (1975). Propagation of apples from adventitious shoots on root pieces. *Rep. Long Ashton Res. Stn for 1974*, p. 39.
Child, R. D., and Hughes, R. F. (1978). Factors influencing rooting in hardwood cuttings of apple cultivars. *Acta Hort.* **79**, 43–48.
Decourtye, L. (1967). Action des rayons gamma sur des variétés de poirier et de pommier en chimère. *Bull. Soc. Bot. France, Coll. Morphol. Exp. 1966*, 48–54.
della Strada, G., Fideghelli, C., Monstra, F., and Quatre, R. (1983). Mutants of the apple Annurca with compact habit and/or more intense fruit colour. *Riv. Frutti. Ortoflori.* **45**, 58–60.
Dermen, H. (1948). Chimeral apple sports and their propagation through adventitious buds. *J. Hered.* **39**, 235–242.
Dermen, H. (1965). Colchiploidy and histological imbalance in triploid apple and pear. *Am. J. Bot.* **52**, 235–242.
Faedi, W., and Rosati, P. (1985). First evaluation of apple mutants induced by gamma ray treatments. *Acta Hort.* **159**, 49–55.
Gilles, G. L., Bormans, H., and Van Laer, P. (1986). Clonal selection of M9 apple rootstock and influence of viruses on pomological value. *Acta Hort.* **180**, 61–67.
Grausland, J. (1986). Observations on clones of apple cultivars Belle de Boskoop, McIntosh and Cox's Orange Pippin. *Acta Hort.* **180**, 45–50.
Lacey, C. N. D. (1982a). Radiation induced mutants of apple trees. *Rep. Long Ashton Res. Stn for 1980*, pp. 192–207.
Lacey, C. N. D. (1982b). The stability of induced compact mutant clones of Bramley's Seedling apple. *Euphytica* **31**, 452–459.
Lacey, C. N. D., and Campbell, A. I. (1979). The positions of mutated sectors in shoots from irradiated buds of Bramley's Seedling apple. *Environ. Exptl. Bot.* **19**, 145–152.
Lacey, C. N. D., and Campbell, A. I. (1980). The characters of some selected mutant clones of Bramley's Seedling apple, and their stability during propagation. *Proc. Eucarpia Fruit Section Meeting on Fruit Tree Breeding, Angers, France, Sept., 1979*, pp. 301–306.
Lacey, C. N. D., Church, R. M., and Richardson, P. (1982). Induced mutants of

Cox's Orange Pippin with apparent increased self-compatibility: 2. Laboratory and orchard assessment and testing. *Euphytica* **31**, 511–518.

Lacey, C. N. D., Goodall, R. A., and Campbell, A. I. (1980). Induction and selection of mutant top fruit plants. *Rep. Long Ashton Res. Stn for 1979*, pp. 22–23.

Lapins, K. O., and Hough, L. F. (1970). Effect of gamma rays on apple and peach leaf buds at different stages of development: II. Injury to apical and axillary meristems and regeneration of shoot apices. *Radiat. Bot.* **10**, 59–68.

Li, Y. Z., Xu, A. B., Li, X. J., Wang, C. X., and Cui, D. C. (1986). Mutation induction in apple. *IAEA, Vienna, Mutat. Breed. Newslett.* **27**, 15.

Maas, V. (1970). Golden Delicious. *In* "North American Apples; Varieties, Rootstocks, Outlooks" (W. H. Upshall, ed.), pp. 69–85. Michigan State University Press.

Pratt, C. (1963). Radiation damage and recovery in diploid and cytochimeral varieties of apple. *Radiat. Bot.* **3**, 193–206.

Pratt, C., Way, R. D., and Ourecky, D. J. (1972). Reversion of sports of 'Rhode Island Greening' apple. *J. Am. Soc. hort. Sci.* **97**, 268–272.

Seth, J. N., Kuksal, R. P., and Joshi, R. P. (1982). The effect of ionizing radiation on the growth and production of young apple compact mutants. 1. Survival, growth, mutation frequency and their pre-selection. *Progr. Hort.* **14**, 94–98.

Silbereisen, R. (1986). Results of mutant tests in apple cultivars under different ecological conditions. *Acta Hort.* **180**, 35–44.

Sparks, T. R., and Campbell, A. I. (1986). The growth, flowering and cropping of four self-fertile clones of Cox's Orange Pippin. *Acta Hort.* **180**, 19–24.

Stewart, R. N., and Dermen, H. (1970). Determinations of number and mitotic activity of shoot apical initial cells by analysis of mericlinal chimaeras. *Am. J. Bot.* **57**, 816–826.

Tilney-Bassett, R. A. E. (1986). "Plant Chimeras". Edward Arnold, London, 199 pp.

Turnbull, J. (1983). Research and the commercial grower. *In* "Apples and Pears" (Elspeth Napier, ed.), *Rep. R. hort. Soc. Conf., Lond.*, pp. 88–96.

Wertheim, S. J., and van Oosten, H. J. (1986). Comparison of virus-free and virus-infected clones of two pear cultivars. *Acta Hort.* **180**, 51–60.

Yoshida, Y., Haninda, T., Tsuchiya, S., Sanada, T., Nishida, T., and Sadamori, S. (1981). Studies on the techniques of apple breeding: V. Induction of a bud sport of Fuji by radiation. *Bull. Fruit Tree Res. Sta., Yatabe, Ibarki, Jpn* **8**, 1–13.

# Part VI Improvement Through Tissue Culture

# 22 Application of Tissue Culture Techniques to Forest Trees

H. E. SOMMER[1], H. Y. WETZSTEIN[2] and S. A. MERKLE[1]

[1]*School of Forest Resources, University of Georgia, Athens, USA*
[2]*Department of Horticulture, University of Georgia, Athens, USA*

In general, tree improvement research involves three fundamental phases: selection, breeding and progeny testing. Tissue culture offers the potential of amplifying a genotype in any of these phases through the rapid production of clones.

At present, intensive efforts in tree improvement are confined to a relatively few conifers. Through tissue culture, it may be relatively easy to capture immediately the superior genotypes of other species, and either rapidly multiply or store them for future use. The same procedure could also be applied to trees with traits not currently included in present programmes and, through somaclonal variation, produce trees with new traits. But first we need to obtain some idea of the state of the art of tissue culture in relation to the propagation of juvenile and mature forest trees.

## 22.1 CRITERIA

Before we can examine to what extent tissue-culture technology is available to implement these and other applications, we need to state some criteria for a successful tissue-culture system. The first prerequisite for the application of tissue culture to a tree improvement scheme is a high-frequency system for organogenesis or embryogenesis. The end-product of the system must be plantlets or mature embryos. Secondly, the plantlets obtained must survive hardening-off, transplanting, and perform in the field in a manner comparable to that of seedlings, cuttings or grafts. Finally, the clones must remain true to type, or maintain introduced variations in a predictable manner.

IMPROVING VEGETATIVELY PROPAGATED CROPS
ISBN 0-12-041410-4

## 22.2 HISTORICAL DEVELOPMENT

The first reports of organogenesis in tissue culture of a forest tree were made by Gautheret (1940a,b). He found adventitious buds in cambial cultures of *Ulmus campestris* and described conditions for their differentiation and developmental anatomy. Later Jacquiot (1949, 1951, 1955a, b) reported further on this phenomenon, using trees that were up to 180 years of age. He determined the effects of exogenous auxins, adenine and inositol on bud differentiation and interpreted his results to indicate that organogenesis *in vitro* was controlled by at least two antagonistic factors. This work is remarkable in that it was done before the discovery of cytokinins. These studies laid the foundation for many of the experimental systems used today to obtain organogenesis in woody angiosperms.

It was not until 1968 that Wolter reported the formation of plantlets from *Populus tremuloides* callus. Subsequently, Winton (1968) cultured triploid *Populus tremuloides* callus, obtained shoots, and rooted the shoots. This was the first instance in which a superior genotype of a forest tree was propagated *in vitro*. This species is hard to root from cuttings. Since then, plantlets of over 120 species and hybrids of forest trees have been obtained *in vitro*. About 80% of these have been hardwoods. We will be able to consider only a few examples, to illustrate the progress that has been made toward their use in tree improvement and cover only a small fraction of the important papers in this field. For a further coverage, one can refer to the many reviews and other publications such as Boxus (1978) and Bonga and Durzan (1982).

## 22.3 TISSUE CULTURE OF HARDWOODS

At the School of Forest Resources in Georgia, we have a particular interest in the improvement of hardwoods as part of a short-rotation, high-intensity coppice forestry research programme. As part of this project, we have been investigating the potential of tissue culture to propagate selections. Much of our work has concentrated on sweetgum (*Liquidambar styraciflua*), primarily because it is a hard-to-root species, and selections are not easily multiplied or placed in a clone bank. It shows some of the problems as well as some of the promise that tissue culture has in forestry.

The first attempts with explants of sweetgum seedlings led to death of the explants. The cause appeared to be the use of Murashige and Skoog (MS) salts. With woody plants it is not unusual that MS medium must be modified or a substitute found (Anderson, 1975; Saito, 1979, 1980; Lloyd and McCown, 1980). Romberger has gone as far as to take 3 years to develop a medium for *Picea abies* shoot meristem cultures (Romberger et al., 1970).

To solve the problem encountered with MS medium, in one experiment we tried two basal salts, Blaydes' (Witham et al., 1971) and a modified Risser and White's (RW) (Sommer, 1983) and 20 combinations of benzyladenine (BA) and α-naphthalene acetic acid (NAA) with hypocotyl sections from six half-sib sources; a total of 2400 cultures. In 4–8 weeks, buds from three bottomland, but only one upland seed lot source, had started to differentiate on one medium: RW with 0.01 mg $\ell^{-1}$ NAA and 0.5 mg $\ell^{-1}$ BA. This gave us a first estimate of a medium for obtaining shoot organogenesis from hypocotyl sections of sweetgum seedlings. The shoots were small, many died after excision, and only some could be rooted. Multiplication rates were low; 2–5 buds per section. Our standard cytokinin had been BA, but in additional trials 2-isopentenyl-adenine (2iP) appeared to be as good or better. When cultures were transferred to a liquid nutrient they produced shoots 2–5 cm long which were harvested and placed on RW medium (Risser and White, 1964; Sommer, 1983). Shoot survival exceeded 90%, and rooting occurred spontaneously within 1 month. The cultures producing shoots could be repeatedly subcultured and in approximately 9 months the following numbers of shoots were obtained: Clone A, 228 shoots; Clone B, 171 shoots; Clone C, 335 shoots. In all cases the shoots exceeded 1 cm in length and rooted spontaneously on RW nutrient.

Shoot tips are another explant source readily available from sweetgum seedlings. However, no multiplication was obtained until we tried a technique developed for teak (Gupta et al., 1980). The use of both kinetin (KN) and BA appears to be particularly important for inducing the outgrowth of preformed buds of teak. With 1 mg $\ell^{-1}$ BA plus 1 mg $\ell^{-1}$ KN or 1 mg $\ell^{-1}$ BA plus 1 mg $\ell^{-1}$ 2iP, 78% and 71%, respectively, of the sweetgum bud cultures showed bud multiplication after 1 month. Again, the shoots produced were unusually small. A few buds gave multiple shoots. This work is continuing.

Another hardwood species which showed promise in earlier studies on short-rotation coppice forestry was yellow-poplar (*Liriodendron tulipifera* L.), which presents seed propagation problems due to low (2%) seed set. Culture establishment from post-dormancy buds have presented some of the same problems as encountered with sweetgum. However, when embryos, removed from seeds 1.5–2 months prior to seed maturity, are placed on a modified Blaydes' medium with 1000 mg $\ell^{-1}$ casein hydrolysate, 2 mg $\ell^{-1}$ 2,4-D and 0.25 mg $\ell^{-1}$ BA, a callus is formed. After subsequent monthly transfers on this medium a pale-yellow nodular callus appears. When this callus is subcultured to the same medium, but without 2,4-D and BA, cultures begin to produce somatic embryos in 1 or 2 months. Within 2 weeks they appear bipolar and within another 2 weeks greening of the cotyledons begins. While most of the embryos fail to germinate normally, some complete germination and form vigorous growing plantlets when transferred to RW medium. This system indicates that it may be possible to mass produce hardwood propagules *via* embryogenesis.

Embryogenic yellow-poplar lines can be grown as suspension cultures in the same Blaydes' medium without agar. These suspension cultures have proven to provide an excellent tissue for the preparation of protoplasts (Merkle and Sommer, unpublished). The purified protoplasts are suspended in callus-inducing medium supplemented with $CaCl_2.2H_2O$ (500 mg $\ell^{-1}$), xylose (250 mg $\ell^{-1}$), fructose (250 mg $\ell$),$^{-1}$ sucrose (40 g $\ell^{-1}$) and 0.5 M mannitol, mixed with agarose and cultured using the high density plating method of Binding and Kollman (1985). Cell wall regeneration is observed after 3 days, 75% of the protoplasts have divided after 7 days, microcalli have formed by day 21. Embryos differentiate after transfer to basal medium and viable plants are obtainable. Thus, yellow-poplar is the first North American forest tree to be regenerated from embryogenic callus following culture of the cells as protoplasts. Such a system opens the possibility of applying many of the techniques of genetic engineering to a forest tree.

Before concluding, we should mention two recent papers. Riffaud and Cornu (1981) have described a method to propagate *in vitro* mature selected *Prunus avium* trees with an average age of 70 years. Using buds from ground suckers and from the crown, bud multiplication was achieved and rooting was accomplished. Seventy trees are in culture, with 20 clones in the nursery. Srivastava and Steinhauer (1981) propagated mature *Betula pendula* from catkins. Initially, callusing and a few shoots were obtained but upon transfer to an extremely high salt medium, 78% of the cultures produced shoots and/or plantlets. Plantlets were transferred to soil. The authors indicate similar results can be obtained from *Alnus*, *Fagus* and *Quercus*. The use of floral parts as explants deserves greater investigation as a means of propagating mature trees by tissue culture. It appears that relatively soon, the propagation of many hardwoods will be possible both from juvenile and from mature trees.

## 22.4 TISSUE CULTURE OF CONIFERS

Conifers are, in general, of greater interest in many areas than are hardwoods. We will consider a few examples of what can be done. For thorough reviews, see Brown and Sommer (1974), Boxus (1978), and Thomas (1979) and David (1982).

*Picea abies* shows both the potential and limitations often encountered in the culturing of conifers. Von Arnold and Eriksson (1978, 1979a,b, 1981a, 1985) have studied organogenesis of Norway spruce using as explants, dormant buds from seedlings and mature trees, needles, and embryos excised from seed. These were cultured on a slightly modified LP medium (Von Arnold and Eriksson, 1977). The process of bud organogenesis was divided into two steps: induction of adventitious bud primordia and development of adventitious buds. For bud induction, $10^{-6}$ to $10^{-4}$ M 2iP was added to the basal LP medium. Primordia were visible after 3 weeks. At the lowest concentration ($10^{-6}$ M), scale-like organs differentiated only on the cotyledon tip, but at $5 \times 10^{-6}$ M, 10–20 scale-like organs differentiated on each of 75% of the embryos. With even higher concentrations ($10^{-4}$ M), tube-shaped protrusions began to appear until they dominated the surface. After 4–5 weeks, the embryos were transferred from induction medium to basal LP medium. Scale-like organs and tube-shaped protrusions differentiated into buds, and shoot formation was observed. These results are typical of what has been observed with embryos of several pines and Douglas fir. Though this study was better conducted than many others, the limitation is that we are at best, multiplying only a seed lot, rather than a selected tree.

Later, Von Arnold and Eriksson (1979a) reported a means for differentiating adventitious buds from dormant buds of Norway spruce. Buds were taken from 5–50-year-old trees and a 75-year-old hedge, at a height of 1 m where possible. Bud scales were removed prior to culture. Adventitious buds were induced on needle primordia by culturing the buds on a basal modified LP medium with $10^{-6}$ to $5 \times 10^{-5}$ M BA or 2iP. In 8 weeks, bud-like structures differentiated and were transferred to half-strength modified LP. After 2 weeks, adventitious buds were found on 10% of the buds from the 75-year-old hedge. In the case of 5–50-year-old trees from a mature stand, about 30% of the buds produced adventitious bud primordia.

The isolation of buds is a rather slow process, so needles were tried as explants. Needles, 1–3 mm long, from the upper half of flushing buds of 5-year-old spruce seedlings, were cut off near the stem and cultured on LP basal medium with $5 \times 10^{-6}$ to $10^{-5}$ M BA. The needles swelled, and bud primordia formed. The addition of 0.5–10 mM IBA or NAA increased the number of primordia. As with the embryos, two kinds of primordia formed: scale-like organs and tube-shaped protrusions. After 6 weeks on induction media, the needles were transferred to half-strength modified LP basal media, and buds developed on 5–15% of the needles. Isolated buds showed shoot growth. These three procedures are quite similar in that bud primordia are induced by cytokinin, buds develop after transfer to a basal medium, and excised buds could develop into shoots. All explants were either capable of starting, or were in a state of, rapid growth. However, only the isolated buds produced evidence for development of adventitious buds on explants from mature trees.

Mentions of rooting of conifer shoots are few. The differentiation of roots on adventitious buds of conifers has been reported in only a few cases, and is usually qualified by a statement that the rate of rooting varies with clone. An extensive discussion can be found in David (1982). Von Arnold (1982) found that about 5% of buds and shoots that differentiated on *Picea abies* embryos cultured on a half-strength modified MS medium rooted. Only one root per shoot was produced when spontaneous rooting occurred. When shoots were treated with $10^{-4}$ M indole butyric acid (IBA) and IAA for 24 hours, washed, and placed on half-strength basal nutrients, up to 25% rooting occurred with up to ten roots produced per shoot. The rooting percentage of shoots varied between shoots from different embryos. Further investigation on plantlet formation indicated that some trade-offs occurred between obtaining high multiplication rates, high-quality buds, successful rooting, as well as genetic influence of the source. Some of these trade-offs centre around the phenomenon known as vitrification and its manipulation to increase the survival of plantlets obtained from vitrified shoots (Von Arnold and Eriksson, 1984). Vitrified shoots or plantlets desiccate and do not survive glasshouse conditions. Adventitious shoots of *Picea abies* exhibit about 70% vitrification, making it a serious problem. Increasing agar concentrations from 0.5% to 2% eliminates vitrification of shoots, but also reduces shoot growth and the percentage rooting of shoots. The solution was to induce adventitious shoots on embryos on 0.5% agar medium and transfer the isolated shoots to 1% agar for growth and rooting.

A potentially more serious trade-off was found when examining genotype and its effect on the production of adventitious buds on needle primordia of *Picea abies* (Von Arnold, 1984). Four 26-year-old selections were rooted and grown in a phytotron for approximately 1 year during which three growing seasons were accomplished. Buds were harvested and adventitious buds produced on the needle primordia. Surprisingly, "trees with the highest ratings in growth and rooting potential gave the lowest yield of adventitious buds, and trees with the lowest ratings gave the highest yield of adventitious buds".

Recently, the development of somatic embryos in cultures of *Picea abies* has been reported (Hakman and Von Arnold, 1985; Hackman *et al.*, 1985). Immature embryos were placed on quarter-strength modified LP basal medium containing $10^{-5}$ M 2,4-D and $5 \times 10^{-6}$ M BA. Three types of callus were found. Two were green, one of which was competent for the differentiation of needles and buds. The third was a white friable callus obtained from the younger embryos. It had polarised structures resembling somatic embryos on its surface. The bipolar bodies were made up of long vacuolated cells, suggestive of the suspensor, that were subtended by clusters of smaller meristematic cells. As these meristematic clumps developed, they turned from white to yellow to green, eventually differentiating embryoids. Plantlets were obtained. In related work, Von Arnold and Hakman (1986) developed a method for obtaining embryogenic callus from mature embryos. When 10 μM 2,4-D and 5 μM BA were added to the media, the sucrose concentration became critical as basal media were changed. When LP medium (Von Arnold and Eriksson, 1977, 1981b) was used, the greatest percentage of embryogenic calli were obtained with 1% sucrose; when Norstog's medium 59 (Norstog and Rhamstine, 1967) was used, the greatest percentage of embryogenic calli was obtained with 3% sucrose. Thus, combined with the work of Simola and Honkanen (1983) and Nagmani and Bonga (1985), it appears that a reliable means of obtaining embryogenesis from a few conifers may soon be possible.

For older conifers, both Boulay (1977) and Ball (Ball *et al.*, 1978) used basal sprouts of coastal redwood as their explant material. According to some theories, these should be juvenile; they are much more responsive in culture than explants from other portions of the tree. Franclet *et al.* (1980) used 11-year-old *Pinus pinaster* and, after intensive fertilisation, were able to regenerate plantlets from needle bundles of the current year's growth. Various other methods including repeated grafting, pruning, and long-term culture with many transfers are other possible approaches to rejuvenation (Franclet, 1979).

## 22.5 ACCLIMATISATION AND PLANTING OUT

An important aspect which is often overlooked in descriptions of *in vitro* culture methodologies is the establishment of plantlets for planting out. Essential to any practical system of *in vitro* culture is not only a high multiplication rate of plantlet production from the explant, but also a high rate of survival during acclimatisation and in the field.

*In vitro* cultured plantlets are grown generally under high humidity and low light conditions. Plantlets removed from culture are very susceptible to wilting and desiccation (Anderson, 1975; McCown and Amos, 1979; Brainerd et al., 1981). Gradual acclimatisation periods of decreasing humidity are necessary for plants, particularly of hardwoods, to survive the transition from culture tube to glasshouse or field conditions.

Plantlets regenerated *in vitro* have a very modified leaf anatomy. We have described the leaf anatomy of cultured and non-cultured sweetgum, *Liquidambar styraciflua*, using light and electron microscopy (Wetzstein and Sommer, 1981, 1982, 1983; Wetzstein et al., 1981). Field-grown leaves showed a typically distinct palisade and spongy tissue with a high cellular density (Fig. 22.1a). Leaves of tissue-cultured plantlets, however, lacked a differentiated palisade parenchyma and had instead spongy parenchyma interspersed with large air spaces (Fig. 22.1b). New leaves from acclimatised plantlets showed an elongation of the upper mesophyll, similar to field leaves, and had fewer intercellular spaces than cultured plantlet leaves (Fig. 22.1c). Differences in stomatal configuration were also seen (Wetzstein and Sommer, 1983). Field-grown leaves (Fig. 22.2a) had ellipsoid guard cells at a level similar to adjacent epidermal cells. In contrast, leaves developed *in vitro* (Fig. 22.2b) had raised circular guard cells. Acclimatised plantlet stomata (Fig. 22.2c) were similar to those in field-grown leaves. Using transmission electron microscopy (Wetzstein and Sommer, 1982), leaves developed *in vitro* were seen to have flattened chloroplasts with an irregularly arranged internal membrane system (Fig. 22.3a), and lacked the organisation into grana and stroma lamellae found in acclimatised leaves (Fig. 22.3b). *In vitro* leaf cuticles were also not well developed compared with acclimatised leaves. Poorly defined mesophyll and/or reduced epicuticular wax has been reported in, for example, plum and apple leaves *in vitro* (Brainerd and Fuchigami, 1981; Brainerd et al., 1981; Fuchigami et al., 1981).

In a study of organogenesis from *Liquidambar styraciflua* callus, three root-like structures were found (Birchem et al., 1981): a typical root form with root hairs, a form without root hairs and a constricted apical region, and a form without a smooth exterior surface or root hairs. We have observed roots from sweetgum plantlets showing at least two morphological forms, and also differences in the number of roots and the number and length of laterals. It remains to be determined which root forms are functional and contribute to the establishment of the plantlet during acclimatisation.

The structural variations seen in the leaves and roots of tissues developed *in vitro* compared with acclimatised tissues suggest that during hardening-off extensive changes in the development of the leaves occur. The

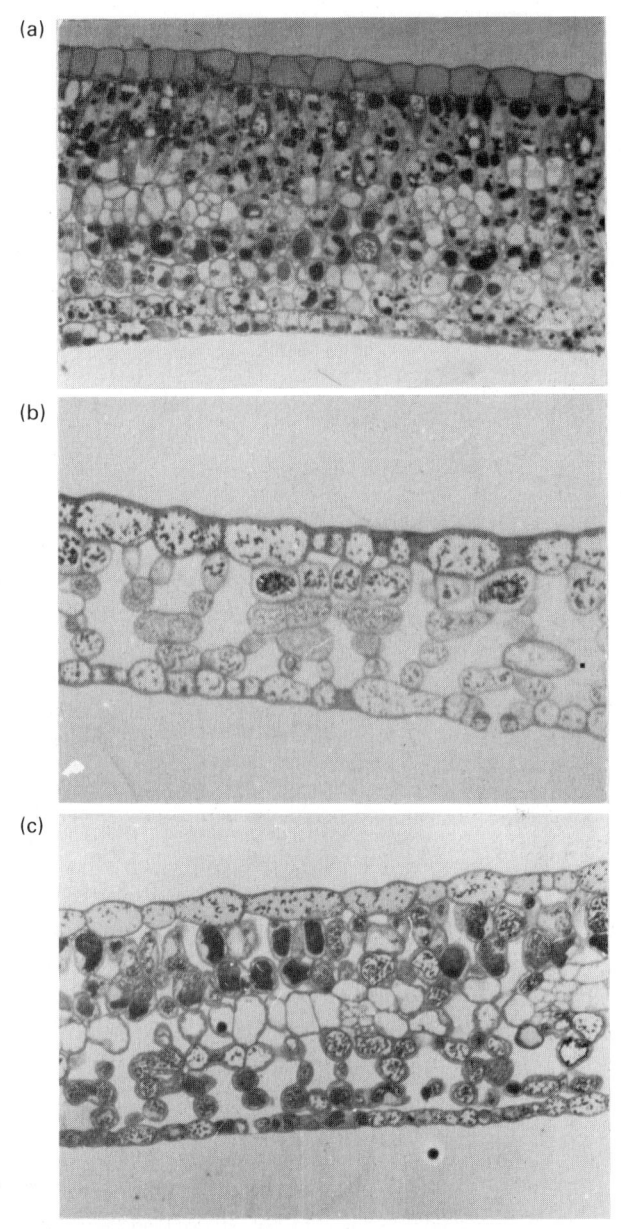

**Figure 22.1.** Cross-sections of *Liquidambar styraciflua* leaves: (a) field-grown leaf ×800; (b) cultured leaf with spongy lacunose mesophyll ×840; (c) acclimatised plantlet leaf showing palisade and spongy parenchyma, ×730.

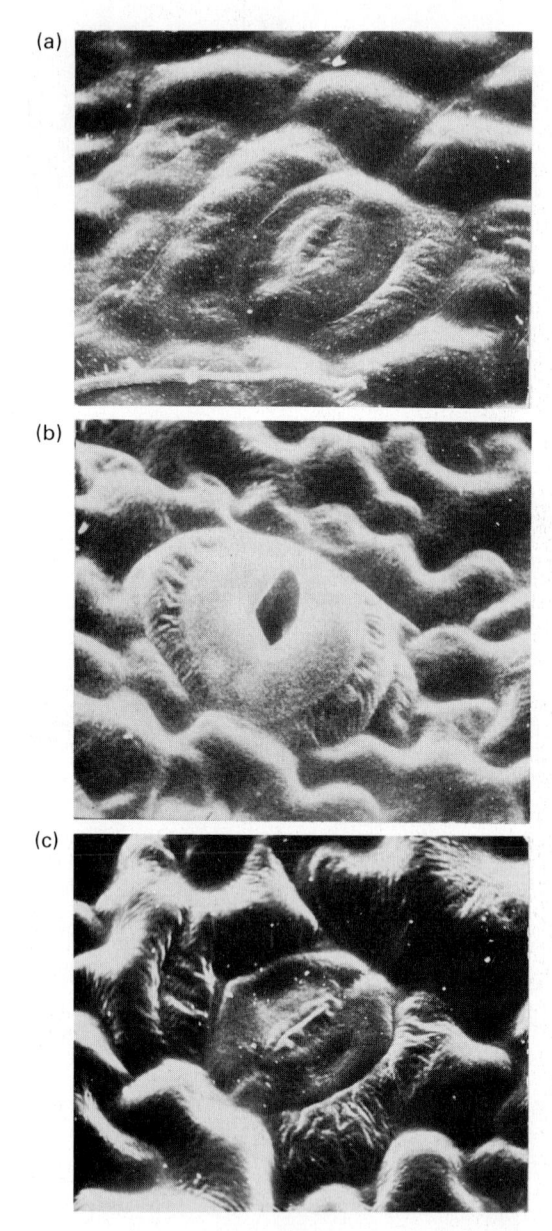

**Figure 22.2.** Scanning electron micrographs of abaxial surface of *Liquidambar styraciflua* leaves: (a) field-grown leaf, ×1550; (b) leaf developed *in vitro*, ×1460; (c) acclimatised leaf, ×1720.

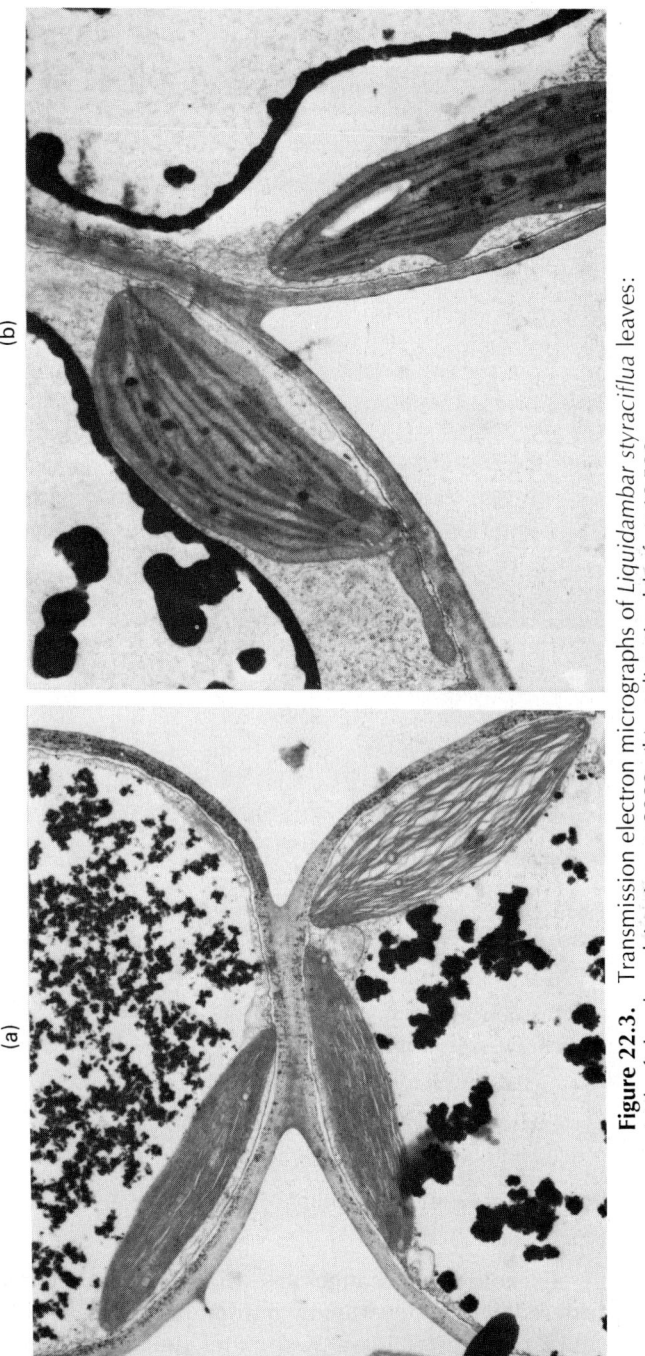

**Figure 22.3.** Transmission electron micrographs of *Liquidambar styraciflua* leaves: (a) leaf developed *in vitro*, ×9900; (b) acclimatised leaf, ×12 500.

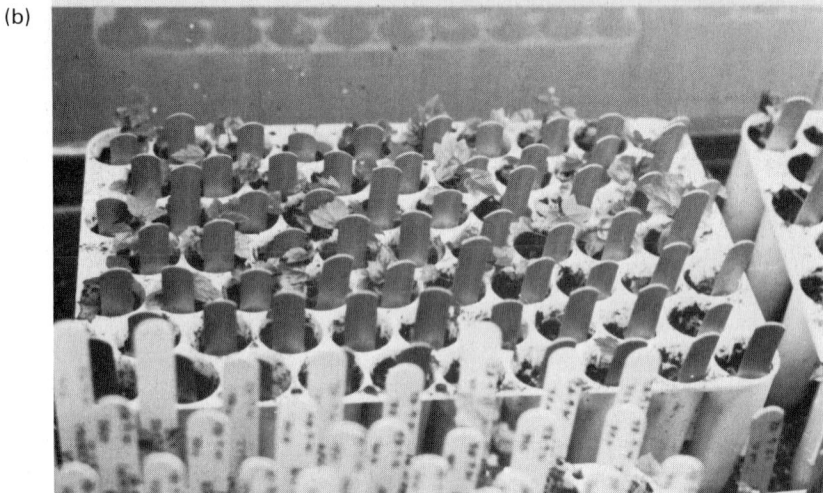

**Figure 22.4.** (a) Acclimatisation chamber, and (b) sweetgum plantlets transferred to soil for hardening-off.

anatomical and physiological changes associated with acclimatisation should be considered when developing hardening-off protocols. These include, the nature of plantlet leaves, stomatal functioning, root physiology, and photosynthetic activity associated with changes from heterotrophic to autotrophic conditions. For example, Skolmen and Mapes (1978) found that roots of *Acacia koa* plantlets formed in agar lacked root hairs and a fully developed vascular system. Their acclimatisation procedures included a stage to allow the roots to become functional. Some plantlets may lack a normal amount of pigmentation (Strode *et al.*, 1979), thus requiring a gradual adaptation to full sun. We have evaluated the effects of quantum flux density on photosynthesis and chloroplast ultrastructure in sweetgum plantlets and seedlings (Lee *et al.*, 1985). Cultured plantlets had appreciably higher photosynthetic rates than non-cultured seedlings grown under comparable light levels. Under our culture conditions, lack of photosynthetic capacity is not a limiting characteristic in plantlet transplant growth, although others have reported lack of autotrophy in cultured plantlets (Grout and Aston, 1978; Donnelly and Vidaver, 1984).

An additional consideration is the condition under which plantlet development occurs *in vitro*. From our studies we can only conclude that the modified leaf anatomy found in cultured plantlets is induced by specific environmental conditions in culture. During the culture period many environmental factors can be easily modified with the goal of producing plantlets which could be acclimatised more rapidly with lower mortality rates. For example, we have found that modification of the medium matrix has a pronounced effect on rooting percentage, rooting rate and shoot : root ratio in *Liquidambar* (Lee *et al.*, 1986).

An example illustrating the effectiveness of modifying *in vitro* and hardening conditions is that described for *Sequoia sempervirens* by Poissonnier *et al.* (1980). Factors for successful plantlet transfer, included an 8–15 day cold treatment (6°C) *in vitro* under light, a hormonal dip, type of potting medium, temperature, photoperiod and mineral fertilisation. Variability among plantlets can also be a major concern in the establishment of acclimatisation protocols (Anderson, 1978).

In the acclimatisation of *Liquidambar*, plantlet size has been found to be an important factor. In our system, 2.5 cm plantlets with two or three leaves survive with less wilting on transfer from tube to soil than 7–10 cm plantlets with more numerous leaves. Shoot : root ratios are greater in larger plantlets and may be a contributory factor. Plantlet age also has affected acclimatisation. Cultures held for longer periods of time prior to soil transfer incurred greater losses than younger cultures with similar development.

There are relatively few papers that report details of acclimatisation procedures or the degree of difficulty in acclimatising plantlets. Two systems most used are mist and humidity chambers. However, there are two major problems with a mist system, uneven distribution of the mist

(Greenwood et al., 1980) and disease problems, particularly in wet areas. Christie (1978) found he could harden-off 90% of his *Populus* plantlets using intermittent mist. High humidity is often maintained by covering the individual plantlets but this method does not adapt well to large production schedules.

*Liquidambar* has been successfully acclimatised using a modified indoor glasshouse chamber (Klima-Gro 5000) inside a temperature-controlled room (Fig. 22.4a). Plantlet trays (Fig. 22.4b) are placed on a bed of water-saturated vermiculite in which are embedded heating cables attached to an aluminium heat grid for even heat distribution. Fluorescent lights supply photoperiodic control, and temperature and humidity are monitored. Through the adjustment of chamber doors, a controlled gradual decrease in humidity is possible. Survival rates of up to 85% have been obtained. McCown (pers. comm.) has described a more sophisticated humidity chamber which he used in the propagation of several woody ornamentals. Poissonnier et al. (1980) have described a bed used in the rooting and acclimatisation of *Sequoia sempervirens*; air temperature and photoperiod are adjustable, soil temperature is controlled by heating coils and high humidity is maintained using polyethylene film tent covers, sometimes with a fogger.

The current status of research on acclimatisation of tissue-cultured plantlets shows an obvious need for more work. Applied studies have not produced a system that adapts satisfactorily to many situations. Critical areas have been recognised, however, such as the need for better humidity control, factors affecting plantlet development *in vitro*, light intensity and quality, photoperiod and plantlet variability. Future research needs to identify critical factors in the differentiation of a continuous functional vascular system, the establishment of autotrophy, and a balance between shoot and root growth.

Once the plantlets are hardened-off, there are still two more steps; field planting and performance. McKeand's (1982) plantings of loblolly pine plantlets grow slower than seedlings. He found that in RL Super Cells in a glasshouse, the root systems of loblolly plantlets were less balanced and less fibrous than those of seedlings. The apparent explanation may be that roots of the cultured plantlets grow down the length of the container and do not develop laterals until they are air-pruned at the bottom of the container. When planted out, the fibrous part of the plantlet root system tends to be deeper below the soil surface and may slow the growth of the plantlets. A glasshouse study of the initial growth of 82 loblolly plantlets compared with seedlings (Leach, 1979), showed that the absolute growth rate of the seedlings was greater in all cases. Moreover, the uniformity of height growth within the clone was no greater than that of seedlings of a half-sib family. These early studies with loblolly are only preliminary results. The North Carolina State Cooperative now has about 2000 plantlets under study to give a far better basis for evaluation (Frampton and Isik, 1986).

Several research groups now have field plantings of tissue culture plantlets. Of particular note are the plantings by AFOCEL of 39 000 *Sequoia sempervirens* representing 134 clones from trees aged 20–100 years. The National Plant Materials Centre at Palmerston North, New Zealand has produced hybrids of *Populus* as part of a rust-resistance programme. Of 19 species and hybrids they have planted from 10 to 14 000 plantlets of each as mother-stock. Other organisations such as the Industrial Forestry Cooperative, Weyerhaeuser Company, International Paper, and Simpson Timber Company each have planted over 1000 trees from tissue culture.

## 22.6 CONCLUSIONS

In many ways, we have presented a relatively conservative view of the role of tissue culture in forest tree improvement for, indeed, the proof of its value is still in the future. In addition, such techniques as the production of haploids *in vitro* or somatic hybridisation are still in their infancy in relation to forest trees. However, the past ten years has seen tissue culture of forest trees progress from a laboratory curiosity to an active field of research, not only in universities, but also within major public and private forestry research organisations.

Acknowledgements

This research was funded by Union Carbide Project 7860-XO2 with DOE, and State and Hatch funds allocated to the Georgia Agricultural Experiment Station and USDA competitive grants in forestry program.

## REFERENCES

Anderson, W. C. (1975). Propagation of rhododendrons by tissue culture: Part I. Development of a culture medium for multiplication of shoots. *Int. Pl. Prop. Soc. Comb. Proc.* **25**, 129–135.

Anderson, W.C. (1978). Rooting of tissue cultured rhododendrons. *Int. Pl. Prop. Soc. Comb. Proc.* **28**, 135–139.

Ball, E. A., Morris, D. M., and Rydelius, J. A. (1978). Cloning of *Sequoia sempervirens* from mature trees through tissue culture. *In* "Round-table

Conference: *In Vitro* Multiplication of Woody Species" (Ph. Boxus, ed.), pp. 181–226. CRA, Gembloux.

Binding, H., and Kollman, R. (1985). Regeneration of protoplasts. *In* "In vitro Techniques: Propagation and Long Term Storage" (A. Schäfer-Menhur, ed.), pp. 93–99. Martinus Nijhoff/Dr W. Junk, Dordrecht.

Birchem, R., Sommer, H. E., and Brown, C. L. (1981). Scanning electron microscopy of shoot and root development in sweetgum callus tissue cultures. *For. Sci.* **27**, 206–212.

Bonga, J. M., and Durzan, D. J. (1982). "Tissue Culture in Forestry". Martinus Nijhoff/Dr W. Junk, Dordrecht, Netherlands.

Boulay, M. (1977). Multiplication rapide du *Sequoia sempervirens* en culture in vitro. *Ann. Rech. Sylvicoles* (AFOCEL) 1977, 37–62.

Boxus, Ph. ed. (1978). "Round-table Conference: *In Vitro* Multiplication of Woody Species". CRA, Gembloux.

Brainerd, K. E., and Fuchigami, L. H. (1981). Acclimatization of aseptically cultured apple plants to low relative humidity. *J. Am. Soc. hort. Sci.* **106**, 515–518.

Brainerd, K. E., Fuchigami, L. H., Kwiatkowski, S., and Clark, C. S. (1981). Leaf anatomy and water stress of aseptically cultured "Pixy" plum grown under different environments. *HortScience.* **16**, 173–175.

Brown, C. L., and Sommer, H. E. (1974). "An Atlas of Gymnosperms Cultured *In Vitro*: 1924–1974". Georgia Forest Research Council, Macon, GA, 271 pp.

Christie, C. B. (1978). Rapid propagation of aspens and silver poplars using tissue culture techniques. *Int. Pl. Prop. Soc. Comb. Proc.* **28**, 255–260.

David, A. (1982). *In vitro* propagation of gymnosperms. *In* "Tissue Culture in Forestry" (J. M. Bonga and D. J. Durzan, eds), pp. 72–108. Martinus Nijhoff/Dr W. Junk, Dordrecht, Netherlands.

David, A., and Thomas, M. J. (1979). Organogenèse et multiplication végétative in vitro chez les gymnospermes. *L'année Biol.* **18**, 381–416.

Donnelly, D. J., and Vidaver, W. E. (1984). Pigment content and gas exchange in red raspberry *in vitro* and *ex vitro*. *J. Am. Soc. hort. Sci.* **109**, 177–181.

Frampton, L. J., Jr, and Isik, K. (1986). Comparison of field growth among loblolly pine seedlings and three plant types produced *in vitro*. *TAPPI Proc. R and D Conf.*, pp. 145–150.

Franclet, A. (1979). Rejeunissement des arbres adultes en vue de leur propagation végétative. *Etude et Recherches (AFOCEL)* **12**, 3–18.

Franclet, A., David, A., David, H., and Boulay, M. (1980). Première mise en évidence de Pin maritime (*Pinus pinaster* Sol.) *C. R. Acad. Sci. Paris* **290**, 927–929.

Fuchigami, L. H., Cheng, T. Y., and Soeldner, A. (1981). Abaxial transpiration and water loss in aseptically cultured plum. *J. Am. Soc. hort. Sci.* **106**, 519–522.

Gautheret, R. (1940a). Recherches sur le bouregeonnement du tissu cambial d'*Ulmus campestris*, cultivé *in vitro*. *C. R. Acad. Sci. Paris* **210**, 632–634.

Gautheret, R. (1940b). Nouvelles recherches sur le bouregeonnement du tissu cambial d'*Ulmus campestris* cultivé *in vitro*. *C. R. Acad. Sci. Paris* **210**, 744–746.

Greenwood, M. S., Marino, T. M., Meier, R. D., and Shahan, K. W. (1980). The role of mist and chemical treatments in rooting loblolly and shortleaf pine cuttings. *For. Sci.* **26**, 651–655.

Grout, B. W. W., and Aston, M. J. (1978). Transplanting of cauliflower plants regenerated from meristem culture: II. Carbon dioxide fixation and the

development of photosynthetic activity. *Hort. Res.* **17**, 65–71.
Gupta, P. K., Nadgir, A. L., Mascarenhas, A. F., and Jagannathan, V. (1980). Tissue culture of forest trees: clonal multiplication of *Tectona grandis* L. (teak) by tissue culture. *Pl. Sci. Lett.* **17**, 259–268.
Hakman, I., and Von Arnold, S. (1985). Plantlet regeneration through somatic embryogenesis in *Picea abies* (Norway Spruce). *J. Pl. Physiol.* **121**, 149–158.
Hakman, I., Fowke, L. C., Von Arnold, S., and Eriksson, T. (1985). The development of somatic embryos in tissue cultures initiated from immature embryos of *Picea abies* (Norway Spruce). *Pl. Sci.* **38**, 53–59.
Jacquiot, C. (1949). Observations sur la néoformation de bourgeons chez le tissu cambial d'*Ulmus campestris* cultivé *in vitro*. *C. R. Acad. Sci. Paris* **229**, 529–530.
Jacquiot, C. (1951). Action du mésoinositol et de l'adénine sur la formation de bourgeons par le tissu cambial d'*Ulmus campestris* cultivé *in vitro*. *C. R. Acad. Sci. Paris* **233**, 815–817.
Jacquiot, C. (1955a). Formation d'organes par le tissu cambial d'*Ulmus campestris* L. et de *Betula verrucosa* Gaertn. cultivés *in vitro*. *C. R. Acad. Sci. Paris* **240**, 557–558.
Jacquiot, C. (1955b). Sur le rôle des corrélations d'inhibition dans les phénonménes d'organogenése observés chez le tissue cambial, cultivé *in vitro*, de certains arbres. Incidences sur les problèmes du bouturage. *C. R. Acad. Sci. Paris* **241**, 1064–1066.
Leach, G. N. (1979). Growth in soil of plantlets produced by tissue culture, loblolly pine. *Tappi* **62**, 59–61.
Lee, N., Wetzstein, H. Y., and Sommer, H. E. (1985). Effects of quantum flux density on photosynthesis and chloroplast ultrastructure in tissue cultured plantlets and seedlings of *Liquidambar styraciflua* L. towards improved acclimatization and field survival. *Pl. Physiol.* **78**, 637–641.
Lee, N., Wetzstein, H. Y., and Sommer, H. E. (1986). The effect of agar vs. liquid medium on rooting in tissue-cultured sweetgum. *HortScience* **21**, 317–318.
Lloyd, G. and McCown, B. (1980). Commercially-feasible micropropagation of mountain laurel, *Kalmia latifolia*, by use of shoot tip culture. *Int. Pl. Prop. Soc. Comb. Proc.* **30**, 421–427.
McCown, B., and Amos, R. (1979). Initial trials with commercial micropropagation of birch selections. *Int. Pl. Prop. Soc. Comb. Proc.* **29**, 387–393.
McKeand, S. E. (1982). Root morphology of loblolly pine tissue culture plantlets. 7th N. Am. For. Biol. Workshop on "Physiology & Genetics of Intensive Culture", July 1982, Univ. Kentucky, Lexington, Kentucky.
Nagmani, R., and Bonga, J. M. (1985). Embryogenesis in sub-cultured callus of *Larix decidua*. *Can. J. For. Res.* **15**, 1088–1091.
Norstog, K., and Rhamstine, (1967). Isolation and culture of haploid and diploid cycad tissue. *Phytomorphology* **17**, 374–381.
Poissonnier, M., Franclet, A., Dumant, M. J., and Gautry, J. Y. (1980). Enracinement de tigelles *in vitro* de *Sequoia sempervirens*. *Ann. Rech. Sylvicoles* (AFOCEL) 1980, 230–253.
Riffaud, J. L., and Cornu, D. (1981). Utilisation de la culture *in vitro* pour la multiplication de merisiers adultes (*Prunus avium* L.) selectionnés en forêt. *Agronomie* **1**, 633–640.
Risser, P. G., and White, P. R. (1964). Nutritional requirements of spruce tumor cells *in vitro*. *Physiol. Plant.* **17**, 620–635.
Romberger, J. A., Varnell, R. J., and Tabor, C. A. (1970). Culture of apical

meristem and embryonic shoots of *Picea abies*—approach and techniques. *Tech. Bull. No. 1409.* US Dept. Agric. Forest Service.
Saito, A. (1979). Effects of inorganic elements on somatic callus culture in *Cryptomeria japonica*. *J. Jap. For. Soc.* **61**, 457–458.
Saito, A. (1980). Effects of inorganic elements in the medium on shoot differentiation from populus callus. *J. Jap. For. Soc.* **64**, 147–149.
Simola, L. K., and Honkanen, J. (1983). Organogenesis and fine structure in megagametophyte phytic callus lines of *Picea abies*. *Physiol. Plant.* **59**, 551–561.
Skolmen, R. C., and Mapes, M. O. (1978). Aftercare procedures required for field survival of tissue culture propagated *Acacia koa*. *Int. Pl. Prop. Soc. Comb. Proc.* **28**, 156–164.
Sommer, H. E. (1983). Organogenesis in woody angiosperms: Application to vegetative propagation. *Bull. Soc. Bot. Fr., Actual. Bot.* **130**, 79–85.
Srivastava, P. S., and Steinhauer, A. (1981). Regeneration of birch plants from catkin tissue cultures. *Pl. Sci. Lett.* **22**, 379–386.
Strode, R. E., Travers, P. A., and Oglesby, R. P. (1979). Commercial micropropagation of rhododendrons. *Int. Pl. Prop. Soc. Comb. Proc.* **29**, 439–443.
Von Arnold, S. (1984). Importance of genotype on the potential for *in vitro* adventitious bud production of *Picea abies*. *For. Sci.* **30**, 314–318.
Von Arnold, S. (1982). Factors influencing formation, development and rooting of adventitious shoots from embryos of *Picea abies* (L.) Karst. *Pl. Sci. Letts* **27**, 275–287.
Von Arnold, S. and Eriksson, T. (1977). A revised medium for growth of pea mesophyll protoplasts. *Physiol. Plant.* **39**, 257–260.
Von Arnold, S., and Eriksson, T. (1978). Induction of adventitious buds on embryos of Norway spruce grown *in vitro*. *Physiol. Plant.* **44**, 283–287.
Von Arnold, S., and Eriksson, T. (1979a). Induction of adventitious buds of Norway spruce (*Picea abies*) grown *in vitro*. *Physiol. Plant.* **45**, 29–34.
Von Arnold, S., and Eriksson, T. (1979b). Bud induction on isolated needles of Norway spruce (*Picea abies* L., Karst) grown *in vitro*. *Pl. Sci. Lett.* **15**, 363–372.
Von Arnold, S., and Eriksson, T. (1981a). Production of adventitious plants from spruce and pine. Symposium on Clonal Forestry, Research Notes 32, pp. 7–31, Dept. of Forest Genetics, Swedish Univ. Agric. Sci., Uppsala.
Von Arnold, S., and Eriksson, T. (1981b). *In vitro* studies of adventitious shoot formation in *Pinus contorta*, *Can. J. Bot.* **59**, 870–874.
Von Arnold, S., and Eriksson, T. (1984). Effect of agar concentration on growth and anatomy of adventitious shoots of *Picea abies* (L.) Karst. *Pl. Cell. Tiss. Organ Cult.* **3**, 257–264.
Von Arnold, S., and Eriksson, T. (1985). Initial stages in the course of adventitious bud formation on embryos of *Picea abies*. *Physiol. Plant.* **64**, 41–47.
Von Arnold, S., and Hakman, I. (1986). Effect of sucrose on initiation of embryogenic callus cultures from mature zygotic embryos of *Picea abies* (L.) Karst. (Norway spruce). *J. Pl. Physiol.* **122**, 261–265.
Wetzstein, H. Y., and Sommer, H. E. (1981). Transmission electron microscopy of cultured and acclimated *Liquidambar styraciflua*. *HortScience* **16**, 405.
Wetzstein, H. Y., and Sommer, H. E. (1982). Leaf anatomy of tissue cultured *Liquidambar styraciflua* (Hamamelidaceae) during acclimatization. *Am. J. Bot.* **69**, 1579–1586.
Wetzstein, H. Y., and Sommer, H. E. (1983). Scanning electron microscopy of *in vitro* cultured *Liquidambar styraciflua* plantlets during acclimatization. *J. Am. Soc. hort. Sci.* **108**, 475–480.

Wetzstein, H. Y., Sommer, H. E., Brown, C. L., and Vines, H. M. (1981). Anatomical changes in tissue cultured sweetgum leaves during the hardening-off period. *HortScience* **16**, 290.

Witham, F. H., Blaydes, D. F., and Devlin, R. M. (1971). "Experiments in Plant Physiology". Van Nostrand Reinhold, New York.

Winton, L. L. (1968). Plantlets from aspen tissue cultures. *Science* **160**, 1234–1235.

Wolter, K. E. (1968). Root and shoot initiation in aspen callus cultures. *Nature* **219**, 509–510.

# 23 Clonal Propagation of Plantation Crops

L. H. JONES

*Unilever Research Laboratory, Colworth House, Sharnbrook, Bedford*

Aseptic culture techniques are helping plant breeders and agronomists in their task of improving crop productivity (Conger, 1981). This chapter considers the use of clonal propagation for multiplication of plantation crops, in particular the commercial applications, and discusses some of the factors to take into account beyond the technical feasibility of plant culture *in vitro*.

## 23.1 USES OF TISSUE CULTURE

Tissue-culture methods can be applied in three main ways:

(1) The direct propagation of selected genotypes to produce clonal plants for sale to growers.
(2) The use of aseptic techniques to introduce some new variation which can be exploited in the form of an improved cultivar. In this case the tissue culture step may be a once-only operation, and subsequent multiplications might be by seed production or by conventional vegetative methods or, indeed, by further aseptic propagation.
(3) Tissue culture may be used to remove viral pathogens. The virus-free stocks must then be multiplied to produce commercially useful numbers, either by conventional methods or by further tissue culture.

Although the second and third approaches may be combined with the first, they are quite different in kind. They require small, specialised laboratories with a few highly trained people, but are not concerned with large-scale propagation. The first application requires the large-scale routine production of plants using relatively untrained labour without loss of hygiene or quality. In this chapter, I shall concentrate on the use of mass propagation techniques to produce plants for direct planting as a productive crop.

I shall not cover the use of *in vitro* techniques for plant breeding, which are reviewed elsewhere (see Vasil *et al.*, 1979; Walbot, 1981; Evans *et al.*, 1983) and in this volume.

## 23.2 SOME COMMERCIAL CONSIDERATIONS

In deciding whether tissue-culture propagation methods can assist in the improvement of any crop, it is essential to evaluate first what benefits, if any, they can provide over existing techniques.

Whatever views are expressed to the contrary, setting up a commercial tissue-culture laboratory is expensive. The evolution of a laboratory protocol for the successful production of test-tube plants is only the first step and does not provide a recipe for guaranteed continuous production of uniformly high quality plants, ready for planting by a grower at a price he can afford. Not only does tissue culture have to provide an advantage, that advantage has also to pay for the extra costs of production and for developing the new technology, which must be reflected in a premium price for the plants. It must also give the power a significant improvement in profitability over traditional planting material.

The first three questions therefore are whether there are significant existing limitations on the availability or quality of planting material, how much improvement could be expected by the application of tissue culture, and whether that improvement would be sufficiently great to command an economic price for the plants.

The size of the market and its structure is also an important early consideration. One of the major advantages of tissue culture is the increased rate of multiplication compared with conventional vegetative methods. It is pointless setting up a tissue-culture laboratory if the demand for plants is only a few thousand per year and cuttings are readily available. On the other hand, annual crops such as legumes planted at densities of, say, 100 plants per m$^2$ (1 million ha$^{-1}$) or cereals, planted at even greater densities, are out of the question for direct vegetative propagation; first, because of their low unit value, and secondly, because of the impracticability of producing such large numbers of plants using current tissue culture methods. For annual crops there is the added problem that the bulk of plants is required for planting over a very short season, whereas the propagation laboratory must spread its production throughout the year. Conversely, long-lived perennial specialist crops such as spices may be of high value, but are required in insufficient numbers to give tissue culture any advantage over rooted cuttings or seed.

A commercial unit is thus limited to crops where there is a market of between $10^5$ and $10^7$ plants per year, and the market has to be accessible. Wide fragmentation of growing areas over different countries with attendant problems of transfer of material across frontiers, extensive smallholder plantings rather than large plantations, and the need for different cultivars in different regions can dramatically reduce the market available to any individual propagator. The market has to be created by convincing the growers that the new clones are a significant improvement on his existing planting material and will repay the premium price. Growers must be in a position to buy and manage the improved material, and the subsistence farmer will need some subsidy to enable him to exploit improved material.

At a meeting on this subject at the Royal Irish Academy (Cassells and Kavanagh, 1983), it was clear in discussion that local strawberry growers were far from convinced that virus-indexed strawberry plants from micropropagation, costing 10–20 pence per plant, could possibly compete with traditional sources of strawberry runners at 3–4 per pence per plant, yet some European strawberry producers now use exclusively micropropagated material, and clearly find the use of virus-indexed, high-quality plants worthwhile.

## 23.3 ADVANTAGES OF CLONES

The advantages of clones over variable seedling progenies are widely recognised and exploited in many different crops. In crops where vegetative propagation is either difficult or impossible by conventional means, tissue culture has a clear advantage, by permitting the development of stable phenotypically uniform clones with well-described varietal characteristics. Provided it is possible to create and identify new genetic recombinants with improved features, clonal propagation can stabilise these unique genotypes in the form of new cultivars, with not only improved yield, but also better product quality and agronomic characters including disease resistance, drought tolerance and fertiliser economy.

## 23.4 SCALING-UP OF THE LABORATORY PROCESS

The two main requirements for successful micropropagation are access to the best genetic sources for propagation (and ability to identify the best individuals) and the ability to propagate those individuals in the required numbers at a competitive price and at uniformly high quality. Translating an initial demonstration that a plant can be micropropagated into commercial reality is both costly and time-consuming. In most cases each clone requires individually optimised media, environment and transfer sequences to obtain a useable multiplication rate, and also optimisation of rooting, hardening and transplanting stages to minimise transfer losses.

As well as biological optimisation, it is also essential to work out a cost-effective operation system for a production unit to maintain efficient flows of appropriate media to the transfer operators, to schedule each day's work and to optimise the rate of plantlet output. Many useful suggestions are made by de Fossard and Bourne (1977). It is essential to maintain adequate hygiene and establish a rigorous quality control system to detect contaminated cultures before they reach the production line. Safeguards are required against mites and insects as well as the more familiar microbial contaminants. It is also necessary to build-in sufficient safeguards against contingencies such as loss of regeneration potential of the plant material, genetic drift, power failures, industrial disputes and impure chemicals. Close attention to detail is essential at all stages if a commercial operation is to be both profitable and provide the grower with the plant material he wants to grow.

It is this R and D work to scale-up a laboratory technique that imposes one of the major constraints on the application of plant tissue-culture methods and limits their effective application to relatively few high-value species lacking alternative propagation methods.

## 23.5 APPLICATION OF PROPAGATION *IN VITRO* TO PLANTATION CROPS

As examples I shall discuss first the application of tissue culture to improvement of the oil palm, and then will examine briefly the prospects of use of tissue-culture methods on some other plantation crops.

### 23.5.1 Oil palm (*Elaeis guineensis* Jacq.)

Until we were able to recover plants from cultures, there was no vegetative propagation method available for oil palm. The use of this technique now enables us to select individual palms of high performance and to create new palm cultivars. The breeding and selection methods required to produce and identify palms for propagation are discussed by Hardon *et al.* (Chapter 3, this volume).

The advantages of clonal material should be reflected primarily in improved yield per hectare, conservatively estimated at a 30% improvement over seedlings in selections from currently available material. In the long term, further oil yield improvements up to an upper limit of perhaps 14 t ha$^{-1}$ per annum can be sought (Hardon et al., Chapter 3, this volume).

Results are now available from the first clones planted in field trials (Corley et al., 1982). These clones were propagated from unselected seedlings, rather than from selected elite palms. Table 23.1 gives the yield data for the first 36 months from planting. There is a wide range of performance, as would be expected, from unselected material. The key information obtained from these trials lies in the confirmation that individual clones have much less variability than seeding populations (Corley, 1982) (Table 23.2). At least two of these unselected clones are promising as potential commercial lines, but emphasis is now being directed at selecting high-quality palms as ortets for propagation. Over 40 clones from selected ortets are now in production for trials and have been planted in extensive clone trials in a number of different countries. From these trials the clones best adapted to growth in each region can be selected for commercial planting.

Table 23.3 (from Corley, 1985) shows the performance of the best clones in several trials in Malaysia relative to mixed seedling controls.

In addition to improvements in yield, we can look for a whole range of agronomic factors, such as drought and disease resistance, ease of harvesting, early fruiting (precocity), regional adaption, economy of fertiliser use, and quality factors, such as the fatty acid composition of the oil, carotene content and triglyceride composition. Hitherto, selection for such characteristics has not been possible and tissue culture in this case will greatly increase the flexibility of the crop both in the field management and end-product use. For example, we now have several years' data from the first fruiting oil palm clones and clearly different patterns of oil composition in different clones are observed (Table 23.4).

The range of variation obtained in this small, unselected sample tells us that oil composition is under close genetic control and suggests that improvements could be obtained by appropriate breeding and selection. Fatty acid composition is also modified by environmental effects, and comparison of three clones grown in Malaysia and Cameroon showed a consistently higher level of longer-chain fatty acids in Malaysian oils. In particular there was a more than proportional increase in linolenic acid (Jones, 1984). Of course, yield cannot be sacrificed unless there is a high premium price for quality of oil. The chance of finding the ideal combination of desirable characters by randomly screening plantation

**Table 23.1.** Yields, up to 36 months after field planting, of clones in 1977 and 1978 plantings[a] at Pamol Estate, Johore, Malaysia (clones are ranked by oil yield).

| Clone | 926 | 905 | 997 | 932 | Seedling control | 924 | 931 | 970 | 907 | 975 | 976 | 949 | 960 | 939 |
|---|---|---|---|---|---|---|---|---|---|---|---|---|---|---|
| Number of palms | 34 | 3 | 14 | | 30 | 10 | 20 | 3 | 5 | 11 | 2 | 10 | 9 | 3 |
| Oil yield (kg per palm) | 9.63 | 8.39 | 7.99 | 6.64 | 5.62 | 5.01 | 4.43 | 3.32 | 3.27 | 2.89 | 2.21 | 1.93 | 0.95 | 0.67 |
| Kernel yield (kg per palm) | 1.53 | 3.41 | 1.36 | 4.24 | 3.26 | 0.86 | 1.12 | 1.45 | 0.29 | 1.24 | 2.75 | 0.83 | 0.41 | 1.21 |
| Oil + kernel (kg per palm) | 11.16 | 11.80 | 9.35 | 10.88 | 8.88 | 5.87 | 5.55 | 4.77 | 3.56 | 4.13 | 4.96 | 2.76 | 1.36 | 1.88 |

*Source:* Corley (1982).
[a] Results from three separate plantings. Clone 926 was included in all three; yields of other clones were expressed as a percentage of that of clone 926 in each planting, weighted mean percentages were calculated, and the overall yield for each clone estimated from this percentage and the mean yield of clone 926. The correlation between the yields of four clones in two different plantings was $r = 0.972$ ($P < 0.05$, 2 $df$).

**Table 23.2.** Analysis of variance for fruit characters of five clones (Pamol Estate, clone trial 3, replicate 1, March–May 1981).

| | DF | Fruit per bunch | Mesocarp per fruit | Dry matter in mesocarp | Oil per dry mesocarp | Kernel per fruit | Shell per fruit | Oil per bunch | Kernel per bunch |
|---|---|---|---|---|---|---|---|---|---|
| *Mean square* | | | | | | | | | |
| Between clones | 4 | 607.46 | 668.89 | 275.38 | 339.08 | 38.39 | 253.85 | 125.75 | 26.05 |
| Between palms within clones | 32 | 58.04 | 9.14 | 79.93 | 16.90 | 1.77 | 2.22 | 13.60 | 0.87 |
| Between bunches within palms | 147 | 54.95 | 5.35 | 54.52 | 21.17 | 1.12 | 1.01 | 12.97 | 0.79 |
| Between seedlings | 7 | 152.73 | 72.58 | 249.10 | 129.77 | 18.33 | 18.35 | 60.06 | 11.31 |
| Between bunches within seedlings | 24 | 33.79 | 5.06 | 78.45 | 61.52 | 1.65 | 0.82 | 17.59 | 1.04 |
| *Variance ratios* | | | | | | | | | |
| Between clones/ between palms | | 10.47*** | 73.22*** | 3.44* | 20.07*** | 21.66*** | 114.20*** | 9.25*** | 29.85*** |
| Between palms/ within palms | | 1.06 | 1.71* | 1.47 | 0.80 | 1.59* | 2.21** | 1.05 | 1.11 |
| Between seedlings/ within seedlings | | 4.52** | 14.33*** | 3.18* | 2.11 | 11.09*** | 22.45*** | 3.41* | 10.91*** |
| Between seedlings/ between clonal palms | | 2.63* | 7.94*** | 3.12* | 7.68*** | 10.34*** | 8.25*** | 4.42** | 12.96*** |

Source: Corley (1982).
*, $P < 0.05$; **, $P < 0.01$; ***, $P < 0.001$.

**Table 23.3.** Performance of the best clones in several trials in Malaysia, relative to mixed seedling controls.

| Trial | Best Clone | Period of recording | Yield as % seedlings Fruit | Oil |
|---|---|---|---|---|
| PCT3 | 997 | 3 years | | 125 |
| PCT5 | 997 | 1 year | 101 | |
| PCT9 | 997 | 1 year | 145 | |
| PCT11 | 115E | 1 year | | 122 |
| PCT14 | 54A | 8 months | 210 | |
| HCT2 | 90A | 1 year | | 161 |
| HCT3 | 90A | 8 months | 127 | |
| HCT4 | 115E | 10 months | 119 | |
| HCT5 | 54A | 9 months | 113 | |
| HCT6 | 92A | 11 months | | 224 |
| HCT7 | 115E | 9 months | | 215 |
| HCT8 | 54A | 11 months | | 159 |

Source: Corley (1985).

palms is extremely low as well as costly. Only by careful selection of the appropriate parents followed by a judicious series of crosses can the desired characters be combined. It then becomes worthwhile to search within the segregating progenies for improved individual phenotypes that can be cloned.

Development of novel types of palm oil is undoubtedly possible. Whether it is economically worthwhile depends on the future price premium for specialist oils, offset against the risk of planting large blocks of a single clones and maintaining separate processing transport and storage facilities for the oil. Oils and fats technology is also advancing rapidly and there is a high level of interchangeability of oils from different sources so that current vegetable-oil prices show relatively little variation for different sources.

We are currently testing a range of clones at different planting densities. Some clones have been planted in both Cameroon and Malaysia to check for genotype-environment interactions. Early results show that the best clone in Malaysia also outyields the others in Cameroon. On the other hand, one of the poorer clones has shown a marked susceptibility to *Cercospora* disease in Africa and has been rejected. We are looking, in African material particularly, for resistance to *Fusarium* wilt, and all clones for potential African use are being tested in wilt-endemic areas.

**Table 23.4.** Composition of oil from 11 clones (2–4 palms per clone, mean of 8 samples per palm).

| Clone | Fatty acids (%) | | | | Iodine value | Triglycerides (%) | | | Carotene (ppm) |
|---|---|---|---|---|---|---|---|---|---|
| | C16:0 | C18:0 | C18:1 | C18:2 | | C50 | C52 | C54 | |
| 905 | 39.9 | 5.8 | 42.8 | 9.8 | 57.2 | 34.6 | 43.1 | 13.3 | 1344 |
| 907 | 42.2 | 4.6 | 35.0 | 16.6 | 62.3 | 41.2 | 40.8 | 9.6 | 357 |
| 924 | 40.3 | 6.2 | 38.3 | 13.0 | 59.0 | 35.6 | 42.0 | 12.2 | 1056 |
| 926 | 42.9 | 5.3 | 36.6 | 13.1 | 57.4 | 40.5 | 39.6 | 9.2 | 640 |
| 931 | 38.9 | 6.6 | 43.0 | 10.0 | 57.7 | 34.4 | 44.1 | 13.9 | 1158 |
| 932 | 42.6 | 5.3 | 40.3 | 9.5 | 54.5 | 39.1 | 40.2 | 9.5 | 1289 |
| 960 | 42.6 | 5.6 | 40.3 | 9.3 | 54.1 | 39.4 | 41.4 | 9.9 | 1268 |
| 970 | 38.9 | 5.8 | 41.9 | 12.2 | 60.7 | 34.5 | 44.8 | 13.7 | 634 |
| 975 | 41.5 | 5.8 | 41.2 | 9.5 | 55.2 | 37.3 | 42.7 | 11.3 | 1093 |
| 976 | 42.9 | 5.1 | 37.4 | 12.7 | 57.6 | 41.3 | 40.2 | 8.8 | 583 |
| 997 | 45.8 | 5.5 | 35.5 | 11.5 | 52.8 | 43.3 | 38.6 | 7.6 | 1204 |
| S.D. | 1.4 | 0.6 | 1.2 | 0.8 | 3.4 | 1.9 | 1.9 | 1.4 | 188 |

Source: Corley (1982).

**Table 23.5.** Mean values of fruit characters for various clones.[a]

| Clone | 905 | 926 | 932 | 939 | 997 |
|---|---|---|---|---|---|
| Fruit set (%) | 56.7 | 46.1 | 78.2 | 47.0 | 48.4 |
| Fruit per bunch (%) | 55.5 | 56.0 | 71.7 | 48.2 | 51.3 |
| Mesocarp per fruit (%) | 79.4 | 88.5 | 74.7 | 54.8 | 90.8 |
| Kernel per fruit (%) | 11.6 | 4.7 | 9.4 | 19.1 | 4.7 |
| Shell per fruit (%) | 5.0 | 5.5 | 13.4 | 14.8 | 3.7 |

Analysis of variance.

|  | Between clones | Between palms | Between bunches |
|---|---|---|---|
| Fruit per bunch (%) | 2006* | 170.5 | 107.3 |
| Mesocarp per fruit (%) | 1223*** | 10.6 | 5.3 |
| Dry per fresh mesocarp (%) | 449* | 88.3 | 45.1 |
| Oil per mesocarp (%) | 32 | 22.2 | 22.9 |
| Kernel per fruit (%) | 109*** | 2.7 | 1.6 |
| Shell per fruit (%) | 802*** | 2.5 | 1.1 |
| Degrees of freedom | 2 | 8 | 22 |

Source: Wooi, Wong and Corley (1983).
[a] Data from three clones, 3 or 4 palms/clone, 3 bunches/palm; 42 months after field planting.
*, $P < 0.05$; ***, $P < 0.001$.

A new clone can only be marketed after being fully proven in the areas where it will be grown, and growers are convinced that their extra planting costs will be quickly paid off to give early returns on their investment.

23.5.1.1 Uniformity, distinctness and stability of clones

Evidence from the field trials to date shows a high degree of within-clone uniformity for many vegetative and fruit characters (Corley, 1982; Corley et al., 1982). The vegetative characteristics of individual clones are readily apparent in their visual appearance, although difficult to quantify. Other characters which are more subject to environmental modification, such as bunch weight, show greater palm-to-palm variation, but in every case the variability observed in seedling controls is greater than that in the clones (Corley, 1982; see also Table 23.2).

Table 23.5 shows the mean values and analysis of variance for various bunch characters; the variation between palms within a clone is not significantly greater than the variation between repeated measurements on the same palm. On the other hand, there are large and significant differences in fruit composition between different clones.

**Table 23.6.** Fruit characters of Clone 932 regenerated from callus culture with two-year interval.[a]

| Date plants regenerated | 1977 | 1979 |
|---|---|---|
| Fruit set (%) | 80.0 ± 22.2 | 81.0 ± 10.5 |
| Fruit per bunch | 72.3 ± 4.9 | 67.8 ± 6.4 |
| Mesocarp per fruit (%) | 73.3 ± 4.3 | 66.4 ± 8.2 |
| Kernel per fruit (%) | 10.0 ± 1.7 | 11.4 ± 1.4 |
| Shell per fruit (%) | 14.7 ± 4.2 | 15.1 ± 2.6 |

Source: Wooi, Wong and Corley (1983).
[a] Data from palms at 30 months after field planting.

There are many reports describing plant variants arising from culture (for review see Krikorian, 1982). The frequency of variation is higher in plants derived from callus than from shoot multiplication. In most callus cultures there are numbers of cells with polyploid and aneuploid and other chromosomal abnormalities, and most plant propagators attempt to avoid the production of callus whenever possible. So far, we have not been able to propagate palms without passing through a callus stage and we might therefore expect to see some abnormal plants arising from culture. Some level of "sport" production is normally found in all vegetative propagation and, indeed, the origins of many improved fruit selections, for example in the apple Red Delicious, lie in such bud sports. So far, in our oil palm work we have found relatively little evidence of somaclonal variation. Established clones show low within-clone variability. Some aberrant plants are produced at the first regeneration stage, but it is not clear whether these result from unbalanced development in culture or from genetic variation. Most such plants are rogued out before the transplanting stage and do not reach the nursery. Further evidence of the stability of plants regenerated from callus was presented by Wooi et al., (1983). Callus cultures from Clone 932, which produced the first clonal palms in 1975 (field planted in January, 1977) later gave rise to further plants which were planted in 1979. These palms came into bearing in 1982, and have retained the growth and fruit characteristics of the first regenerants (Table 23.6). The distinctive features of high fruit to bunch and high shell and kernel content of the fruit are shown by both sets of material. Recently, however, we have observed some flowering abnormalities in young palms of a few clones which have been in continuous production for several years (Corley et al., 1986). The cause may be physiological (disturbed hormone balance), epigenetic or genetic, although the last possibility seems unlikely because several clones show the same abnormality. Work is in progress to identify the nature of the disorder.

Although regenerant plants appear to be genetically stable, at least for the first few years in culture, the cultures from which they are derived do show some cytologically abnormal cells, and in some cases numerous polyploid and aneuploid cells are present (Smith and Thomas, 1973; Jones et al., 1983). Krikorian and Kann (1981) made a detailed karyotype analysis of *Hemerocallis* cultures and plants derived from them, and also found that regenerant plants retained normal karyotype in spite of the presence of abnormal cells in the cultures.

So far, the plants regenerating from culture appear to be normal diploids. Perhaps in this species embryogenesis is such a rare event that any cytological unbalance precludes it. Hence, evidence of cytological instability in some cells in a culture does not necessarily imply that variability will be observed in plants regenerated from those cultures.

### 23.5.1.2 Risks of clonal planting

Two questions are frequently asked concerning the risks associated with clonal propagation.

The *first* concerns the risks of susceptibility to spread of disease through clonal plantings. This requires careful evaluation. We do not know to what extent the heterogeneity of genotypes in an outbreeding species such as oil palm protects it from disease. As a safeguard, we would not recommend planting large monoclonal blocks, and we expect to provide a choice of perhaps ten or more good clones which can be planted in an area. There are now many years experience with the use of clonal rubber plantations and, with appropriate selection of clones for specific local conditions, there seem to be few real problems.

Whether oil palm clones should be kept in small block plantings or interplanted in mixed stands remains to be determined. If block plantings are to be used we need to determine the optimum block size and spacing for each clone. Trials comparing mixed plantings with monoclone blocks were planted in 1984.

Our early trials work (Corley et al., 1982) indicated a high degree of flowering and sexual synchrony within clonal plantings. Interplanting may ensure that some palms in an area will be in a male phase at any given time, thus maintaining adequate pollination. On the other hand, if a clone proves susceptible to a disease it is easy to clear and replant a block, but impossible to replace trees in a mixed planting. All these, and other questions of clone management have yet to be resolved in field trials.

The *second* question implies that by planting clonal material rather than seedlings we shall be decreasing the size of the gene-pool.

The gene-pool consists of the wild groves of native palms which still exist but must be dwindling, and will do so whatever strategy is adopted in commercial plantings. This population is also represented in a number of collections in different countries (Hardon, 1976). At the next level, and probably most important is the working gene-pool used in palm breeding programmes. The present commercial oil palm stock is derived from a very narrow genetic base (see Hardon *et al.*, Chapter 3, this volume). For conventional seed production, the introduction of unimproved stock from the genetic collections would involve impractically long periods of backcrossing and reselection.

Many breeding programmes, particularly in Africa, have sought to increase homozygosity of parent lines in order to produce uniform hybrid seed. Each inbreeding generation restricts the parental heterozygosity and hence the size of the gene-pool. Breeding to produce ortets for cloning requires a highly heterozygous and heterogeneous population from which to select the best recombinants. Thus, there is now more stimulus to plant breeders to use a wider range of genetic material in their breeding programmes.

### 23.5.2 Coconut (*Cocos nucifera* L.)

Clonal propagation of coconut would provide advantages comparable with, or perhaps exceeding, those for oil palm in terms of improved yield. The relatively recent introduction of hybrid seed, produced from crosses between a dwarf cultivar with high nut number and local tall types with large nuts, have provided greatly improved yield. Unfortunately, numbers of nuts produced by a mother-palm are small and there is a shortage of hybrid seed.

Consequently, a method of propagation from hybrid embryos would have commercial benefit in coconut whereas, in oil palm, unselected hybrid seeds are of no significance. Propagation of monocotyledons has proven easier from embryonic tissues than from mature plants, and this may well become the first practical application. Of course, as in the oil palm, the greatest benefits will come from propagation of the best hybrid material once it has been proven in the field.

Coconut, like oil palm, is a clear candidate for tissue-culture propagation since there is no existing vegetative propagation method.

There have been several recent developments in tissue-culture propagation of coconut. In 1983, Branton and Blake reported successful plant regeneration *in vitro* but were unable to establish them in soil. We have now obtained a viable plant, regenerated from callus derived from immature inflorescence tissue, which has been planted in the Solomon Islands, where it is putting on new growth (Hughes, 1985). Several plants have recently been established in pots at the Hindustan Lever Research Laboratory in Bombay by Baskharan (pers. comm.). They developed from structures similar to those described by Pannetier and Buffard-Morel (1982) on young leaf tissues *in vitro*. The Wye College team have now obtained more plants *in vitro*, with every indication that they will be successfully transferred to soil (Branton, pers. comm.). There are also recent reports from India (Raju *et al.*, 1984) of coconut regeneration *in vitro*.

Although it cannot yet be said that the method is now available, the feasibility of tissue-culture propagation of coconut is established.

### 23.5.3 Date palm (*Phoenix dactylifera* L.)

In the date palm, clonal propagation using offsets is the normal method used. The clones are well characterised and the limitation is the availability of offsets for planting. In this particular crop, political and economic problems have conspired to create a desperate shortage of plants. The rate of offset production declines as palms age. Factors such as the oil boom and labour shortages in traditional date-growing countries have reduced the rate of replanting. Now, increasingly aged plantings are becoming debilitated. *Fusarium* (Bayoud) disease is taking its toll, and new planting material is increasingly scarce. Thus, a tissue-culture method for propagation of date palm could be of great value in the re-establishment of old date gardens.

During the past few years, propagation methods have been worked out by Reynolds and Murashige (1979) and by Tisserat (1980), and others, including ourselves, have successfully regenerated date palm plants from culture and planted them in soil. Several laboratories have now developed a routine production system capable of turning out uniform quality plants, established in soil ready for planting by the grower at a price he can afford. A comparative review of the culture of oil palm, coconut and date palm was provided by Blake (1983).

### 23.5.4 Rubber (*Hevea brasiliensis* Muell. Arg.)

As yet there is no reliable tissue-culture method for propagation of rubber, although steady progress is being made.

Some success was reported by Carron and Enjalric (1982) and by Mascarenhas *et al.* (1983); more recent reviews of progress and methodology were given by Chen (1984) and Wang and Wu (1985).

Much effort has been directed at the use of tissue-culture methods for crop improvement by use of pollen haploids, and Chen *et al.* (1982) described the regeneration of rubber plants from embryoids derived from the culture of anthers. A mixture of haploid, diploid and aneuploid plants was obtained with continuing chromosome instability in the regenerant plants.

Little success was reported in true clonal propagation using traditional explants, but Wan Abdul Rahaman *et al.* (1985) reported considerable success in obtaining embryogenic cultures from anther wall tissues. Wang and Wu (1985) described Chinese work comparing the performance of plants of clones Haiken 1 and Haiken 2 with their ortets. A promising observation was that the clonal plants retained the parent phenotype but gave taller and higher-yielding plants, with the advantage of both buddings (trueness-to-type) and seedlings (vigorous tap-root system). As yet the multiplication rate is low and the methodology will require much further development, but it is now clear that clonal propagation of rubber will be feasible.

The more conventional micropropagation methods using nodal cuttings is also possible with some clones (Carron *et al.*, 1985) but, again, multiplication rates are low, and in this case the plants lack the advantages of the embryogenic route.

At present, rubber clones are produced by bud-grafting selected scions on to seedling rootstocks. The method is quick, simple, reliable and cheap. To be competitive, the cost of production of tissue-culture plants would have to be less than 3 or 4 pence at the pre-nursery stage, unless their improved performance warrants a significant premium price. There is, at present, no good evidence of the effects of rootstock on performance, although such an effect would be expected. If beneficial stock–scion interactions are found, it may be worth developing clonal rootstocks, perhaps *via* tissue culture.

The possibility now clearly exists that tissue culture methods can be developed. To compete with existing clonal propagation methods, a high multiplication rate will have to be coupled to an efficient plantlet handling and rearing system, but could convey some benefits in terms of yield and vigour. There are also considerable possibilities in the use of tissue-culture methods to develop novel clones in advanced plant-breeding programmes.

### 23.5.5 Cocoa (*Theobroma cacao* L.)

Cocoa propagation provides perhaps the most intriguing challenge. Clones are less well developed than in rubber, but are being increasingly used. In Malaysia, approximately 20% of new cocoa plantings use bud-grafted clones. However, most cocoa plantings use hybrid seed which show great variability. One Malaysian observation showed that 40% of the seedlings in one progeny collection contributed 75% of the total yield (Ramadasan and Arasu, 1980).

The production of clonal cocoa is limited by the dimorphism of the cocoa tree. The initial growth of a seedling is orthotropic, forming an upright stem or chupon. Indeterminate, plagiotropic fan branches develop from the terminal bud of the chupon. The process of bud grafting in cocoa is even easier than with rubber, because the ages of the stock and scion material are less critical. Bud grafts and cuttings will produce orthotropic growth only if taken from chupon material. Buds from fan growth give plants with a low habit that require careful pruning and management which adds to the cost of clonal plants.

Suitable buds from chupon material are naturally limited in number and fan buds are normally used. A tissue-culture technique could clearly overcome this limitation, and the extra cost of staking and pruning fan plants allows some latitude in the price of plants from tissue culture that could allow them to compete profitably.

As yet, there is no effective method for propagation of cocoa *in vitro*. Hall and Collin (1975) were able to initiate callus and induce root morphogenesis. Since then, Pence et al. (1979) demonstrated asexual embryogenesis from immature zygotic embryos. So far, no regenerants have been obtained from mature tissues, and major problems have been experienced in disinfection of such material (Passey and Jones, 1983). Tamin and Ghazali (1985) reported multiple shoot production from stem node sections of bud-grafted orthotropic shoots. Multiple bud proliferation from axillary buds was obtained. The resulting shoots were easily rooted but were vitrified and incapable of establishment in soil.

Although micropropagation of chupon buds would be the initial method of choice, it is interesting to speculate on the morphogenetic responses of cells from fan and chupon sources. Is the morphogenetic potential different in fan- and chupon-derived cells, and would either somatic embryogenesis or adventitious bud formation restore fan cells to chupon growth? Stoutmeyer and Britt (1965) found that adult and juvenile characteristics were maintained in cultures of *Hedera helix*. Roots were formed more frequently in cultures from juvenile origins but they did not obtain embryogenesis. Banks (1979), also working on *Hedera helix*, found that callus from juvenile tissues regenerated shoot buds, giving plants of juvenile habit. Callus from adult tissues underwent somatic embryogenesis, again producing juvenile growth. Thus, it seems likely that if morphogenesis could be obtained from either fan or chupon, the outcome might well be chupon growth. The answer to these problems could be of great interest in a range of other species with dimorphic growth, e.g. *Eucalyptus*, and in many conifer species, where there are marked transitions from juvenile to mature growth which limit conventional cloning methods.

Dimorphism is only one of the problems of clonal cocoa. There is a considerable degree of both self- and cross-incompatibility in many populations, and care will be needed to ensure that either self-compatible clones are produced, or that cross-compatible clones are interplanted.

Resistance to black-pod disease is perhaps the most important requirement, worldwide, as it is responsible for 30% of crop loss (Gray, pers. comm.). In West Africa, there is a need for clones resistant to swollen-shoot virus. The selection and cloning of resistant individuals would be of inestimable value.

Clearly there is great potential for application of tissue-culture methods to cocoa improvement. As with other crops, it will have to be coupled with a good breeding programme, with careful ortet selection and with thorough field testing of the clones.

### 23.5.6 Cassava (*Manihot esculenta* Crantz)

Cassava is one of the world's major food crops, providing nourishment for perhaps 300 million people in tropical countries, and is a major source of starch for industry and animal feeds. It is traditionally a smallholder crop, vegetatively propagated, and with a large number of different varieties. Many of the stocks are infected with Cassava mosaic virus, and the first objective was to produce virus-free planting material using tissue-culture methods. The work was technically very successful (Kartha and Gamborg, 1977). There is now widespread international exchange of virus-free cassava germplasm (Roca *et al.*, 1983). Cassava propagation was, however, almost impossible to apply commercially because of the fragmented and impoverished nature of the market. The advent of the new interest, mainly in Brazil, in renewable resources for energy production has aroused fresh interest in cassava as a source of carbohydrate for fermentation to alcohol to be used as fuel (Goldemberg, 1977). There is now a large demand for plants for plantation use, providing an ideal market for tissue-culture production of virus-free plants. This enables bulk propagation to be combined with a quality improvement providing added market value, but commercial viability of cassava propagation requires a lower production cost than is currently possible.

The point once again is that tissue-culture propagation can only be used if it is possible to recoup the added value of improved plants by a premium price that pays for their production.

### 23.5.7 Sugarcane (*Saccharum officinarum*)

Sugarcane is already normally vegetatively propagated and is easily and cheaply multiplied. As pointed out by Heinz *et al.* (1977), there is little point in considering mass propagation by *in vitro* methods unless they can be coupled to selection for useful traits such as disease resistance. Thus, the major application of tissue culture has been to exploit the genetic variability revealed by *in vitro* methods, rather than to look for clonal uniformity (Ming Chin-Liu, 1981).

## 23.6 CONCLUSIONS

For plant tissue culture to be used effectively for bulk propagation of plantation, or indeed any other crop plants, exacting criteria have to be met. These concern the price premium which must be applied to the plants to make the process worthwhile, and the benefit derived by the grower if he buys plants from an *in vitro* propagation unit. Unless government agencies intervene to subsidise the process, all other factors are subservient. In each crop the analysis will be different. It may be that clonal propagation is clearly worthwhile, as in oil palm and coconut, or that, circumstances change to make it so, as in cassava. Alternatively, the technical problems may make the research too long-term or costly, as in the case of rubber. In crops such as sugarcane, mass propagation *in vitro* cannot compete in price or numbers with traditional propagation methods. In these crops, however, tissue culture is making a major contribution to the production of improved cultivars (Ming Chin-Liu, 1981). Similarly, great advances have been made in coffee breeding using tissue culture methods (Sondahl and Sharp, 1977).

In easily propagated crops the contributions to plantation crop improvement will come by coupling tissue culture with selection *in vitro*, or the by-passing of sexual mating barriers by any of the methods of somatic fusion, organelle transfer, chromosome manipulation or recombinant DNA technology. Subsequent propagation of the improved cultivars may well be better carried out by traditional methods.

Acknowledgements

I am indebted to my many colleagues in Unilever who made constructive comments on the drafts of this paper. In particular, I thank Dr R. H. V. Corley for help and for permission to use the data from his experiments. I would also like to thank Mr C. H. Teoh and Mr B. B. Ang of Harrisons and Crosfield Ltd., Rubber and Cocoa Research Station in Malaysia for information on vegetative propagation methods in these crops.

# REFERENCES

Banks, M. S. (1979). Plant regeneration from callus from two growth phases of English ivy. *Hedera helix* L. *Z. Pflanzenphysiol.* **92**, 349–353.

Blake, J. (1983). Tissue culture propagation of coconut, date and oil palm. In "Tissue Culture of Trees" (J. H. Dodds, ed.), pp. 29–59. Avi, Westport, Conn.

Branton, R. L., and Blake, J. (1983). Development of organised structures in callus derived from explants of *Cocos nucifera* L. *Ann. Bot.* **52**, 673–678.

Carron, M. P., and Enjalric, F. (1982). Studies on vegetative micropropagation of *Hevea brasiliensis* by somatic embryogenesis and *in vitro* microcutting. In "Proceedings of Vth International Congress of Plant Tissue and Cell Culture", (A. Fujiwara, ed.), pp. 751–752. Maruzen, Tokyo.

Carron, M. P., Deschamps, A., Enjalric, F., and Lardet, L. (1985). Vegetative *in vitro* propagation by microcutting of selected rubber trees: a widespread technique before 2000? *Proc. 1st Int. Rubber Tissue Culture Workshop, Kuala Lumpur, Malaysia.* Rubber Research Institute, Malaysia.

Cassells, A. C., and Kavanagh, J. A. eds. (1983). "Plant Tissue Culture in Relation to Biotechnology". Royal Irish Academy, Dublin.

Chen, Z. (1984). Rubber (*Hevea*). In "Handbook of Plant Cell Culture, Vol. 2, Crop Species" (W. R. Sharp, D. G. Evans, P. V. Ammirato and Y. Yamata, eds), pp. 546–571. Macmillan, New York.

Chen, Z., Qian, C., Qin, M., Xu, X., and Xiao, Y. (1982). Anther culture of *Hevea brasiliensis*. *Theor. appl. Genet.* **62**, 103–108.

Conger, B. V. (1981). Agronomic crops. In "Cloning Agricultural Plants *via in vitro* Techniques", (B. V. Conger, ed.), pp. 165–215. CRC Press, Boca Raton, Florida.

Corley, R. H. V. (1982). Clonal planting material for the oil palm industry. *J. Perak Plrs' Assoc.* 1981, 35–49.

Corley, R. H. V. (1985). Agriculture factors affecting yield and supply of palm kernel and coconut oils. FRA/FOSFA Symposium, Palm Kernel and Coconut Oils.

Corley, R. H. V., Lee, C. H., Law, I.-H., and Wong, C. Y. (1986). Abnormal flower development in oil palm clones. *Planter, Kuala Lumpur* **62**, 233–240.

Corley, R. H. V., Wong, C. Y., Wooi, K. C., and Jones, L. H. (1982). Early results from the first oil palm clone trials. In "The Oil Palm in Agriculture in the Eighties", Vol. 1, (E. Pushparajah and Chew Poh Soon, eds), pp. 173–196. Incorporated Society of Planters, Kuala Lumpur.

de Fossard, R. A., and Bourne, R. A. (1977). Reducing tissue culture costs for commercial propagation. *Acta Hort.* **78**, 37–44.

Evans, D. A., Sharp, W. R., Ammirato, P. V., and Yamada, Y. (1983). "Handbook of Plant Cell Culture: Vol. 1. Techniques for Propagation and Breeding", Macmillan, New York, 970 pp.

Goldemberg, J. (1977). Alcohol from plant products: A Brazilian alternative to the energy shortage. In "Plant Cell and Tissue Culture: Principles and Applications" (W. R. Sharp, P. O. Larsen, E. F. Paddock and V. Raghavan, eds), pp. 55–65. Ohio State University Press, Columbus, Ohio.

Hall, T. R. H., and Collin, H. A. (1975). Initiation and growth of tissue cultures of *Theobroma cacao*. *Ann. Bot.* **39**, 555–570.

Hardon, J. J. (1976). Oil palm breeding—introduction. In "Oil Palm Research"

(R. H. V. Corley, J. J. Hardon and B. J. Wood, eds), pp. 89–108. Elsevier, Amsterdam.
Heinz, D. J., Krishnamurthi, M., Nickell, G., and Maretzki, A. (1977). Cell, tissue and organ culture in sugarcane improvement. *In* "Plant Cell, Tissue and Organ Culture". (J. Reinert and Y. P. S. Bajaj, eds), pp. 4–17. Springer-Verlag, Berlin.
Hughes, W. A. (1985). Unilever success in coconut tissue culture. *Coconis* **19**, 1.
Jones, L. H. (1984). Novel palm oils from cloned palms. *J. Am. Oil Chem. Soc.* **61**, 1717–1719.
Jones, L. H., Barfield, D., Barrett, J., Flook, A., Pollock, K., and Robinson, P. (1983). Cytology of oil palm cultures and regenerant plants. *In* 'Proceedings of Vth International Congress of Plant Tissue and Cell Culture", (A. Fujiwara, ed.), pp. 727–728. Maruzen, Tokyo.
Kartha, K. K., and Gamborg, O. L. (1977). Cassava tissue culture—principles and applications. *In* "Plant Cell and Tissue Culture; Principles and Applications" (W. R. Sharp, P. O. Larsen, E. F. Paddock and V. Raghavan, eds), pp. 710–725. Ohio State University Press, Columbus, Ohio.
Krikorian, A. D. (1982). Cloning higher plants from aseptically cultured tissues and cells, *Biol.* Rev. **57**, 151–218.
Krikorian, A. D., and Kann, R. P. (1981). Plantlet production from morphogenetically competent cell suspensions of Daylily. *Ann. Bot.* **47**, 679–686.
Mascarenhas, A. F., Hazara, S., Potdar, U., Kulkarni, D. K., and Gupta, P. K. (1983). Rapid clonal multiplication of mature forest trees through tissue culture. *In* "Proceedings of Vth International Congress of Plant Tissue and Cell Culture", (A. Fujiwara, ed.), pp. 719–720. Maruzen, Tokyo.
Ming Chin-Liu (1981). *In vitro* methods applied to sugar cane improvement. *In* "Plant Tissue Culture—Methods and Applications in Agriculture" (T. A. Thorpe, ed.), pp. 299–323. Academic Press, New York and London.
Pannetier, C., and Buffard-Morell, J. (1982). Production of somatic embryos from leaf tissues of the coconut *Cocos nucifera* L. *In* "Proceedings of Vth International Congress of Plant Tissue and Cell Culture", (A. Fujiwara, ed.), pp. 755–756. Maruzen, Tokyo.
Passey, A. J., and Jones, O. P. (1983). Shoot proliferation and rooting *in vitro* of *Theobroma cacao*, type Amelonado. *J. hort. Sci.* **58**, 589–592.
Pence, V. C., Hasegawa, P. M., and Janick, J. (1979). Asexual embryogenesis in *Theobroma cacao* L. *J. Am. Soc. hort. Sci.* **104**, 145–148.
Raju, C. R., Prakash Kumar, P., Chandramohan, M., and Iyer, R. D. (1984). Coconut plantlets from leaf tissue cultures. *J. Plant. Crops* **12**, 75–78.
Ramadasan, K., and Arasu, N. T. (1980). Vegetative propagation of *Theobroma cacao* L. and related problems. *Planter, Kuala Lumpur*, **56**, 49–59.
Reynolds, J. R., and Murashige, T. (1979). Asexual embryogenesis in callus cultures of palms. *In Vitro* **15**, 383–387.
Roca, W. M., Rodriguez, J., Beltran, J., Roa, J., and Mafla, G. (1983). Tissue culture for the conservation and international exchange of germ plasm. *In* "Proceedings of Vth International Congress of Plant Tissue and Cell Culture", (A. Fujiwara, ed.), pp. 771–772. Maruzen, Tokyo.
Smith, W. K., and Thomas, J. A. (1973). The isolation and *in vitro* cultivation of *Elaeis guineensis*. *Oléagineaux* **28**, 123–127.
Sondahl, M. R., and Sharp, W. R. (1977). Research in *Coffea* spp. and applications of tissue culture methods. *In* "Plant Cell and Tissue Culture—Principles and

Applications" (W. R. Sharp, P. O. Larsen, E. F. Paddock and V. Raghavan, eds), pp. 527–584. Ohio State University Press, Columbus, Ohio.

Stoutmeyer, V. T., and Britt, O. K. (1965). The behaviour of tissue cultures from English and Algerian ivy in different growth phases. *Am. J. Bot.* **52**, 805–810.

Tamin, M. S. M., and Ghazali, H. (1985). Preliminary results on organ cultures of *Theobroma cacao* L. Simposium Kultur Tisu Tumbulian Kebangsan KeII. Malaysian National Tissue Culture Conference, Serdang, Selangor, Malaysia. 15–17 October.

Tisserat, B. (1980). Propagation of date palm (*Phoenix dactylifera* L.). *J. exp. Bot.* **30**, 1275–1283.

Vasil, I. K., Ahuja, M. R., and Vasil, V. (1979). Plant tissue cultures in genetics and plant breeding. *Adv. Genet.* **20**, 127.

Walbot, V. (1981). The application of genetics to plants. *In* "Impacts of Applied Genetics—Micro-organisms, Plants and Animals", pp. 137–164. Congress of the USA, Office of Technology Assessment, Washington D.C.

Wan Abdul Rahaman, W. H., Cheong, K. F., and Mohd. Ghouse, W. (1985). *Hevea* Tissue Culture. Prospect and Retrospect. Prospect of the First International Rubber Tissue Culture Workshop. Serdang, Selangor, Malaysia. RRIM/IRRDB. 15–17 October.

Wang, Z., and Wu, H. (1985). Inheritable characters of anther somatogenic plants of *Hevea brasiliensis*. Proc. 1st Intl. Rubber Tissue Culture Workshop. Kuala Lumpur, Malaysia. Rubber Research Institute, Malaysia.

Wooi, K. C., Wong, C. Y., and Corley, R. H. V. (1983). Genetic stability of oil palm callus cultures. *In* "Proceedings of Vth International Congress of Plant Tissue and Cell Culture", (A. Fujiwara, ed.), pp. 749–750. Maruzen, Tokyo.

# Index

Abscisic acid
  in embryogenesis suppression, 94
  in growth restriction, 308
*Achimenes sp.*
  adventitious bud techniques, 343
  apex formation, 279
  dominant mutations, 341
  mutation-breeding, 336
Adventitious bud techniques, *in vitro*/*in vivo*, 343–344
Ageing, in plants, 267
*Alnus*, tissue culture, 368
*Alstroemeria spp.*, 162
  mutation breeding, 336, 344
  *see also* Flowers, cut
*Amphorophora spp.*, 138
Apex formation, in adventitious buds, 279–282
Aphid, rosy apple, 117
Apical meristems
  cryopreservation, 308–311
  culture, 305–306
  grafting, 265
Apomixis, induced, 193
Apple(s)
  assessment of natural mutations, 349
  benzyladenine, and selection for compactness, 346
  breeding priorities, 119–120
  chimeras, 277, 279, 280
  conventional propagation
    mutant clones, 353–355
    after re–irradiation, 355–357
  selection, 351–360
  Cox's Orange Pippin, drawbacks, 120–122
  disease and pest resistance, 114–117
    apple scab, 114
    mildew, 115
    rosy apple aphid, 117
  homohistant mutants, 353
    tests through root cuttings, 357–360
  mutation breeding, 349–362
  new cultivars, 120–122
  periclinal chimerism, preliminary tests, 353–354
  phase change, 268
  quality testing, 118–119
  root cuttings, and homohistant clones, 358–360
  storage, 118–120
  yield, 117–118
Auxins
  'habituation', 265
  induction of bud primordia, 369–371
  rooting of tree cuttings, 230–231

Barley, mutation breeding, 318–319, 325–326
  pleiotropic effects, 326
Benzyladenine, in selection systems, 346
Bermuda grass
  cold-resistance, 326, 336, 341

mutation breeding, 344
*Betula sp.*, tissue culture, 368
Blackcurrants
  breeding, 135–147
  frost tolerance, 141
  juice quality, 142
  resistance to pests and diseases, 142–144
  reversion virus, 143
  yield, 141–142
  see also Ribes
Blayde's medium, tissue culture, 367–368
*Botrytis cinerea*, 138
*Brevipalpus sp.*, 97

Callus, cryopreservation, 308–311
Callus culture, 165
  embryos, genetic stability, 304
  'habituation', 265–266
Carnation(s)
  chimeras, 163, 275, 277–278, 281
  wilt, defence mechanism, 173
Cassava, see *Manihot sp.*
*Cecidophyopsis sp.*, 143
Cell determination, phase change, 266
Chimera(s), 271–284, 341
  advantages, 273, 342
  in apple, 277, 353
  in carnation, 163, 275
  in *Chrysanthemum*, 345
  in *Citrus*, 89
  in *Crataegomespilus*, 274
  disadvantages, 276, 335, 342
  and fungal resistance, 275
  induction, 271, 341
  in *Laburnocytisus*, 274
  mutagenic induction, 271
  in ornamentals, 162
  in *Pelargonium*, 274, 278
  in potato, 273, 277, 280
  in rice, 325
  in *Rhododendron*, 274, 281
  in sugarcane, 24, 337
  in tomato, 275
*Chaetosiphon sp.*, 162
*Chondrostereum sp*, (silver leaf), 130
Chimera(s)
  '244' and '422', 276
  advantages, 273–276, 342
  in chrysanthemum, 164–165
  in citrus, 89
  different types, 353
  disadvantages, 276–278, 335, 342
  and fungal resistance, 275
  induction, 271–273, 341–343
  periclinal, 272–273, 353
  reversion, 351
  solid mutants, 278–282
*Chrysanthemum sp.*
  irradiation, 345
  market value, 160
  mutation breedng, 163–164, 337, 340, 345
  pest resistance, 172
  short duration crop, 170–171
CIP, see Potato, International Centre
*Citrus*, species and cultivars, 86–101
  breeding, 83–110
*Citrus*
  aneuploidy, 87
  bridging crosses, 103
  chimeras, 89
  cold-hardiness, 96, 98
  'crossability', 102
  diploid megagametophytes, 85
  disease resistance, 97, 99–100
  germplasm, 104
  grafting of apical meristems, 265
  incompatibility, 87–88
  juvenility, 90–92
  'Mal seccho', 97
  mutation breeding, 88
  new cultivars, 103–104
  nucellar embryony, 84, 92–95
  pollen storage, 102
  polyploidy, 83–87
  rootstock breeding, 98–100
  salt-tolerance, 99
  seed storage, 102
  thornlessness, 98
  tissue culture, 100–102
  trifoliate leaf character, 98
  Tristeza virus, 97
  zygotic seedlings, 92–95
Clone trials
  apple, 351
  forest trees, 203, 207–208
  oil palm, 69–78

# Index

*Picea sp.*, 251–257
potato, 186–190
rubber tree, 28, 31–32, 47
sugarcane, 10–11
*Triplochiton sp.*, 231–240
Clones, advantages, 387
Cocoa, tissue culture, 399–401
Cocoideae, 63
*Cocoa nucifera* (coconut), tissue culture, 397–398
Colchicine, induced polyploidy, 84, 167
*Colletotrichum sp.*, leaf disease, 31, 33, 40, 46
Conifers, tissue culture, 368–371
Conservation
  gene parks, 295
  *in situ*, 294
  of seeds, 290–294
  of vegetatively propagated crops, 294
*Corticium sp.*, 33
*Crataegomespilus*, chimera, 274
Crossing systems
  biparental crosses, 15, 21
  polycrosses, 15–16, 21
  repetition, 16
Cryopreservation, 303–313
*Cynodon sp.*, *see* Bermuda Grass
Cytokinin
  habituation, 265
  induction of bud primordia, 369–370

*Dahlia*, mutation breeding, 345
*Dasyneura sp.* (blackcurrant midge), 144
Date palm, tissue culture, 398
*Deuterophoma sp.*, in citrus disease, 97
*Didymella sp.* (spur blight), 139
*Dioscorea sp.*, restricted-growth storage, 308
*Diplocarpon sp.* (black spot), 161
Disease resistance
  in apple, 114
  in blackcurrant, 142–144
  in *Citrus*, 97–100
  in mutants, 325
  in oilpalm, 74
  in ornamentals, 172
  in pear, 115
  in plum, 130

  in *Poncirus trifoliata*, 83, 95, 98
  in potato, 181, 185, 192, 295
  in raspberry, 138
  in rose, 161
  in rubber, 31, 33, 46, 50
  in strawberry, 151
DNA, adult and juvenile tissues, 266–267
*Drepanopeziza sp.* (blackcurrant leaf spot), 142
*Drysaphis sp.* (rosy apple aphid), 117

*Elaeis guineensis*, *see* Oil palm
*Elsinoe sp.* (raspberry cane spot), 139
EMLA fruit tree scheme, 350
Enviromax Planting Recommendations, 48
*Eremocitrus sp.*, 96, 98
*Erianthus sp.*, hybrids, 4
*Erinnyis sp.*, pest of rubber, 33
*Erwinia sp.*, fireblight, 117
*Eucalyptus sp.*, phase change, 268

*Fagus sp.*, tissue culture, 368
Fasciation, induction, 322
Fireblight, 117
Flowers, cut
  commercial mutants, 339
Forestry, *see* Trees
*Fragaria ananassa*, 157
  *see also* Strawberry
Freesia
  polyploidy, 167
  seed-propagation, 168
Freezing, *see* Cryopreservation
*Fusarium spp.*
  carnation wilt, 173
  *Narcissus* basal rot, 174

Gene banks, *see* Genetic resources: activities
Gene parks, 295
'Genecology', 329–332
Genetic resources, 285–301
  *see also* under crop names
  activities, 287–300
  conservation, 290–295

data management, 297
evaluation, 295–297
exploration, 287–290
international co-operation, 299–300
training, 298–299
utilisation, 298
duplicates, elimination, 296
'exotic germ plasm', 298
genetic diversity, 285–287
genetic erosion, 286
International Board for Plant Genetic Resources, 299
pre-breeding, 298
Genetic variability
in forest trees, 204
in Norway spruce, 252
in tropical trees, 222
in temperate trees, 248
in Centres of Origin and Diversity, 286
in *Citrus*, 104
in sugarcane collections, 3
in rose, 161
in potato, 190
*Geranium sp.*, polyploidy, 167
*Gerbera sp.*
flower longevity, 174
seed propagation, 168
Germplasm, storage techniques, 303–311
*Globodera sp.*, 181, 184
Gooseberry mildew, 142
Grapefruit, see *Citrus paradisi*
*Gypsophila sp.*, 162

'Habituation', callus cultures, 265
Haploid induction, 100
*Hedera sp.*, see Ivy
Heterozygosity, see Genetic variability
*Hevea brasiliensis*, see Rubber tree
*Hordeum vulgare*, see Barley
Hormones
flower induction, 49
growth restriction, 308
'habituation', 265
induction of bud primordia, 369–370
rooting of cuttings, 230–231
see also auxin, cytokinin

Hybridisation, interspecific, 161–162

Incompatibility, *Citrus*, 87–88
International Board for Plant Genetic Resources, 297, 299
International Institute of Tropical Agriculture, 299
International Potato Centre (CIP), 296
International Rice Research Institute, 299
Irradiation
chimeras, apple, 276
potato, 281
detached leaves, 343
by gamma rays, 351
mutant production, 318–319
programmes, 336–337
Ivy
micropropagation, 400
phase change, 264–266
rooting, 267–268

Juvenility, 263
in apple, 113
in *Citrus*, 90–92
in *Cacao*, 400
in Redwood, 371, 377–379
in pear, 113
in plum, 130

*Kalanchoe blosfeldiana*, 169

*Laburnocytisus sp.*, chimera, 274
Latex, 27
Leaf blight, South African, 27, 33, 41, 46–47
Leaf structure, *Liquid ambar* explants, 372–377
Lemon, see *Citrus limon*
*Leptosphaeria sp.*, raspberry cane blight, 135, 139
*Liquidambar sp.*
acclimatisation, 378
leaf structural variations, 372–377
organogenesis, 372
tissue culture, 366–367
*Liriodendron sp.*, tissue culture, 367–368

*Liryomyza sp.*, 172

Male sterility, induction, 17
*Malus robusta*, 115–117
*Malus zumi*, 115–116
  see also Apple(s)
*Manihot sp.*, tissue culture, 401
*Mentha sp.*, wilt-resistance, 325, 337
  irradiation method, 345
Meristems, *see* Apical meristems
*Microcyclus sp.*, disease of rubber, 33, 46
Micropropagation
  *see* Tissue culture
Mildew, American gooseberry, 142
  see also *Podosphaera*
*Miscanthus spp.* (sugar cane), 4
*Monilinia sp.* (fruit rot), 130
Mutagens
  chemical, 347
  list, 322
  in production of chimeras, 271–273
  sodium azide, 326
Mutant genes
  advantages/disadvantages, 342–343
  adventitious bud techniques, 343–344
  cultivar list, 319, 323
  desired mutations, 325–326
  and flower colour, 341
  'genecology', 329–332
  homozygosity, multiple, 319–325
  improvement of existing genotypes, 338–340
  and parental stocks, 340–341
  induction, 335
  pleiotropic action, 326–329
  recombination, 319, 324–325
  selection value, 317–319
Mutation breeding, 335–347
  in *Achimenes spp.*, 338
  in *Alstroemeria spp.*, 162, 336, 344
  in apple, 349–362
  applications, 335–347
  in barley, 318, 325
  in Bermuda grass, 344
  in *Chrysanthemum*, 163, 337, 340, 345
  in *Citrus*, 88
  in cut flowers, 339
  in dahlia, 345
  homohistant mutants
    tests for, 357
  in ornamentals, 162
  periclinal chimeras
    tests for, 353
  in *Pisum*
    pleiotropism, 327
  polyploidy, induced, 166, 341
'Mutator' genes, 324

*Narcissus sp.*, 'basal rot', 174
*Narenga sp.* (sugarcane), 4
*Nauclea sp.*, 241
Nematodes, 99, 181, 184–185
*Nicotiana sp.*, apex formation, 279
Norway spruce, *see Picea*
Nucellar embryony, 92–95

*Oidium sp.*, disease of rubber, 33, 46
Oil palm
  breeding, 63–81
  breeding for seed production, 64–68
  *Cereospora sp.*, 392
  clone trials, 78, 396–397
  defoliation, effect, 73
  Deli *dura* population, 64–68
  disease resistance, 74
  *dura* x *dura*, 77
  *dura* x *pisifera*, 70, 75
  *dura* x *tenera*, 64–67, 75, 77
  fertiliser, use, 75
  fruit characteristics, 391–395
  harvesting, 75
  'ideotypes', 78
  interspecific hybridisation, 78
  life cycle, 63
  oil: bunch ratio, 72
  oil, composition, 74
    palm oil: kernel oil, 74
  oil content, variation, 390–393
  ortet selection, 72
  palm kernel oil, 63
  *pisifera*, 65–68
  selection for vegetative propagation, 68–78
    objectives, 71–75

populations, 75–78
progress, 64–68
shell thickness, 64
*tenera* x *tenera*, 75
tissue culture, 63, 388–397
wilt resistance, 392
yield
   expected improvement, 65, 69
   variation, 72–73, 77
Orange, *see* Citrus
Ornamental plants
   autopolyploidy, 166–167
   breeding objectives, 159–179
   commercial mutants, 339
   generation of genetic variants, 161–167
   longevity, 174–175
   low-temperature tolerance, 171–172
   mutation breeding, 162–166
   pest and disease resistance, 172–174
   seed propagation, 168
   year-round production, 169–171
   *see also* Flowers, cut, *and* specific names
*Otiorhynchus sp.*, vine weevil, 144

Palm oil, *see* Oil palm
Pear(s)
   breeding, 113–123
   disease and pest resistance
     fireblight, 117
     mildew, 115–116
   new cultivars, 122
   rootstock, 114, 116
   yield, 118
Peas, *see Pisum sp.*
Pedicel culture, 165–166
*Pelargonium sp.*
   chimeras, 274, 278
   polyploidy, 167
   seed propagation, 169
*Peperomira sp.*, apex formation, 279–280
Pest resistance
   in apple, 114
   in blackcurrant, 142
   in *Chrysanthemum*, 172
   in *Citrus*, 97
   in forest trees, 207

in ornamentals, 172
in pear, 115
in potato, 181, 185, 192
in raspberry, 139
in strawberry, 162
Peppermint, *see Mentha*
Phase change, 263–268
   in apple, 268
   in *Eucalyptus*, 268
   in ivy, 264
   in Redwood, 268
   rooting ability, 267
*Phoenix sp.*, see Date palm
*Phragmidium sp.*, 139
*Phytophthora spp.*, 33, 40, 46, 151, 153
*Phytophthora infestans* (potato blight), 185
*Picea abies*
   adventitious bud culture, 369–371
   genetic variation, 251–257
   somatic embryos, 370
   tissue culture, 367, 369
*Pinus spp.*
   genetic variability, 204
   late-juvenile arrest, 206–207
   loblolly, plantlet performance, 378
   plantlet production, 371
*Pisum sp.*
   fasciation, 322–325
   'genecology', 330–332
   mutant genes
     pleiotropic effects, 327–329
     specific interactions, 317–322, 324
Plant growth substances, *see* Hormones
Plum(s)
   breeding, 125–133
   flowering time, 127–129
   juvenile phase, 130–133
   limitations on improvement, 130–132
   species, 125–126
*Podosphaera sp.*, apple mildew, 115–116
Pollen storage, *Citrus*, 102
Polycrosses *vs* biparental crosses, 15
Polyploidy,
   in *Citrus*, 83–87
   induction, 166–167, 341
   *see also* Chimeras
*Poncirus trifoliata*, 83, 95–97, 98
   disease resistance, 100
Poplar, yellow, *see Liriodendron*

*Populus sp.*
  rust resistance, 379
  tissue culture, 366
Pot plants, top ten, 160
Potato
  apomixis, 193
  breeding, 181–196
  chimeras, 273, 277–278, 280–281
  conventional breeding programmes, 186–189
  cultivars, 181–182
  cytogenetics, 183–184
  future strategies, 191–193
  historical perspective, 183–184
  International Potato Centre, 296
  leafroll virus, 185
  non-conventional approaches, 190–191
  pest and disease resistance, 181–182, 185–186, 192
  restricted growth storage, 307, 308
  true potato seed *vs* tubers, 193
*Potentilla spp.*, in strawberry breeding, 157
'Provenance' selection, trees, 249
*Prunus spp.*, 125–126
  antigenic determinants, 266
  tissue culture, 368
  *see also* Plum(s)
*Pseudomonas mors-prunorum* (bacterial canker), 130
*Puccinia sp.*, 172–173
*Pyrus spp.*, *see* Pear

*Quercus*, tissue culture, 368

*Radopholus sp.* (nematode), 99
Raspberry
  breeding, 135–147
  cane blight, 135
  disease resistance, 138
  fruit quality, 137
  fruiting season, 137
  pests, 139
  primocane cultivars, 137
  yield, 135
  *see also* Rubus

Recording of field collections, 292–293
Redwood
  cuttings, 268
  juvenile shoots, explants, 371, 377, 379
  phase change, 268
Resistance,
  to cold,
    Bermuda grass, 326, 336, 341
    blackcurrant, 141
    *Citrus*, 96
    ornamentals, 171
  to diseases, *see* disease resistance
  to pests, *see* pest resistance
  to salt, *Citrus*, 99, 101
  to wind, rubber, 41, 50
  *see also* Pest; Disease resistance
*Resseliella sp.*, cane midge, 139
Restricted-growth storage procedures, 307–308
Reversion virus, blackcurrant, 143
*Rhododendron* chimera, 274, 281
*Ribes spp.*, 141–144
  *see also* Blackcurrant
Rice,
  International Rice Research Institute, 299
  mutation-breeding, 325
Root-cuttings, apple, and homohistant clones, 358–360
Root development, *Liquidambar* explants, 372
Rooting ability, and phase change, 267–268
Rootstock breeding, *Citrus*, 98–100
*Rosa* (rose)
  *R. foetida*, 161
  *R. rugosa*, 161
  black spot, 161
  long duration cropping, 170
  market value, 160
  pest resistance, 172
Rubber tree
  biometrical genetics, 44–45
  breeding, 27–61
  breeding objectives, 49–51
    selection cycle, 43–44
  breeding problems, 28–34
  crown budding, 47
  clonal improvement, 28

conservation, 36–37
disease resistance, 31, 33, 46, 50
flower induction, 34, 48
fruit-set, 34, 49
GCA (general combining ability), 44
genotype-environmental
    interactions, 33–34
international co-operation, 35–38,
    51–52
length of breeding and selection
    cycle, 32, 49, 50
multiple characters, 32–33
new genetic introductions, 35–38
nodal cuttings, 399
pedigrees, Malaysian clones, 30–32
polycross progenies, 45
polyploidy, 38
'primary clones', 29
recommendations, 47–48
seasonality, 34
selection techniques, 39–47
shortening of breeding cycle, 38–44
tissue culture, 38, 398–399
Wickham collection, 29, 36
wind-tolerance, 41, 50
yield
    productivity, 50
    testing, 39–41
    theoretical maximum, 52
*Rubus spp.*, 136–138
    *see also* Raspberry
RW (Risser and White) medium,
    367–368

*Saccharum spp. see* Sugarcane
*Saintpaulia sp.*
    apex formation, 279
    low-temperature tolerance, 171
SALB (South American leaf blight),
    27, 33, 41, 46–47
Salt-tolerance, *Citrus*, 99
Sampling
    recording sheets, 292–293
    seeds, 287–288
    vegetatively propagated plants,
        288–290
*Sclerostachya sp.* (sugarcane), 4
Seed(s)

commercial production, tropical
    trees, 216
conservation, 290
orthodox, 290
propagation
    *Gerbera*, 168
    ornamentals, 168
    *Pelargonium*, 169
    strawberry, 156
recalcitrant, 291
sampling, 287
storage, 102
true potato seed, 193
*Sequoia sp.*, *see* Redwood
*Solanum spp.*, 184–190
    *see also* Potato
Somaclonal variation, 346
Sorghum, crossed with sugarcane, 4
*Sphaerotheca sp.*, 142
Storage, apples, 118–120
Storage, vegetative germplasm
    cryopreservation, 308–311
    *in vitro*, 304–307, 311
    restricted-growth, 307–308
Strawberry
    apical meristems, cryopreservation, 311
    breeding, 149–158
    and ionising radiation, 157–158
    recurrent selection programme, 155
    seed propagation, 156–157
    virus-indexation, 387
*Streptocarpus sp.*, apex formation, 279
Sugarcane
    breeding, 3–26
    disease testing, 22–23
    economic importance, 24
    farm trials, 23–24
    flowering time, 8–9, 17
    genetic resources, 3–24
        base collections, 3–5
        commercial breeding
            programmes, 6, 9–14
        crossing systems, 15–21
        selection systems, 9–15, 21–24
        utilisation, 5–9
        working collections, 9–10
        World Collections, 3–8, 22
    male sterility, 17
    mutation breeding, 24, 337

noble canes, 4
tillering, 22
tissue culture, 24, 401–402

Sweetgum, *see Liquidambar*

Teak (*Tectona*), 229
  tissue culture, 367
*Terminalia sp.*, 240, 241
*Tetranychus sp.* (mite), 162
*Thanatephorus sp.*, disease of rubber, 33
*Theobroma cacao*, *see* Cocoa
Tissue culture (*see also* callus culture)
  acclimatisation, 378
  *Alnus*, 368
  auxins and habituation, 265
  *Betula*, 368
  callus, 165
  *Citrus*, 100
  cocoa, 399
  coconut, 397
  commercial (scaling-up), 386–388
  conifers, 368
  date palm, 398
  embryos
    *Citrus*, 85, 101
    cocoa, 400
    genetic stability, 304
    *Picea abies*, 370
  *Fagus*, 368
  forest trees, 365–383
  ivy, 400
  *Liquidambar*, 366
  *Liriodendron*, 367
  *Manihot*, 401
  oil palm, 63, 388–397
  pedicel culture, 165
  *Picea abies*, 367
  plantation crops, 385–405
  protoplasts, poplar, 368
  rubber, 38, 398
  storage, 304–311
  sugarcane, 24, 401
  teak, 367
  tropical trees, 213
  *Ulmus*, 366
  vitrification, 370, 400

Tobacco, vector-induced tolerance gene, 346
Tomato, chimeras, 275
Trees
  alternative strategies, 200–203
  clonal forest
    advantages, 203
  clonal forest *vs* seedlings, 205–207
  clonal trials
    management, 207–208
    Norway spruce, 251–257
  forest, 199–209, 245–261
  genetic variability, 204, 206
  gene conservation, 208
  hybrids, 201
  mixtures *vs* pure stands, 207
  pest resistance
    clonal *vs* seedlings, 207
  seed production areas, 201–203
    *vs* clonal option, 205–206
  temperate species
    levels of variation, 246–251
    Norway Spruce, 251–257
  test extent, and duration, 203–204
  tissue culture
    acclimatisation, 371–372
    conifers, 368–371
    hardwoods, 366–368
    planting out, 378
    structural variations, 372–377
  tropical, 211–227, 229–242
    conservation and the genetic base, 222
    cyclophysis, topophysis and periphysis, 214
    experiments, environmental effects, 223
    genetic information, 215–216
    *Nauclea diderrichii*, 241
    seed production, 216
    *Terminalia sp.*, 240, 241
*Triplochiton scleroxylon*
  branching habit, prediction, 236–240
  clonal trials, 231–240
  commercial methods, 217–221
  vegetative propagation, 212–214
  wood quality, 240
Tristeza virus (in citrus), 97
*Tylenchulus sp.* (nematode), 99

*Ulmus sp.*, organogenesis, 366

Variation, genetic basis, *see* Mutant genes
*Venturia sp.*, 114
*Verticillium* wilt, 149
Virus elimination,
 apical meristems, 305
 fruit trees, 350
 *via* tissue culture, 385
 tropical trees, 213

Vitrification, 370

Wickham collection, rubber, 29, 36

*Xanthomonas albilineans*, 4
*Xanthomonas campestia*, 97
*Xanthosoma*, restricted-growth storage, 307
X-ray irradiation, *see* Irradiation